Stefan's problems

Stanislav Južnič

Title: Stefan's problems

ISBN: 979-8-88676-027-9

Author: Stanislav Južnič

Cover image: Wikimedia Commons

Publisher: Generis Publishing
Online orders: www.generis-publishing.com
Contact email: info@generis-publishing.com

Table of Contents

Summary

Stefan's mentor Marian Koller used to be the very-best astronomer from Carniola besides Augustin Hallerstein. Today, his work connected with Dalton minimum is again in limelight as the alternative explanation of disputed global warming phenomena.

Koller decisively supported Josef Stefan, a Slovene from Klagenfurt, who won his blitzkrieg pedagogical path from a high school professor to the leading scholar in the Habsburg Monarchy. His greatest helper was his fellow Slovenian Marian Koller, a Benedictine erudite from Bohinj as a leading counsellor at the Ministry of Education. The successful cooperation of both leading Slovenian experts promoted the Habsburgian sciences worldwide. Koller's international connections and pedagogical-scientific ideas enabled early success of young Stefan. In his turn, Stefan made his atomistic kinetic theories mandatory in Habsburgian monarchy except for Ernst Mach's Prague which traditionally opposed Viennese ideas already during the prevailing influences of Jesuit Rudjer Bošković's sciences a century earlier.

Already in 1842, Koller's astronomical measurements read for the Londoner *Monthly Notices of the Royal Astronomical society* were noted in *Phil.Mag.* Stefan widely used that prestige of his benefactor Koller. It was previously not widely known that Josef Stefan published more than twenty papers in London, Edinburgh, and Dublin *Philosophical magazine*.

Primary Mathematics Subject Classification Number (Following the 1991 Mathematics Subject Classification compiled by Mathematical Reviews and Zentralblatt fur Mathematik): 35R35, 80A22

Secondary Mathematics Subject Classification Number: 35B40, 35C05, 35C15, 35Kxx, 35R30, 46N20, 49J20, 65Mxx, 65Nxx, 76R50, 76S05, 76T05, 93C20.

Primary Keywords (Palabras claves primarias): Josef (Jožef) Stefan, Stefan Problem and Stefan Number, *Philosophical Magazine*, collaboration between literati of London, Paris, Geneve and Vienna, history of physics in middle Victorian era, Marian Koller.

Secondary Keywords (Palabras claves secundarias): Boltzmann, lectures on kinetic atomistic heat, Lamé-Clapeyron's elasticity, Fresnel's wave optics, Weber's electrodynamics of Faraday's induction in Vienna, Dalton minimum, Sunspots, Bohinj, Kremsmünster, History of Observational Astronomy, Geomagnetism, and Meteorology, Central European Studies, History of Science and Technology, Melting and Solidification of Metals for metallurgy of continuous ingots as the continuous casting methods for ingot processing, Melting and Solidification of Arctic Ice as well as agricultural land and its crops, Global Warming Hypothesis, oil industry oilfield-petroleum mechanics, melting of nanoparticles.

Introduction

Stefan also published in French language in Paris and in Geneve as well as in Berlin and Leipzig. Boltzmann learned English after his teacher Stefan gave him English vocabulary to enable his readings of Maxwell's publication and Boltzmann then used to send his personal letters to the "Gentleman" of *Phil.Mag.*

English and French language translations of contemporary Stefan's papers are used for the first time in this study, as well as the relevant Habsburgian 19th century archival material involving Stefan and Koller. The final goal is the explanation: how the dying Stefan's Habsburg monarchy topped the worldwide erudition by its exceptionally productive swansongs. The mainstream success of Stefan-Boltzmann Law and Stefan Problem are the examples of the lasting Habsburgian legacies.

The common people wonder how Josef Stefan managed to advance from the initially illegitimate son of a suburban single mother Maria Startinik to a leading Central European academic. The answer is called: Koller.

First Part

The Mentor Koller

Leading Mid-European Astronomer: Koller

Introduction to Koller's Merits

Compared to the initially poor Josef Stefan, Marian Koller was much better off as the firstborn of the manager (Bergwerkeverwesser, Mines Corrector) of baron Zois' Upper Carniola ironworks. After two decades of lecturing physics and managing of the observatory at Kremsmünster, Koller became a leading Habsburgian reformer of technical schools.

Marian was baptized as Wolfgang Koller on 31 October 1792 at 3 o'clock in the morning in Wochein Feistritz (Oberfeistritz, Bohinjska Bistritza) no. 78 (87) in Carniola. His godparents were the neighboring (Nachbare) locals from Bohinjska Bistrica, Simon Ravnik and Marija Sodja the wife of the farmer (hubler) who was also a godmother of Marian's elder brother Ignatz Koller on 29 July 1791(Bufon, 1970, 40; Book of Births and Baptisms of the Parish of Bohinjska Bistrica, page (folio) 19 (27) last row; Wurzbach, 1864, 12: 346-347; Strakosch-Grassmann, 1905, pp. 227 230 (with picture)). Both surnames are still common in Bohinj today. Already on 4 August 1792 the future Koller's teacher and Zois' protegee Valentin Vodnik OFM of Ribnica parisha asked for a newly established local parish in Koprivnik in Bohinj; the application was resolved on 4 December 1792, and he became the pastor of the Koprivnik curacy after January 1793, a couple of months after the birth of the future Benedictine Marian Koller. In 1793, because of his progressively damaging gout Zois visited his Bohinj blast furnaces and ironworks for his last time. Zois was also a supporter of the new Koprivnica parish. To make it easier to use Vodnik's talents in his cultural circle, Zois helped him move as appointed cooperator to the parish priest Josef Pinhak (Jožef, * 1755 Zweckendorf (Účel (Účely) vesnice, Namenska vas, Purpose village) in Moravia; † 20 August 1814 Ljubljana) at the former Jesuit church of St. Jakob in Ljubljana on 23 June 1796. Therefore, Vodnik was Koller's neighbor for three years and a half in Bohinj areas in Koprivnik 4 km northeast of Bohinska Bistrica. Vodnik conducted at least two Alpine research tours for Zois, financed by treasury supervised by Koller.

In early August 1895, Vodnik went to the hills in the company of count Franc Jožef Hanibal Hohenwart, mountain guides and a priest of the Benedictine grammar school in Ljubljana, Jožef Pinhak, who meet them at Javornik. They climbed Belščica, Veliki and Mali Stol; through Bohinjska Bistrica they went to the source of Bistrica. After visiting Karl Zois, Francophiles Jožef Pinhak and Vodnik climbed towards Triglav on 20 August 1895. In memory of the expedition, Vodnik wrote an ode to the 2194-meter-high Vršac near Kanjavec, which has since been called Vodnikov Vršac. After leaving Bohinj, Vodnik became Pinhak's chaplain at St. Jakob in Ljubljana.

The senior Andreas Koller was the general manager of the baron Sigismund Zois' ironworks. His wife Josepha, born noble Seethal belonged to the old Carniolan-Croatian family, five generations earlier still surnamed Matežič. For centuries, her ancestors managed the Upper Carniolan and Lower Carniolan blast furnaces.

Koller family was always associated with mining, as the original spelling of their surname in the Slovenian lands of Inner Austria was Kohler, therefore the Charcoal-burner.

In the first half of the 16th century, the furnaces in the Bohinj area were in Bohinjski kot, Bohinjska Bistrica, Pozabljeno, and Stara Fužina. In the middle of the 17th century, they were owned by the Locatelli brothers who started to modernize them. It was known as Locatelli brother's enterprise, which Zois' family took over in the middle of 18th century. In written sources it is mentioned that Locatelli received his privilege in 1670 as "Janez Locatelli (Johann, Giovanni), who was also a co-owner of the furnace in Plavž, Rudno cave by Kamnik and Rovte". One of the mines was also the Klinar's shaft at the mountain called Lakes (Jezerca). Josef Kalasanc Erberg noted Johann Joseph Locatell(i) among five Carniolan mathematicians.[1] Janez Jožef Locatelli (Giovani Giuseppe, * Trieste?) wrote about meteorology at Plavž (Furnace) by Jesenice.

---

[1] Murko, 1974, 34; Vidmar, 2009, 356, 358-359; Uršič, Erberg, p. 138

Zois' manager in Bohinj was Ignac Potočnik (Jakob Ignac Pototschnig, Potozhnig, * 24 July 1756 Kropa no. 32, today's no. 56; † 1816) from Kropa where he later served as Napoleonic maire. In 1788, he resigned because the planned reforms of Žiga (Sigismund) Zois seemed too difficult for his taste. Therefore, Andrej Koller took over the management. As a probable native of Idrija with its mercury mine, he was not a great expert in ironworks in the first place, but he used to be a careful and meticulous executor of Zois's measures including the erection of a new blast furnace in 1791. He emerged from a new generation of non-Jesuitical experts, mostly trained informally or outside the Habsburgian areas. They worked as engineers before Andrej Koller's son Marian developed modern real technical schools in Habsburgian Monarchy after the Spring of Nations. Just like Andrej Koller, in his era the experts managed the mines in Idrija and Banská Štiavnica, including Franc Anton von Steinberg and Samuel Mikovíny. Similar officials were Balthasar Hacquet in Idrija or Trieste-based freemasonic builder Vincenc Jurij baron Strupi (Struppi, * 1733; † 1810) who achieved Carniolan title of baron on 18 June 1790 as a relative of mother of Wolfgang Marian Koller's professor J. Jenko.

A. Koller planned Zois's Bohinj road with bridges in 1783 and completed it in six years until 1789 through the right banks of the Sava river over several bridges. The previous route through Štenge by the left banks of the Sava was not passable for a horse-drawn carriage, so they had to open their carts there and continue their journey on foot (Cundrič, 2002, 42). Andrej's younger son, the forester Jožef, certainly advanced his father's ideas by designing the Solkan-Trnovo road three quarters of a century later between 1855-1860.

The father Andrej Koller was of small stature, so he was nicknamed Kolarček, meaning the small Koller. Even his son Marian Koller grew up to be a comparatively small man. Andrej Koller was mayor under French rule. In fact he decided about everything in Bohinj; so, people respected him and asked for his advices. In Zois' ironworks manor on the western edge of Bohinjska Bistrica not far from the central castle of the counts Attems, A. Koller personally experienced the peak and fall of Zois's furnaces. During the economic crisis of 1812 caused by Napoleonic blockade of trade, Koller temporarily stopped the plant and dismissed his workers as caretaker of Zois's metallurgy in Bohinj and later in neighboring Javornik 20 km northeast of Bohinjska Bistrica near the eastern suburbs of Jesenice city. According to his boss baron Zois' notes of abt. 1816, Koller and simultaneously in 1788 newly appointed mining scribe-supervisor Anton Žerovnik as a local from Bohinj had the most complete topographic map of Bohinj and Triglav areas (Die Wochein, Sestavil P. Richter (professor Franc Janez Ksaver Richter (1783 Osoblaha in Moravia-1856 Vienna)) po spisih iz zapuščine pokojnega Sigmunda Zoisa barona Edelsteina Od točke 3 do točke 5, *lllyrisches Blatt*, 25 June 1821, pp. 82-83, Translated by Franc Ceklin (* 1911 Stara Fužina in Bohinj; † 2004 Črni vrh): Triglav in Bohinj, *Planinski vestnik*, 1977, 77/1: 17, 77/2: 143, 144, first part July 77/7: pp. 427, 429, second part August 1977 77/8: 503-510).

A. Koller's eldest son, the future astronomer W. Marian Koller, received his first upbringing and education at his parents' house. He listened to his first lessons at his hometown school. W. Koller went to study on Zois' initiative. Of course, his younger brothers, a forester and a lawyer, also studied successfully.

Koller's Studies

In 1802 Koller matriculated at the k. k. Normal school in Ljubljana and in 1803 passed to the local high school there, which he completed in 1808 with very good progress; in the latter year he also took the exam to enable his private tutoring. His juvenile acquaintance Valentin Vodnik taught him Slovene language (Črnivec, 1999, p. 348; Bufon, 1970, 40). Koller's classmates were the son of Ribnica landlord Kopitar's protégé Jožef I Anton Rudež (* 1793 Ribnica; † 1846) and Ljubljana native the future physician Frančišek (Fran) Weber. In July-August 1862, Jožef I Anton Rudež's second son and heir Jožef II Rudež (* 1819 Ribnica; † 1871 Ribnica) and his younger brother Karel Dragotin Rudež (* 1833 Ribnica; † 1885 Hrastje on Gracarjev Turn in municipality Šentjernej) joined Gregor Mendel (* 1822; † 1884) onboard the second pleasure train (Vergnügungszug). They travelled to Paris, visited the International Exhibition in London in August, and took some sightseeing of Stuttgart. On 2 August 1862, the *Vereinigter Laibacher Zeitung* wrote about twelve travellers from Ljubljana: three large landowners, five businessmen, a wallpaper decorator, a calligrapher, a photographer, and an engineer. Their company included the large landowner S. or J. Bals. That person was Sigmund Bals' (* 1785) and Aloisia's son Sigmund Bals (* 11 March 1813 Capuchin suburb no. 18 in Ljubljana; † after 1869) who studied his second class of humanities in Ljubljana in 1830 but still lived unmarried in 1869. Possibly the person who boarded the train was Sigmund Bals' brother Josef. Maybe the traveller was Sigmund Bals' niece or sister-in-law, both named Josefa Bals, who later owned former Sigmund Bals's house at Ljubljana Capuchin suburb no. 18 (after 1867 the inn Pri Balsu (By Bals) with a drugstore, renamed to Theater-Gasse Gledališka ulica no. 1 after 1877, todays Wolf's street no. 1). The younger Sigmund Bals' (* 1785) and Aloisia's son Joseph (* 14 March 1814 Capuchin suburb no. 18 in Ljubljana) married Josefa (* 1817) and lived with their daughters Josefa (* 1839), Albina and Augusta in 1837 and afterwards at neighbouring Capuchin suburb no. 25 in Ljubljana (todays Kongresni Trg no. 8). That building Capuchin suburb no. 25 and neighbouring house was later bought by Franc Kollman (* 1805 Zgoša 32 pri Tavčarju; † 1876/1900 Ljubljana) or by his son Franc Kollman (Franjo, * 1839 Zgoša by Begunje near Radovlji□a; † 1908 Ljubljana). That same Sigmund Bals was also registered alone at Capuchin suburb no. 78 in 1837.

Among Mendel's Slovenian travel companions were also a landowner Ludwig von Kuchtich (Kučtić de Oskocz) of Razwor by Zidani Most in Lower Styria, the merchant Joseph Friedrich Sennig, and the landowner mayor of Cerknica in Carniola in 1860-1870 Adolf Obreza (* 1834 Gori□a; † 1886 Cerkni□a). Koller and Stefan were too busy to join them.

In the years 1809 and 1810 Koller attended both philosophical classes at the k. k. Lyceum of Ljubljana with outstanding success. In 1809 he passed the exam as a private and domestic-house instructor using Latin language for all classes of the high schools.

In 1810/11, during the French Occupation of Carniola, Koller studied general chemistry, natural history, and mathematics at the Central School. He finished with the grade "l'optime" at the faculty of medicine in Ljubljana (École centrale de médecine de Laibach). Additionally, he also studied Italian and French language, which he used a lot later. In 1810 the Prague Jew Gunz taught him his bellowed mathematics, while Idrija native Hladnik lectured on Natural History. J. Jenko taught physics and mathematics until 14 June 1810. Until Kersnik replaced him in 1811, Giovan Maria Zendrini (Giovan Battista, Giovanni, Gianmaria, 1783 Breno in Valcamonica 90 km north of Brescia-22 January 1858 Torbiato (di Adro) 25 km west of Brescia) taught chemistry in Ljubljana. Zendrini began as a substitute professor in Chemistry, Mineralogy and Natural History in the Liceum/High School Department of Mella 14 km south of Brescia in 1806. Later, he was a professor of Chemistry at the

University of Ljubljana in 1810-1811. Again, in Brescia he taught Botany and Agriculture in the Lyceum High School Department in 1811-1816. He lectured as a substitute professor of special Natural History at the University of Pavia (1816-1819), then as a full professor (1819-1851/52); meanwhile he was also director of the Natural History Museum (1817-1852) and Rector of the University (1847/48). He was a great admirer of the young novelist Alessandro Manzoni (1785 Milan-1873). While he taught in Ljubljana, from 5 August 1810 Zendrini was an active member of the Academy of the Department of Mella (Accademia di scienze, lettere, agricoltura ed arti), then the Ateneo of Brescia (Accademia di scienze, lettere, ed arti) founded in 1802. He was also referred to as "Honorary Member. Zendrini published *Filosofia zoologica, ossia Prospetto generale della struttura, funzioni e classificazione degli animali del dott. Giovanni Fleming; traduzione dall'inglese* (2 voll.; Pavia 1829, translation from English in 2 volumes, *The philosophy* of *zoology; or, A general view of the structure, functions, and classification of animals*. Edinburgh, 1822, by Free Church of Scotland minister, vitalist, and fan of Abraham Gottlob Werner's (and Zois') Neptunism geology John Fleming (1785 Kirkroods farm near Bathgate-1857 Leith)) and *Rudimenti mineralogici compilati ad uso degli incipienti lo studio della mineralogia da G.Z. Conuno* (Pavia: Fusi e comp, 1840, seconda edizione, *Mineralogical rudiments compiled for the use of beginners the study of mineralogy*, second edition signed by initials G.Z. First edition was published anonymously as *Frammenti mineralogici* by the same Pavia: Fusi e comp, 1834). Zendrini was certainly a great expert, but the university of Ljubljana was irregularly financed in his era: so, he left for his home areas.

The Benedictine Rafael Zelli came to Ljubljana from Zadar. He supervised all schools in Illyrian provinces and even instructed Marmont in his chemical experiments which may have inclined Koller to the Benedictine order.

Koller's special patron, baron Zois, wanted to have the young man trained as a mining engineer at the mining Academy in Banská Štiavnica (kk Bergakademie zu Schemnitz), but his plan was thwarted due to the unfavorable conditions. Napoleon sieged Bratislava in summer 1809 as a part of Campaign against the Habsburgian Empire while the French took over Vienna and fought two major battles at Aspern and Wagram. Also, the father A. Koller had probably already figured out that Zois's ironworks were irrevocably sinking under the blows of Napoleon's blockades and better Swedish steel.

Instead, Koller went to the university in Vienna. In 1811/12, he studied mathematics there under the famous astronomer Johann Tobias Bürg (* 1766 Vienna; † 1836 Wiesenau Castle 54 km northwest of Klagenfurt). Koller got the best superior marks (cum nota "eminentiae"). In 1812/13 he attended the additional similar lectures of applicative mathematics (ex mathesi forensi) with the grade next to first prominence ("primam cum eminentia"). In the mid 18th century and in Koller's times, "mathesi forensi" dealt with arithmetic, geometry, architecture, mechanics, hydrostatics, chronology political sciences and juridical uses connected with statistics at all higher and lower courts. He excellent Gunz's mathematical lectures in Ljubljana were a great foundation of those additional Koller's studies.

Koller's Carniolan and later Viennese neighbor, the Upper Carniola Repenje near Vodice native leading linguist Jernej Kopitar, liked to make fun of Zois' protégé Koller, especially regarding Koller's juvenile sympathies for Viennese girls. Kopitar reported from the Imperial Vienna to Zois's Ljubljana: "In four days Koller will took over the post of a home teacher with a salary of 200 florins in the Viennese currency, with food and the like at the house of economic councilor of the count Fries. Gran fortuna per lui anyway, because I suspect he likes too much his lovely persona inside his current households. He attends the natural science lectures of the chemist-botanist Johann Baptist Andreas knight Scherer (* 1755 Prague; † 1844 Vienna), but he still has a lot of free time to instruct. I am just having a little fun while mocking him. There is only one Vienna in the world, but there is also only one Vienna for the damage of young people. He passed the mathematics exam as the first among the excellent students (Matheseos attestatam habet primam cum eminentia, Letter from Jernej Kopitar to Žiga Zois from Vienna on 29 April 1812, when Koller was nineteen and a half years old).

Kopitar continued: "Koller is in excellent condition in all respects, for which he has to thank his compatriot Jugovic, to whom I recommended him. He often travels to the country with his boss's family on Sundays and other holidays. His employer is the economic director of Count Moritz Christian Johann Fries (* 1777; † 26 December 1826 Paris). As Alsatian-Swiss Protestant banking director, Fries played a role of Maecenas, like Zois did in Ljubljana. Indeed, Fries inherited the monarchy's wealthiest Viennese merchant house. Fries was the artistic patron of Beethoven, Haydn, and F. Schubert as a collector of natural history books until his bankruptcy in 1826. At least as early as 1838, his Viennese palace was taken over by another benefactor of Ljubljana based astronomers, Simon Sinas Baron Hodos-Kizdia (* 1810; † 1876).

Kopitar continued: "It is good to see Frančišek (Fran) Weber from Ljubljana as well". Weber was the first prize winner as a classmate of W. Koller and J. Rudež in 1808 in the 2nd grade of Ljubljana Humanities. There, Weber used a scholarship from the Čemšenik priest Jurij Tomaž Rumpler's Fund. With Knafelj's scholarship, he studied medicine in Vienna in 1814 (Črnivec, 1999, pp. 347-349). He then served as physician in the newly conquered Habsburgian Dalmatia, Postojna, Kamnik, Ljubljana, Rijeka (Fiume), Pazin, Zadar and Trieste as a descendant of the Ljubljana doctor of medicine Anton von Weber († 22 February 1757 Ljubljana), Zois's relative through marriages in the Kappus von Picheistein's family.

Kopitar added that "it was high time for Koller to get more work done: as idleness had already begun to show its detrimental impact. He did his math exams with distinction (eminentia), but I did not see his certificate with marks received at exams of science. I once warned him fatherly and I hope that everything will be fine now" (Letter from Jernej Kopitar to Žiga Zois signed in Vienna on 13 June 1812).

Librarian and later an administrator at the Viennese Court Library Kopitar then reported that "Koller is not at sight in those days. Koller's professor, former Klagenfurt teacher astronomer Johann Tobias Bürg, was relieved of his duties for three years; he practiced his astronomy in anti-Jesuit Franz Xaver Zoch's (1855 Pest-1932 Paris) Seeberg by Gotha, briefly taught as a professor of physics at Lyceum in Klagenfurt in 1791/92 until the ex-Jesuit Francis Treisnecker (1745 Mallon in Lower Austria-1817) picked up Bürg as professor of higher mathematics at the University of Vienna Treisnecker's Viennese assistant and finally a replacement in 1817. In 1802 and later Bürg was a teacher Baroness Elisabeth Maria Josepha von Matt (née Edler von Humelauer, 1762 Vienna-1 March 1814 Vienna), a wife of Ignaz baron Matt (baron in 1793, 1740 Konstanz-19 July 1814 Vienna), whose observatory west of the university observatory near St. Stephan and geodetic instruments for the measurements in West Bohemia they used in 1804-1814. Elisabeth was the only female astronomer whose observations of the asteroids Pallas and Juno were personally published in European journals including Zach's (and his assistant Bernhard von Lindenau's) *Monatliche Correspondenz*. Her home was also a fashionable saloon where Schiller's and Goethe's works were played and read with the writer Karoline Pichler, * 1769 Vienna; † 1843 Alservorstadt of Vienna) attending (Peter Brosche, Klaralinda Ma-Kircher, Elisabeth von Matt (1762-1814), An Enlightened Practitioner of Astronomy in Vienna, Journal of *Astronomical History and Heritage* 13 (2010) 3: pp. 187-193; Peter Brosche, Klaralinda Ma-Kircher, Die Wiener Astronomin Elisabeth von Matt (1762–1814). *Wiener Geschichtsblätter* 67, Heft 3 (2012), pp. 259–273). Bürg recalled that Matt was the only person who cared fir him while he was deaf, which was probably the reason why Bürg got his free three years after 1813, probably also for his and Mott's geodetic measurements.

During his absence, Bürg was replaced by the professor of commercial accounting at Real School, after 1815 at new Johann Joseph von Prechtl's Viennese Polytechnic, Joseph Hantschl (* 1769 Cvikov (Zwickau) in northern Bohemia; † 2 June 1826 Vienna). Hantschl was an acquaintance of Ž. Zois' nephew-godson Anton Bonaczy von Bonazza (1785-1853), the owner of the ironworks in Mislinje. Bonaczy lived in Croatia after marrying the niece of the Zagreb bishop Vrhovec, Franjica Novosel (* 1791) in Zagreb on 1 November 1809, until Franjica died and Bonaczy remarried in Trieste on 13 May 1814. Weber complained that Koller owed him 50 florins in Viennese currency already for many months. Kopitar described Weber's accusation as "Clodius accusat Moechum" by Latin saying against Cicero's enemy Claudius, a famous womanizer. Clodius accused other as the adulterers while trying to divert attention from himself by blaming his own sins on another. That populist politician Publius Clodius Pulcher (93-52 BC) allegedly even had an affair with Cicero's bride. Kopitar was therefore making fun of Weber's complaint about Koller, as Weber also borrowed money: "Since May, Koller has also owed 30 florins in Viennese currency to Kopitar. In summa (If I add it all up) it seems to me that the young gentleman would like some extra presents. Sed haec interea inter nos. Nil nocebit illi, aliquid expertum esse (But this for now between us. Nothing will hurt him anyway if anyone finds out (He must learn from that experience)). Koller must not fall deep, because of his situation in the house of the economic councilor of count Fries. From 3 May 1812 to 1814 he served there as a home teacher, therefore he had to pay attention to his reputation." Kopitar there mocked again: "Anyway, he is from Bohinj (where folks used to be (in)famously frugal like the Scots)! However, I would not want to involve you; forget it! Zois's copyist Vincenc Karnof should - as far as I am concerned - write to his father A. Koller again. Weber and Koller (W. & K.) are students et punctum!" The last idea means trat the students are somewhat allowed to perform some naughty immoralities.

Kopitar added the remark "Jenko?" He meant the Upper Carniolan native professor Jožef Jenko, whom Kopitar repeatedly recommended to Zois (Letter from Jernej Kopitar to Žiga Zois signed in Vienna on 13 February 1813). In those days Jenko was still a professor of mathematics in Linz, but he soon transferred to Graz and later to the Viennese university with Zois' help.

As a student of mathematical sciences, Koller from Bohinj taught privately in Vienna. Zois mentioned to Kopitar his willingness to cover Koller's debts himself if they were not too expensive. Like Fries, Zois was almost financially ruined by Napoleon's continental trade blockade.

During that time, Koller passed several exams needed to gain a Viennese chair for mathematics. But he was put off for the future because older professors were believed to be better choice.

Sometimes small, apparently insignificant circumstances become decisive for a person's future. That has proven itself very well in Koller's case just before the Congress of Vienna, which shaped the fate of the world between 1 November 1814 and 9 June 1815. There, a balloonist from Carniolan Bloke Gregor Kraškovič (* 1767; † 1823) excelled himself with his flights above Prater. One evening, a great fan of the game of chess Koller sat in a coffee house. He was trying to checkmate one of greatest Viennese masters. That Viennese café was probably Zur Goldenen Krone pub, where often played the main chess players of the capital, including Anton Witthalm, Count Johann Somssich and Johann Baptist Allgaier (1763-1823).

Next to Koller were several gentlemen at a table. One of them mentioned that he was looking for a teacher of French and Italian languages for his children already for a long time. Unfortunately, in vain. He did not know whom to contact to achieve his purpose. Koller, who heard this, started a conversation with the gentleman Joseph Ludwig after Koller's own chess game was completed. Koller inquired about all the circumstances and conditions, and finally declared, "I am offering myself for that position". He already decided that he should wait for the university chair of his favorite subject of mathematics somewhere outside Vienna for some time, as he needed to achieve good connections and reputations. Therefore, Koller joined the households of ironworks handler Joseph Ludwig in Steinbach an der Steyr in Upper Austria, where he soon gained the affection and love of his pupils, but also the respect of parents and friends and acquaintances of the family. Joseph Ludwig, along with Anton Ludwig, traded in Steinbach an der Steyr in Upper Austria also in 1843-1848. In 1772, the manufacturer of knives Johann Anton Ludwig Johann run his manufacture in Bad Hall 10 km north of Steinbach, while Johann Georg Ludwig had the similar business on Hochgasse no. 9 in Steinbach, and Matthias Ludwig traded at Grünburg no. 64 in Steinbach (Provinzial-Handbuch von Osterreich ob der Enns und Salzburg, 1848, 354-355; Kieweg, 1998, 87; Reslhuber, 1866, 354).

During his two years spent in Steinbach, Koller also developed a friendship with the family of Joseph Redtenbacher in Kirchdorf near Krems 15 km southwest of Steinbach. Joseph Redtenbacher married Koller's cousin Joseph Maria Theresa born Koller while at that time a branch of the Koller family also lived in Steinbach. The Redtenbacher's family highly valued the young Koller, while they were blessed with many children who became Koller's students. Joseph Redtenbacher's three sons, Joseph, Ludwig, and Wilhelm, together with their cold cousin and brother-in-law Ferdinand Redtenbacher, enthroned the modern science in Central Europe. As a professor of Chemistry at the University of Vienna, Joseph Redtenbacher founded the Viennese Academy with Koller among the founding members. Dr. Ludwig Redtenbacher became director of the Imperial Zoological Court Cabinet. Dr. Wilhelm Redtenbacher, on the other hand, became a general practitioner in Vienna as a former physics student of Marian Koller at the Institute of Philosophy in Kremsmünster.

Astronomer in Kremsmünster

During his time in Steinbach from 1814 to 1816, Koller visited the nearby abbey Kremsmünster several times. He learned all about their scientific institutes and the work of the Benedictines. Since he found some satisfaction in his literary endeavors here, Koller decided to become one of them, the Benedictine. By his excellent testimonials with the assurance of his current use scientific studies he was gladly granted acceptance. Koller entered the novitiate on October 5, 1816 and got the monastic name Fr. Marianus. After a probationary period, Koller studied theology from 1817 to 1821 at Lyceum of Linz with outstanding success. He filled his spare hours with a zealous study of the mathematics he had become so fond of.

On June 13, 1821, Koller forever connected himself to that abbey by the solemn vows. The Benedictine fourth wow granted the connection of a monk to his abbey; later as a Viennese official Koller resided in the Viennese palace of his Benedictines. He was ordained a priest on 18 August. On 8 September 1821 in the parish church of his birthplace, as a new priest he sang and preached his personal first mass in the church of St. Nicholas (Nikolaj) of his native Bohinj, at the parish of Bohinjska Bistrica founded in 1788 (Klemenčič, 1981, 16). His younger brother Andrej initially made a similar decision: on November 30, 1823, the outgoing bishop of Ljubljana, Avguštin Gruber, granted him his first tonsure with four lower orders. But Andrej soon changed his mind by marrying the nice maid of the wealthy Zollner (Colnar) family in Ljubljana.

In November 1821 newly promoted pater Marian Koller became a cooperator in the pastoral care of the parish Sipbachzell 5 km south of Kremsmünster, which was operated by the Kremsmünster monastery. He gained the general love and respect of the parishioners through his excellent work as a pastor, catechist and a great pulpit speaker. Consequently, he received the highest commendation in 1824.

Three years of his pulpit preaches, elaborated in his manuscript, testify to Koller's zeal in this profession. His teachings of the Christian religion are as comprehensible to the mind as to the heart. He used simple, noble language.

Professor and Custos of Physics cabinet 1826-1839

P. Thaddäus Derfflinger (19 December 1748 Mühlwang by Gmunden-18 April 1824) was the director of the Kremsmünster Observatory famous for his observations of Sunspots. After his death, the leaders of Philosophical Studies and the Directors of Convict and Observatory called Koller back to the Abbey to become a professor of natural history at the Philosophical Institute. Koller started his lecturing in the school year 1824/25. At the beginning of November 1824, he provisionally took up this professorship. Additionally, in 1826 he took over the professorship of physics after proper preparation. On 17 February 1827 he passed the strict exam (Rigorosum) for both subjects in one day at the k. k.

Lyceum in Linz. Koller received his excellent results, after which the court confirmed him as a full professor of both fields.

From 18 to 26 September 1837, Franz Unger from Graz, Andreas Ettingshausen, Koller and professor of botany with mineralogy at Kremsmünster Basilius Schönberger (1807 Kirchdorf by Krems-1850) attended Prague meeting of the naturalists and physicians (15th Versammlung Deutscher Naturforscher und Ärzte). Koller reported about the sad state of the observatory in Prague to his colleague Augustin Reslhuber on 15 September 1837 (Fellöcker, 1864, 266). Martin Alois David (1757-1834) was previously a director of the Clementine Observatory in Prague.

For almost a decade and a half, Koller took care of the physics laboratory needed for lectures in his abbey (Fellöcker, 1864, 258-261). Among other things, he purchased several devices related to vacuum techniques:

In 1833 Koller acquired the Windbox (*Windbüchse*, air rifle, wind rifle) as an airgun designed by clockmaker ironmonger Bartolomeo Girardoni (Girandoni, 1729 Cortina d'Ampezzo in what was then Habsburg Tyrol-1799 Penzing in western Vienna) around 1779. From 1780 it was also used by the Habsburg army. The device was donated to Koller by Friedrich Häusler, a monastery economist. Also, Koller acquired a device proving Mariotte's law for compressed and diluted air (Fellöcker, 1864, 259).

Koller used Plateau's anorthoscope and on 8 December 1835 he additionally received the similar stroboscope designed by Koller's friend Stampfer for 4 fl (Fellöcker, 1864, 260).

In 1827 Koller acquired a device for the presentation of all important electro-magnetic experiments of Hans Christian Oersted, Ampère, and others, the so-called Ampère's frame. It was designed according to Baumgartner's description communicated in the first volume of Baumgartner's and Ettingshausen's Viennese Journal *Zeitschrift für Physik und Mathematik* in 1826. Koller bought it at a price of 200 fl (Fellöcker, 1864, 261).

Director of the Observatory 1830-1847: Beginnings

After the unexpected death of Thaddäus Derfflinger's successor, the astronomer P. Bonifacius Schwarzenbrunner (Bonifaz, baptized as Jakob, 25 June 1790 Garsten in Upper Austria-29 April 1830), Koller took over the management of the observatory. He consequentially gave up his professorship of natural history at the end of the school year 1830. He retained the chair of physics until the end of the academic year 1839. Consequently, Koller devoted most of his energies to the service of astronomy.

During the election process of a new Board of Directors on 23 September 1840, after the death of the abbot Joseph Altwirth (May 4, 1840), Marian Koller achieved a significant number of votes. The major part of voters did not choose him, because they believed that Koller's strictly scientific direction might bother his supreme administration, although in his work up to date they had no evidence either for or against such claim. Some also worried that, as abbot, he would devote all his care only to the university and the observatory, while neglecting other branches of monastic activity. Therefore, Koller did not get the abbot's dignity, because the monks wanted to save him for a position in which he should better use his knowledge and experience.

Koller's character proved itself on this occasion in the noblest way. Without any spitefulness against those who did not support his election, he remained the man as he always was, and thereby achieved even higher respect of his fellow monks.

The newly elected abbot, Thomas Mitterndorfer (* 1797 Sierning by Steyr; † 1860), who has always been an intimate friend and admirer of Koller, approached him with sincere cordiality. The fraternal relationship between the two remained unchanged for the most benevolent influences on the spiritual harmony of the whole monastery.

After the death of the director of the k. k. Observatory in Vienna, Joseph Johan Edler von Littrow (* 1781 Horšovský Týn in Bohemian Sudetenland; † 30 November 1840 Vienna), the authorities searched for his replacement. The abbot Thomas reminded and emphasized the knowledge and merits of Marian Koller whom he recommended as director. After long negotiations, his request was "rejected," but the proposal itself fully honored Koller's personal merits and the quality of his observational astronomy. In 1842, Littrow's son Karl edler Littrow (* 1811 Kazan in Russia; † 1872 Veni e) was ele ted; of ourse, he fully collaborated with Koller.

On 1 March 1843, the abbot Thomas appointed Marian Koller as deputy director of the Philosophical Study Institute, and Director of the k. k. Convict. Certainly, Koller maintained the directorship of the observatory.

With his excellent specialist knowledge, Koller headed his educational institution with such caution that his work was highlighted by the authorities several times. Even the emperor Ferdinand praised him on 13 December 1846. Ferdinand's tutor used to be the baron Josef Kalasanc Erberg from the Carniolan manor of Dol by Sava river.

Koller's knowledge and achievements as a professor and astronomer, his experience, and merits as head of a well-attended famous educational institution created a great impact on the broadest circles. So, it happened that after the death of the court counselor (Hofrath) physicist-astronomer Cassian Hallaschka (František Ignác Kassián Halaška, * 1780 Budišov nad Budišovkou (above Bautsch) in Moravian Silesia; 1799 Piarist; † 12 July 1847 Prague) they searched for his successor. The anatomist Jan Evangelista Purkyně's (1787–1869) friend Halaška used to be a member of the Government Commission for Higher Education (*Studienhofcommission*). As such, he headed all philosophical institutes, technical schools, naval schools, mining schools and forestry academies throughout the monarchy. At the k. k. Study Court Commission attention was directed to Koller. The son of the former director of the Idrija mine Karl Borromäus Rudolf count Inzaghi von Kindberg (Inzaghy, * 1777 Idrija; † 1856 Graz) visited Koller in Kremsmüster on 22 September 1847 (Fellöcker, 1864, 242). From (8 December 1817) 9 May 1818 to 26 April 1819, Inzaghi served in Ljubljana as the fourth Land-Governor of the kingdom of Illyria established after Napoleonic debacle. Therefore, he knew Koller's family very well. In 1842, Inzaghi became the upper court chancellor of interior ministry as well as president of Koller's court commission for studies.

Inzaghi's selection committee's attention was thus focused on Koller and on 30 October 1847 he was appointment as governmental councilor, speaker for the philosophical courses and technical studies at the Study Court Commission (Studien-Hofkommission). Koller worked under the supervision of Karl Borromäus Rudolf count Inzaghi and other two presidents of Study Court

Commission, with Andrej Mešutar as Koller's colleague there. While residing in his abbey's Renaissance Benedictines' palace Kremsmünsterhof at Annagasse 1003 (now no. 4) south of Stephansdom in Viennese Innere Stadt, Koller also became the President of the Philosophical Faculty of the k. k. Viennese University of the emperor Ferdinand I (Bufon, 1970, 40; *Hof- und Staats-Handbuch* 1848: pp. 228, 236, 284, 1853: pp. 72, 148, 180, 1866: pp. 48, 144, 151, 237).

While celebrating that Koller's political success, the bishop of Linz, Koller's fellow Benedictine Gregorius Thomas Ziegler (* 1770 Kirchheim in Schwaben near Augsburg; OSB 1788 Wiblingen Abbey south of Ulm; † 1852 Linz), made him a real consistorial councilor for his services to the religious-moral education of the student youth. Ziegler also reestablished the Jesuits at their settlement on the Freinberg near Linz.

Koller took up the honorary office of governmental councilor at the beginning of December 1847 and soon found his ways in his new influential sphere.

To recognize and honor Koller's services to the sciences, the Imperial Academy of Sciences in Vienna chose him as their full member on 26 January 1848, which was confirmed by the emperor on 1 February 1848, just before the March revolution almost ruined Habsburgians. Karl Inzaghi and archduke Ludwig were also honorary members of that Viennese Academy of Sciences.

When he left Kremsmünster, Koller had no idea how fateful the year 1848 would be for Habsburgians, especially for Vienna. In all the dangers and turmoil, he persevered bravely in his place and in office, and only retired to the Melk Abbey for a few days at the end of October. From there he returned to Vienna at the beginning of November. After he banned fourth song from Byron's "Childe Harold", Inzaghi became the confidante of the unmarried archduke general artillerist Ludwig (Louis Joseph, 1784 Florence-1864 Vienna) and member of interior minister count Franz Anton Kolowrat's moderate liberal politician party. Ludwig was a head of the State Conference (from 1836 to 1848) which controlled all government offices on behalf of Emperor Ferdinand I. Metternich and Kolowrat were other members of that State Conference, while Ludwig favored Metternich's politics and supported absolutism. Ludwig retired after the revolution of March 1848. Kolowrat also did not praise Inzaghi much after Metternich's downfall. Therefore, Karl Inzaghi resigned and retired after the Viennese March revolution on 21 March 1848, a day after

Kolowrat became the first constitutional Minister-President of Austria for one sole month in office. Inzaghi's official support was over, but Koller was clever enough to retain his political position in every incoming governments of Habsburgian Minister-Presidents who quickly replaced each other in those dangerous revolutionary years.

In 1849 after the March revolution, the k. k. Supreme Study Court Commission dissolved and a Ministry of Worship-Religion and Education (Kultus und öffentlichen Unterrichtes) was provisionally established in its place. Koller was appointed there as the section counsellor (Sectionsrath) responsible for the secondary schools, polytechnic, nautical and astronomical institutes. Therefore, he just continued his previous works under different frame. His collaborators were A. Mešutar and philosopher-Herbartianist grammar school & philosophical studies reformer in 1845-1847 Franz Serafin Exner (* 1802 Vienna; † 1853 Padua). The father of the physicists, Franz Serafin Exner, reformed the other part of secondary schools besides Koller's technical schools. Their collaborator was also court counsellor Johan Kleeman as section counselor (Rat). They worked under the leadership of former Exner's Prague university student the minister Leopold count Thun-Hohenstein (Leo, 1811 Děčín in Czech Sudetenland-1888 Vienna) until the downfall of Bach's absolutism in 1860. Thun-Hohenstein was the Old-Czech party's nationalist founder of political Catholicism whose home teacher was Bedřich Smetana (* 1824). As the opportunist Kleeman then already quarreled with his former collaborator the leader of Old Slovenian party Janez Bleiweis. So, Kleeman probably had troubles with Thun-Hohenstein's nationalist politics, as Thun-Hohenstein was a close friend of the leader of Federalist Old Czech Party, the Protestant historian František Palacký (Palacky, 1798-1876). They promoted Austroslavism which may also have been M. Koller's secret option even if M. Koller's nephew later supported the liberal Germanization (Bufon, 1970, 40; *Hof- und Staats-Handbuch* 1853: pp. 72, 148, 180).

On 29 October 1849, by the provisional law on the organization of the academic university authorities, Koller's employment as the president of the philosophical faculty of the k. k. University of Vienna came to an end. The doctoral college did not want to lose Koller from their board of directors. Therefore, "in recognition of the many and important services to science and education, with the expression of the deepest veneration for his personal characteristics" Koller became a full member of the doctoral college of the philosophical faculty of the Viennese University, appointed by acclamation. They gave him the honorary doctorate for his great merits for erudition ("de republica literaria optime meritus").

When the Ministry of Culture and Education was definitively organized in 1851, Koller was elected to the Ministerial Council (Bufon, 1970, 40). On 27 May 1859, the emperor Franz Joseph I praised Koller's merits, as he signed in his diploma, He stated: "Koller worked as the professor of natural sciences and director of the observatory, the philosophical institute, and the convict Kremsmünster for twenty-three years, then since 1847 he was employed in the civil service as a government councilor and speaker at then k. k. Study Court Commission. After its dissolution he joined the k. k. Ministry of Education as section council. Since 1851 he operated as ministerial councilor and reformer of the secondary schools, technical, nautical, and astronomical institutes. He was especially helpful for the rapid upswing of secondary schools in Habsburgian monarchy in his brilliant way." Through award of Knight's Cross of the Imperial Habsburgian Leopold Order in 1859, Koller was also advanced to the higher salary at his Ministerial Council in November 1859. He worked in this position until the unexpected end of his life on 19 September 1866. With his excellent intellectual abilities and tireless enthusiasm, Koller had acquired thorough in-depth knowledge of all the subjects of the natural history, mathematics, physics, astronomy, meteorology, terrestrial magnetism, and modern languages.

As a professor of natural sciences, he distinguished himself by clear, precise lectures, elegant language, strict logical consistency, sharpness of judgment and exhaustive thoroughness; he used to be a well-trained theorist and excellent experimenter. He eagerly took up every recent discovery and invention, exploited it in his lectures, and showed its importance in the theoretical or practical relationships. So, to mention just one case, Koller's students memorized his lecture about physics in the last years of the third decennium of 19th century: "When the professor told us about the experiment of the then very new discovery, the distraction of the magnetic needle by electric current, he said: Gentlemen, this discovery has a great future! Several of the students laughed, not realizing how this apparently small thing should ever become meaningful to them. Indeed, its application to telegraphy followed a few years later, carrying the thoughts of people to the most distant points with lightning speed."

Koller procured the necessary equipment and aids for his demonstrations. Thus, he became the actual benefactor of physics cabinets and natural science collections in Kremsmünster, the wealth and completeness of which many universities envy.

When Koller took over the management of the observatory in 1830, it was his first concern to activate and maintain the two-feet wide circle of meridians under the guidance of the director J. J. von Littrow and the professor S. Stampfer. They had been measuring it already in the times of Emperor Franz I, but their results had not yet been completely successful due to the difficulty of that peculiar construction of the observatory. Koller finished it to the complete satisfaction, so that the regular observations could start in July 1831. They have been continued until now without interruption, such as the determinations of the position and course of the standard clock, the location as well as the errors of the meridian circle as instrument for timing of the passage of stars across the local meridian. In 1829, he determined the geographical longitude of his Kremsmünster observatory by observing the star Alpha Centauri and measured a difference of 11.79 with respect to the observatory in Paris. His results were confirmed by ten other observatories while the today's value of a difference is just slightly higher: 14.133°E - 2.3365° E = 11.7965 (Fellöcker, 1864, 244 erroneously noted Koller's value as 10.79). Koller measured the position of the moon, the apparent locations of stars relative to the moon, the planets, and many fixed stars.

In 1831, Koller designed a rotunda with a stone pillar and a rotating dome on the top of the observatory to accommodate the portable equatorial, acquired by the grace of abbot Joseph Altwirth. Equatorial was a measuring instrument for the accurate determination of stars in 19th century. It is an equatorially mounted, long-firing telescope with large, finely divided circles and reading microscopes. In contrast to the meridian circle, it also enables the measurements of direct rectification and declination differences outside the meridian. Koller's equatorial's lens had the aperture 28 lines, 9 inches hour circle, and declination circle 12 inches in diameter. The hour circle, together with declination and distance from the planet's centre of mass, determines the location of any celestial object. It is the great circle through the object and the two celestial poles as a higher concept than the meridian defined in astronomy, which takes account of the terrain and depth to the centre of Earth at a ground observer's location.

Koller got his equatorial from Simon Stampfer's Viennese polytechnic institute manufactured there by Stampfer's workshop Georg Christoph Starke (Christopher Stark) who also produced Georg Friedrich von Reichenbach's (* 1771 Durla h; † 1826 Muni h) meridian ir les. The chronometer measuring seconds for meridian-circle was produced by the son of the Danish royal watchmaker Copenhagen master Urban Jürgensen (1776 - 1830) who specialized his craft in Paris as an apprentice to Abraham Louis Breguet and Ferdinand Berthoud and in London with John Arnold. Danish king gave it as the gift of a general-cartographer Ludwig August baron Fallon (1776 Namur; † 4 September 1828 Vienna) and Koller received it after Fallon's death. Fallon was the director of astronomical-trigonometric surveying, advisor to the real estate tax court commission, and director of the cadastral recording and as head of the topographical institute of the general staff. He merged civil and military surveying and promoted the introduction of Senefelder's lithography. Fallon recommended assuming sea level as a comparison plan, refined the barometric altitude measurements and constructed a new catoptric instrument in addition to a reflex distance meter. His main work consists of the map produced on the orders of Schwarzenberg "The Austrian Empire with considerable parts of the bordering states" in 1822 (Koller, 1833. Abhandlungen „über den Meridiankreis und das neue transportable Äquatoreale" der Sternwarte zu Kremsmünster. *Astronomische nachrichten* volume X, no. 225: 137-138, sent to the editor on 22 April 1832).

In the same year 1831, together with his friend professor Stampfer, Koller completed the setup and rectification of the instrument, with which they carried out observations outside the meridian, especially of comets. That telescope with lenses from the immortal Fraunhofer was characterized by precision. Schuhmacher's "astronomical news" in volumes VIII to XXV contain the numerous results of Koller's work.

Koller used the observations of his subordinates in Kremsmünster including Haslberger and Lettmayr, as well as the mineralogist Theodor Fuchs who later taught at Viennese University. From 1831 to 1834 Koller supported the astronomical observations of the professor of mathematics at the philosophical institute P. Wolfgang Danner (baptized Josef, * 1792 Sierning by Steyr river in Upper Austria; OSB 1816; † 1854), to whom he gave lessons in theoretical and practical astronomy and mathematics. By Koller's instigation Fr. Augustin Reslhuber studied theology in 1831 and 1832 at the University of Vienna with professor of higher mathematics Andreas von Ettingshausen. Under the director

J. J. von Littrow, Reslhuber studied astronomy in 1833. In November 1834, Reslhuber became Koller's adjunct at the observatory, and in 1840 P. Sigismund Fellöcker joined the group as an assistant. That increase in the number of workers and enabled the expanded range of work on the observatory.

Almost every year during the holiday months from 1831 to 1847, the professor of practical geometry at Viennese Polytechnic Simon Stampfer (1792 Slovenska Mataria (Windischmatrai, today Matrei in East Tyrol)-1864 Vienna), visited Koller's observatory as a loyal sincere friend and counselor. They jointly carried out the most varied examinations and determinations, including the Meridian Circle:
a) valuation of the divisions of the libel measurement site;
b) Construction of a new Meridian measurement site on the south side of the observatory at a distance of 918 Viennese Klafter (1.9 m), which was set up for night observation illuminated with a lamp;
c) Trigonometric determination of the distance of the new measurement sites from the center of the meridian circle;
d) Attaching curved threads to the micrometer of the meridian circle; one from both sides at fairly equal distances from the central thread and, if possible, the same inclination angles to the same. They needed it to determine the azimuthal changes in the tube by adjusting the calibrations to the crooked threads and reading from the vertical circles;
e) investigation of the bending (curvature) of the telescope;
f) Investigation of the shape of the cones of the axis of rotation of the meridian circle and determination of the corrections to be applied to the observations based on the determined cone shape;
g) attaching a feeler device to the alidade of the height circle for convenient and precise determination of the changes in the position of the alidade. It is a turning board which allowed one to sight a distant object and use the line of sight to perform a task;

All those investigations and determinations were carried out using mostly new methods devised by the professor Stampfer. In 1837, Stampfer first implemented his idea of a light point micrometer in the middle of the field of view for easier observation of faint objects through the portable equatorial, which is useful for comet observations.

The expected return of Halley's comet (with a orbital period of 75.3 years) caught the general attention of astronomers in 1836. During the night of August

21-22, Koller and Stampfer as the first in German speaking areas observed the comet almost at the same minute. Later, they continuously measured its orbit until their optical aids were no longer sufficient due to the vanishing comet's light.

Director of Observatory 1830-1847: Worldwide Window Opened in Summer 1838

In the spring of 1838, the professor Andreas von Ettingshausen invited Koller to a journey through northern Germany, across then significantly larger Denmark via Hamburg to London and Paris. As desirable and highly attractive as this opportunity seemed to him to expand and enrich his knowledge, and to meet the men of science, with whom he was already in the learned correspondence, there was a problem of raising money to cover the cost of such an important trip. The propriety was happily resolved when the abbot Joseph Altwirth granted a significant travel sum. The confreres who were close to Marian and other friends made their savings readily available. Koller, unlike seasick Hallerstein or J. Stefan, just loved to travel, especially because he was able to replace Valvasor's horses with a more comfortable train. J. Stefan, of course, grew up in a much humbler environment, where extensive travel was an unattainable luxury.

Baumgartner initially wished to join them, but in 1833 he fell ill with his throat and limited his Viennese lectures as his voice failed him. He finally canceled his participation in the trip (Folk, Holovatch, 2020, 4). On 18 July 1838, Ettingshausen and Marian Koller left Vienna for Prague together with Dr. August Kunzek von Li hton (* 1795; † 1865), a professor of physics at the university in Lvov (Lemberg). In Prague they visited the university professor of physics with applied mathematics Ferdinand Hessler (* 1803 Regensburg; † 1865 Vienna), the professor of mechanics and physics at Polytechnic Karl Wersin (1803 Sokolov (Falkenau an der Eger) in Czech Sudetenland-1880 Prague) and a professor of mathematics at Polytechnic Christian Doppler (1803 Salzburg-7 March 1853 Venice).

Figure 1: The most important Carniolan scientist Koller among other members of mathematical-natural historical class of Viennese academy in 1853, probably after both founding members of opposition fraction died in Venetian areas of lung diseases, Doppler in March and Exner on 21 June 1853. Staying from left to right: founding member plant chemist Friedrich Rochleder (1819 Vienna-1874), Doppler's teacher professor of mathematics and engineering at Viennese Polytechnic later baron Adam Burg (1797 Vienna; † 1882 Vienna), Karl Ludwig von Littrow, Koller's student Josef Redtenbacher, Gregor Mendel's teacher the botanist Eduard Fenzl (1808 Krummnußbaum-1879 Vienna) as father-in-law of J. Stefan's critic the mineralogist Gustav Tschermak and the astronomer Edmund Weiss, the anatomist Josef Hyrtl (*1810 Eisenstadt; † 1894 Perchtoldsdorf by Vienna), Koller's best friend Simon Stampfer, founding member the establisher of Viennese Polytechnic Johann Josef Prechtl (1778-1854 Vienna), Koller, founding member the pathologist later baron and president of Academy Carl Rokitansky, Mendel's Darwinist professor the paleontologist Franz Unger (1800 manor Amthof by Lučane (Gut Amthof by Leutschach, Gradišče no. 57) in Upper Styria-1870 Graz). Siting from left to right: Koller's travel companion the secretary Andreas baron Ettingshausen,

Koller's friend the president Andreas baron Baumgartner, the mineralogist Ettingshausen's son-in-law and scientific consultant of Tegetthoff's Austro-Hungarian North Pole expedition (1872-1874) Anton Schrötter von Kristelli (1802 Olomouc-1875), and Josef Maximilian Petzval. Among the missing founding members on this picture was the mineralogist Franz Xaver Maximilian Zippe (1791 Kytlice (Kittlitz) in Děčín District of Bohemian Sudetenland-1863). Koller obliviously inherited a small stature from his father and even used a great disproportional head like his older Carniolan contemporary Gabriel Gruber (Source: reference code 106.I.2146 of Archive of images at Viennese university (Archiv der Universität Wien, Bildarchiv Signatur), picture taken in 1853).

After those visits to important Doppler's Prague, Koller and his companions boarded a train for Berlin. In Berlin, Koller and his companions stayed at the hotel De Russie. On 22 July 1838 and next day they enjoyed the company of a young Encke's Swiss student, Johann Rudolf Wolf (* 1816 Fällanden near Zuri h; † 1893 Züri h), later a hief resear her of sunspots. Former Ettingshausen's Viennese student, Wolf, visited them after Joseph edler Würth (Josef, * 1817 Vienna; † 1855 Vienna) told him about Ettingshausen's arrival. Therefore, Wolf and Würth paid a respectful visit which pleased Koller. Wolf's Viennese friend Würth became a lawyer-politician of the Frankfurt Parliament in 1848, which might have been too liberal position for the priest M. Koller, but not for Koller's nephew (Wolf, Larcher, 1993, 72).

Koller first visited the Berliner observatory. Unfortunately, he did not meet the famous director J. F. Encke. The Berliner astronomer Johann Gottfried Galle (1812 Radis by Kemberg in Saxony–1910 Potsdam) was the friendly guide to Koller, his travel companions, Wolf and Würth. He accompanied them through the rooms of the excellently equipped observatory. Later in 1846 Galle made the first planned observations of the planet Neptune, which Urbain Jean Joseph Le Verrier (Leverrier) predicted by calculation. Galle's success astonished all European astronomers including Koller. Koller's predecessor, the Benedictine Dom Placidus Fixlmillner (1721 Achleuthen by Kremsmünster-1791 Kremsmünster), among the first astronomers computed the orbit of Uranus: Koller would be glad to do the same for Neptune, but Galle happened to perform that job quicker.

Koller and his fellow travelers used their days up to the 25 July 1838 to visit the local mechanicians such as the former Alexander von Humboldt's optician August Daniel Oertling (1803-1866) with his apprentice half-brother later

Londoner optician Ludwig August Friedrich Oertling (1818 Schwerin in Mecklenburg-1891). They also went to see the microscopes of the manufacturer Friedrich Wilhelm Schiek (Schick, * 1790 Herbsleben 25 km northwest of Erfurt; † 1870) and Wilhelm Hirsːhmann the senior (1777-1847). They were even received by Encke's Berliner astronomer famous for his maps of Mars and the Moon, Johann Heinrich von Mädler (1794 Berlin-1874 Hannover), by Magnus' PhD supervisor the chemist-crystallographer Eilhard Mitscherlich (1794-1863) and by the first Jewish member of the Berliner Academy Peter Theophil Riess (Gottlieb Rieß, * 1805 Berlin; † 1883). Earlier on 23 April 1834, Riess brought Ettingshausen's letter to Koller's Kremsmünster although Riess and Ettingshausen advocated the opposite theories of electricity. At Heinrich Gustav Magnus' (1802 Berlin-1870 Berlin) richly equipped cabinet Koller and his companions performed several experiments including the compression of water and various gases. They tested the sound of the Syren under water which Koller also demonstrated in his lectures in Kremsmünster.

During the night on 25 July 1838, they left Berlin and went to Hamburg, where they arrived on 27 July 1838. While walking through the downtown Hamburg, the hikers noticed a building that they immediately recognized as an observatory; it was the building of the navigation school. Its board director, the former British astronomer Carl Ludwig Christian Rümker (1788-1862), received the Habsburgian travelers in a very friendly manner and readily instructed them how to set up their own observatory.
At 11 o'clock in the morning in then Danish Altona they visited the conference counselor Heinrich Christian Schumacher, the famous editor of the "Astronomische Nachrichten". After the warmest greeting, he told the travelers that he was expecting Sir John Herschel's visit next day. He invited them to postpone the departure to London for few days, which they willingly accepted.

In the meantime, Koller and his companions met the naval clockmaker Hendrik Johan Kessels (Heinrich, 1781 Maastricht-1849 Claverham by Bristol) in then Danish now German Altona. They also visited the designers of optical instruments and syringes the brothers Adolf Repsold (1806 Hamburg-1871 Hamburg) and Georg Repsold (1804 Hamburg-1885 Hamburg) in Hamburg. They examined Schumacher's excellent collection of instruments where Koller saw the planimeter for his first time.

On 29 July at a friendly dinner at Schumacher's headquarters the travelers joined Sir John Herschel, Kessels, later Schumacher's replacement at Altona and as the editor of *Astronomische Nachrichten* (*Astronomical Notes*) Christian August Friedrich Peters (1806-1880), Rümker and the brothers Repsold.

The next day, Koller and his Habsburgian friends visited the workshop of the brothers Repsold again. There they manufactured the large meridian circle for the observatory at Pulkowa of Friedrich Georg Wilhelm von Struve (1793 Altona-1864) and Peters who joined Pulkowa staff later in 1839. Sir John Herschel detailly inspected its objective made at the famous Fraunhofer Institute in Munich. Fraunhofer died a dozen years earlier, but his optical instruments were still extremely famous. Koller soon ordered many devices from the experts he met along those journeys.

Herschel travelled to meet Gauss in Göttingen. He certainly also went to the neighboring Hanover to see his famous Jewish aunt the astronomer Caroline Lucretia Herschel (1750 Hannover-9 January 1848 Hannover) of a small stature and big heart.

After a pleasant farewell from friends in Altona and Hamburg, the travelers boarded a steamer for London in the company of Sir John Herschel on 1 August 1838 (Fellöcker, 1864, 269). The ship had to face headwinds, so the travelers did not arrive in London until 6:00 pm on 3 August.

On 4 August 1838, they visited the Adelaide Gallery, the institute of practical sciences for the exhibition of objects with many amusements founded in 1834. It offered numerous objects, painting, and drawing from the fields of physics, natural history, chemistry, and technical sciences. The Irish optician and electrical inventor Edward Marmaduke Clarke (* 1805/1806 Ireland; † 26 January 1859 Birchin Lane 4, London), was their friendly guide through the gallery. In 1837, he helped to establish the Electrical Society of London.

On 6 August, Koller and his accompanying gentlemen were at Sir John Herschel's home for breakfast, which the president of Royal astronomical society Francis Baily (1774-1844), Charles Wheatstone (* 1802; † 1875) and Charles Babbage also attended. Babbage led the guests into his house and explained to them his projected computing machine a decade after Babbage visited the Carniolan Postojna cave to get his samples of Proteus. On 17 July 1828 he signed himself at the book of Postojna guests on his way from Florence to Vienna and Berlin. From 30 August until 6 October 1828, Humphry Davy and the physician John J. Tobin were in Ljubljana and afterwards crossed

Postojna area on their way to Trieste; on their way back, they stayed in Postojna again until 13 October 1828 (Južnič, 2005, 215-216). Babbage or Zois' friend Davy might have met Koller on that occasions as Koller also loved to visit his native areas. On 22 October 1825, Koller acquired a human fish (Proteus anguinus) during his trip to Venice and donated it to his domestic Kremsmünster monastery collection; he had it captured in Magdalena Cave near Postojna (Adelsberg) in Carniola and preserved it until August 1828. After that proteus died, he brought another proteus to the monastery in October 1828; that one lived until July 1829 (Fellöcker, 1864, 145). Koller somehow inherited the habit of observation of proteus from his patron Zois, who used to acquire his human fish from his estate near Stična. The Cistercians of Stična used proteus for their weather forecasting and Koller might have done that too. In addition to many other associations, Koller as a devoted Bohinj native become a member of the Agricultural Societies (Landwirthschafts Gesellschaften) of Upper Austria, Lower Austria, and a corresponding member Carniolan Agricultural Society. He was as well a member of the Museum society (Landes-Museums Vereinein) of Upper Austria and Carniola; he joined that Carniolan society probably already in late 1839, but he was no longer a member after a considerable reduction of membership in 1856. From 6 October 1852, by the recommendation of the botanists deputy presidents Ludwig knight Heufler zu Rasen und Perdonegg (from 1865 baron Hohenbühel, * 1817 Innsbru□k; † 1885 Altenzoll by Hall in Tirol) and the Director of the Imperial Botanical Cabinet professor Dr. Eduard Fenzl (1808 Krummnußbaum by Melk in Lower Austria-1879 Vienna), Koller became a full member of the Viennese zoological-botanical society established by the researcher of Diptera (flies) Georg knight Frauenfeld (1807 Vienna-1873 Vienna) on 9 April 1851. As the leading scientist Frauenfeld sailed around the world by Novara Expedition of 1857-1859 before the frigate SMS Novara was put under the command of the Maribor born admiral Wilhelm von Tegetthoff in October 1862. The geological part of Novara Expedition of 1857-1859 was described by Franz Unger in *Wiener Zeitschrift für Kunst, Literatur und Mode* (Zoologisch-Botanischen Gesellschaft in Wien, Fellöcker, 1864, 253; Reslhuber, 1866, 381; *Verhandlungen des Zoologisch-Botanischen Vereins Wien Jahrgang 1852*, 1853 2: 89).

The following day, the Habsburgian guests visited Wheatstone's cabinet at Somersethouse, where the small shy genius Wheatstone presented his electro-magnetic telegraph, his speech machine, the device for determining the speed of light, a device for interference of sound, and many more. Then they paid a visit to the meteorologist John David Roberton (1800-1843), the assistant secretary of the Royal Society, to whom Koller gave the meteorological observations made for Herschel in Kremsmünster. From 2 April 1835 until 30 November 1835, the vacancy in the offices of Assistant Secretary and Librarian has been supplied by the appointment of Roberton as assistant secretary, at a salary of 160l. per annum, with the use of a bedroom, sitting-room, coals, and candles. His whole time was at the service of the Society. The guests visited the Society's library, where Newton's manuscripts and his first telescope were kept.

Already in 1837 and 1838, Koller's observatory accepted invitation of the president of Royal British astronomical society Baily to participate in the redefinition of the positions of the number of fixed stars to improve Baily's star catalog of the "British Association for the advancement of science". Therefore, after he left Roberton, Koller visited Baily. On that occasion Koller gave to Baily some of Koller's corrections of Baily's tables of 1826. Four months later, Koller handed over to Baily his finalized notes about 208 fixed stars newly observed at the meridian circle of Kremsmünster. In his French accompanying letter addressed to Baily, Koller excused himself for his four months long delay because he was busy in the meantime by observations of Encke's comet, 2P/Encke, which completes an orbit of the Sun once every 3.3 years as the shortest period of a reasonably bright comet. It was seen from August to December 1838 with its perihelion on 18 December 1838 Greenwich time. Koller published his observations of Encke's comet in *Astronomische Nachrichten* in 1838. Koller used additional several days to make the reductions according to Friedrich Wilhelm Bessel's first formula which Koller published in that same *Berliner Astronomisches Jahrbuch* on pages 188 and 196 in 1838. Bessel obtained his function of the first and second kind by the study of a problem of Kepler for determining the motion of three bodies moving under mutual gravitation in 1824. Koller approved Baily's opinion that the stars 337 Fornacis, 42 Virginis and 2460 Capricorni could not be retrieved anymore ("confirmérent ce que vous avez avancé; c'est à dire, que les étoiles 337 Fornacis, 42 Virginis, et 2460 Capricorni, ne peuvent plus être retrouvées"). Baily suggested the same already before Koller's visit in 1837 in his Londoner booklet entitled *An Address* on page 3 as: "337 and 2460 cannot now be found; nor is 42 Virginis sufficiently identified yet, although it was seen by previous observers".

In 1842 Koller published a list of 208 new stars observed by the Meridian district of Kremsmünster, which were reduced to the beginning of the year 1838 and to 1840.0. His results were published in the 12th volume of the *"Memoirs of the British Royal astronomical Society"* (Klemenčič, 1981, 16).

Under Koller's direction, the assistant of the observatory, P. Sigismund Fellöcker, edited and completed a table of data of stars as session VII of the Berliner academic star catalogues of Johann Franc Encke (* 1791 Hamburg; † 1865 Spandau), which had been approved by the commission set up for that purpose.

After his examination of Baily's rooms, Koller visited Babbage again, and also stayed with John Dollond's grandson George Dollond (1774-1852). Dollond presented to Koller many beautiful optical and other physical instruments, especially a new polarizing instrument designed by George Biddell Airy (1801-1892).

On 9 August 1838, the Habsburgian travelers visited Slough in Berkshire 20 miles west of central London, Windsor, and Eton. In Herschel's garden at Slough, they observed the famous giant telescope of the late Sir William Herschel. On 10 August, they admired large battery of a great fan of Slovenian Alpine tours Humphry Davy in Michael Faraday's headquarters. On 11 August, Koller visited the observatory in Greenwich.

The following days the travelers toured the curiosities of London and visited the Adelaide Gallery again, where they experimented with the microscope filled by the Hydro-Oxygen gas; it was a oxyhydrogen microscope, variously spelled with or without a hyphen, or as the hydro-oxygen or gas microscope. That oxy-hydrogen (limelight) microscope projected microscopic subjects onto a screen using limelight, magnifying lenses, and a magic lantern. It entertained audiences from 1825 through the first decades of cinema. Oxyhydrogen microscope was a sister to the magic lantern. The optician and lecturer Edward M. Clarke designed his own oxyhydrogen microscope and gave the credit to Charles Woodward. Clarke set up his instrument business close to the Adelaide Gallery, giving him exposure to those who visited the Gallery. He set up a small theatre in his establishment where he could exhibit the oxyhydrogen microscope and other instruments. He marketed his own model of the oxyhydrogen microscope, which offered relatively minor "improvements" to the work of others. In 1840, he was appointed lecturer at the Adelaide Gallery itself, giving him an opportunity to entertain audiences with his oxyhydrogen microscope and dissolving views while advertising his wares. He even developed his own dissolving view. The Adelaide Gallery was the first to purchase one of Cary and

Cooper's microscopes. A few years later, the Polytechnic ordered a larger model, also built by Cary's firm. The Adelaide Gallery's chemistry lecturer, William Maugham, was put in charge of exhibiting the "Grand Oxy-hydrogen Microscope," which was housed in the Microscope Room of the Gallery. He delighted audiences with enormously enlarged fleas, insect wings and eyes, and the microscopic inhabitants of Thames River water. Maugham also demonstrated various chemical experiments, including the melting of platinum with an oxy-hydrogen blowpipe, supposedly of his own design. Another exhibitor of the oxyhydrogen microscope at both the Adelaide Gallery and the Polytechnic was John Frederick Goddard (1795-1866). At the Royal Victoria Gallery in 1840, the distinguished physicist William Sturgeon (1783-1850) lectured on electricity, galvanism, and optics, with a demonstration of the Gallery's own oxyhydrogen microscope. Sturgeon had lectured at the Adelaide Gallery in the 1830s and was appointed Superintendent of the Royal Victoria Gallery in 1840 (Edward M. Clarke. 1837. Improvement in the mechanical arrangement of the hydro-oxygen blowpipe. *The Annals of Electricity, Magnetism, & Chemistry; and Guardian of Experimental Science* 1: 303-305).

On 15 August 1838, Koller visited the first lieutenant William Samuel Stratford (1791-1853), the superintendent of the *Nautical Almanac and Astronomical Ephemeris* for the years 1831-1853. That journal was printed in London by William Clowes and Sons, Stamford-Street; and sold by John Murray, Albemarle-Street; H.M. Stationery Office. In his volumes, Stratford published many works of Koller up to date as notes from Koller's yearbooks of the observatory at Kremsmünster. Certainly, Koller also promised to send his new observations which were indeed published in London in following years.
On 17 August 1838, Koller went to Greenwich again to visit the famous director Airy, who warmly received him. After visiting most remarkable Londoner places of interest, the scientific institutes and museums and making personal acquaintances with the experts of then natural sciences, the travelers left the metropolis of England extremely satisfied with their sixteen-day stay. On 19 August they sailed to Antwerp, after which they continued their journey to Brussels.

On 21 August 1838 in Brussels, they visited the director Adolphe Quetelet's (1796-1874) observatory and the museum; the next day, they carried out the experiments with the radiation and with the anorthoscope of Quetelet's doctoral student Joseph Plateau (1801 Brussels–1883 Ghent). Wolf tried to find Quetelet several weeks later, on 14 September and again on 16 September 1838, but with no avail. Quetelet was an authoritative boss who did not wish to play a host to greenhorns like then still unknown Wolf. After his Postdoc studies in Vienna from 29 September 1836 until 8 April 1838, Wolf travelled at the same time as Koller from late July to 31 December 1838 but in somewhat different direction: Berlin-Potsdam-Brandenburg-Magdeburg-Halberstadt-Göttingen-Gotha-Bonn-Brussels-Paris-Geneva-Nyon-Lausanne-Payerne-Bern-Zürich (Wolf, Larcher, 1993, 80; Lutstorf, 1893, 10, 12, 16, 22-24).

On the evening of 22 August, Koller and his companions left for Paris, where they arrived at 8:00 am on 24 August 1838. A month later, on 24 September 1838, Rudolf Wolf arrived there too, Wolf stopped at Parisian western suburb Saint-Cloud by St. Germain station together with Aeppli, Schmid, Kubli, two Viennese, and the other man named Koller (Wolf, Larcher, 1993, 82). Later in 1849 Wolf proved the Parisian count Buffon's idea by throwing the needle to the chessboard squares to measure $\pi = 3.1596$ after more than 5000 experiments, which one of the best Carniolan mathematicians Ivo Lah liked a lot a century later.

On 27 August, after a visit to the clockmakers Henri-Prudence Gambey (1787-1847) and Louis François Clément Breguet (* 1804 Paris; † 1883), Koller went to the meeting of the Parisian Académie des sciences, where he met François Arago, Bouvard, Félix Savary, baron Siméon-Denis Poisson, the mathematician who took over Ampère's academic post in 1836 the Protestant Jacques Charles François Sturm (1803 Geneve-1855 Paris), the chemist Michel Eugène Chevreul (1786–1889), the zoologist André Marie Constant Dumeril (Duméril, 1774–1860), the geologist Alexandre Brongniart (1770–1847), the volcanologist-botanist Jean-Baptiste Geneviève Marcellin Bory de Saint-Vincent (1778 Agen-1846 Paris), Babinet, the researcher of phase transitions baron Charles Cagniard de la Tour (1777–1859), and Arago's friend Alexander von Humboldt (1769 Berlin-1859 Berlin) who happened to be in Paris at diplomatic mission of the king Louis-Philippe. Half of a century later Koller's protegee Josef (Jožef) Stefan upgraded Charles Cagniard de la Tour's research of solid-liquid phase changes in his six treatises between the years 1889 and 1891, related to the broader context of his interest in transport phenomena, particularly liquid-gas phase changes and chemical reactions in from 1873 to 1889. Stefan's work crossed the gap between the modern data and the first experimental and analytical attempts to describe the solid-liquid phase change by the pioneers Georgius Agricola (1494–1555), J. Kepler, and Joseph Black's (* 1728; † 1799) latent heat required for phase transition in the 18th century. Black's data were published by his successor at the university of Glasgow the anti-Jacobine John Robison as *Lectures on the Elements of Chemistry, Delivered in the University of Edinburgh by the Late Joseph Black* in 1803. Their works were continued by Fourier, Thomas Andrews, Lamé, Clapeyron and Franz Ernst Neumann's (* 1798; † 1895) speᵢifiᵢheat of ᵢompounds in the 19th century. In honour of Stefan's work involving moving and free boundaries of multiphase systems at changeable borders of melting ice, the concepts of Stefan problem and the nondimensional Stefan number are widely used nowadays in treating the multiphase systems (Šarler, 2011, 10, 21; Šarler, 1995, 83; Stefan, 1889, 965). Stefan in Carinthia and Koller in Bohinj certainly knew all about ice.

On 28 August Koller and his companions visited the Parisian observatory, where the director Alexis Bouvard (1767-1843) kindly presented his instruments. On 3 September 1838 Koller spent his evening at headquarters of Jacques Babinet (1794-1872) where they carried out various interesting optical experiments focused on crystallography, rainbow, and masses of planets.
On 4 September, Félix Savart (* 1791; † 1841 Pariz) presented to the travelers his labs at the Museum du College de France. They were accompanied by Jean Claude Eugène Péclet (1793-1857). Koller also visited Claude Servais Mathias Pouillet (1790-1868) in the Conservatoire des arts et manufactures. Koller and Stefan used Pouillet's textbooks a lot. Later in 1879, Koller's protegee J. Stefan in his famous paper determined the temperature of the Sun by using the measurements of Pouillet and the experiments of Charles Soret, a professor at Geneva. Stefan needed them to calculate the first meaningful value for the temperature of the Sun amounting 5580°C.

During the following days, the travelers met Arago, newly promoted baron Augustin-Louis Cauchy, Charles Chevalier (1804-1859 Paris) and others. Cauchy had just returned from his two years of voluntary exile in Gorizia, and Ettingshausen had already consulted him about optics earlier while Cauchy was in Prague in September 1834. After those more official talks, Koller and his friends made excursions in the areas of Paris.

On 12 September, they began their return trip via Freiburg im Breisgau, where the meeting of German natural scientists and doctors was held in the next few days. They arrived to Freiburg im Breisgau on 16 September 1838. Afterwards Koller and his fellow gentlemen attended some meetings and they befriended the Berliner geologist Leopold von Buch, the Oxford geologist William Buckland (1784–1856), the Munich botanist Dr. Karl Friedrich Philipp von

Martius (1794 Erlangen–1868 Munich), Basel researcher of electrical polarization Dr. Schönbein, Heidelberg physicist acting as president of the physics-astronomy-physical geography section of the meeting Georg Wilhelm Muncke (Munck, Munke, 1772 Hilligsfeld-1847 Großkmehlen), and the Würzburg Chemist as secretary of the physics-astronomy-physical geography section of the meeting Gottfried Wilhelm Osann (1796 Weimar-1866 Würzburg). Osann was the professor of physics and chemistry at the university of Würzburg and onetime lover of Adele, the sister of philosopher Schopenhauer. The Habsburgian travelers also met the optician Wilhelm Eisenlohr (1799-1872) from the Karlsruhe Polytechnic and several others.

On 20 September Koller and his friends traveled to Augsburg, where they visited the Augustinian Augustin Stark (* 1771 Augsburg; † 8 March 1839 Augsburg), and then they went to Munich. Stark was well versed in astronomy and meteorology, but he died son afterwards.

In Munich they first visited the professor Carl August von Steinheil (1801 Alsace–1870 Munich), who presented them the telegraph he had constructed, a centrifugal machine, the heliotrope, a magnetometer, and the idea of a new heliostat. Koller and his friends enjoyed some of Munich's sightseeing during the rest of day. On 24 September, Ettingshausen and Koller attended a debate about the lightning rods at Freiburg meeting (*Bericht über die Versammlung Deutscher Naturforscher und Ärzte*, Freiburg 1838, 1840, p. 63).
On 27 September morning their host was Johann Lamont (* 1805 Corriemulzie in Scotland; † 1879 München). In addition to the beautiful optical instruments from Fraunhofer's famous workshop, Lamont presented to Koller and companions his instruments for determining the hourly averages of pressure of air and its temperature, a machine for calculating the refraction, and another tool for reducing the mean locations of the stars to apparent ones. Lamont's new micrometer on the large refractor attracted special attention as Koller needed those aids for his own research. Lamont excelled as an early observer of Neptune seven years later.

On the evening of that same day the travelers left Munich; Koller returned from Linz to his Kremsmünster, while Ettingshausen and Kunzek continued their trip to Vienna.

Director of the Observatory 1830-1847: Back Home with a Great Knowledge

Koller's journey brought the most gratifying fruits. As an excellent natural scientist himself, he has so far been left in the narrow circle. Although he already had some useful teaching aids, he was largely dependent on himself in his comparatively isolated working place. He was mostly an autodidact in theory and practice. His assistant used to be a trained mechanic who was only the mediocre; Koller had to teach everyone else how to handle the instruments that he obtained from faraway.

On his European trip, Koller saw and learned a lot of new things. As far as the funds were available, his new instruments were immediately bought, some others ordered. According to Koller's instructions, many tools were manufactured in his observatory's workshop, so that the physics cabinet soon received equipment that was appropriate to the state of the art and the progress made in the natural sciences worldwide.

Koller's personal acquaintances with the most excellent professional authorities was followed by a lively correspondence involving many of them. Most recent inventions and discoveries were discussed, the mutual doubts resolved. Koller brought home many donated books, and his foreign publishers promised to print further series of his observations.
Koller himself was extremely satisfied with the success of his excursion, and those who provided him with the funds congratulated themselves on having invested their capital at such attractive rates.

An amazing spectacle for observers in a broad path up to 204 km wide from Iberian Peninsula, Arago's southern France, Northern Italy including the central line passing south of Milano and less than one minute total eclipse observed from Venice. It was also seen in Slovenia, Austria, Southern Russia, and Northern China, but not in the USA night. The total solar eclipse lasted for 4 minutes and 5 seconds. Maximum eclipse was at 07:06:21 UT in the earliest hours of the morning. Koller's friends Airy and Francis Baily observed the total solar eclipse from northern Italy. Baily focused his attention on the solar corona and prominences and identified them as part of the Sun's atmosphere although the attempted daguerreotypes were not the best. In 1842, the Danish conference councilor C. H. Schumacher made his trip to Vienna to observe the total solar eclipse on July 8. The central line was south of Vienna and north of Buda. He paid his return visit to Koller in Kremsmünster and stayed in the monastery for several days. His sojourn in monastery headquarters became particularly important for Koller's observatory because Schumacher brought eight chronometers with him to determine the difference of the geographic longitudes between Vienna (todays value: 16.3062 = 16° 10' 58" to 16° 34' 43" E) and Kremsmünster west of Vienna (14.1315 = 14° 7' 53" E). Because of the earth's rotation, there is a close connection between longitude and chronometric time. For the difference of geographic latitudes between Vienna (48° 12' 30" N = 48.1441 N) and Kremsmünster west of Vienna (48° 3' 13" N = 48.0537 N) they measured 8' 59" while todays value is somewhat higher. That result with the other previous determinations harmonized down to the small error of 0.05 s.

Schumacher also carried a normal barometer, which was exactly balanced to that of the Parisian observatory. A multiple comparison with Koller's main barometer showed the needed correction of + 0.432 Paris lines (ligne = 2.2558 mm), which has been applied to all observations since then. The older observations were also improved.

When he took over the management of the observatory, Koller paid particular attention to the local meteorological observations. For this purpose, he purchased new instruments. Their average corrections were arranged according to then best methods. His apparatuses which had previously been used were compared with the new ones to enable the improvement of older observations. His number of daily records was increased, and the system of observations expanded, so that air pressure, temperature, vapor pressure, humidity of the air, types of clouds, velocity of clouds, density of clouds, wind direction and strength, types as well as amount of precipitation, notes of special phenomena and a brief characteristic of the everyday weather were included after 1831.

The obtained results were not left unused, because the daily, monthly, and annual averages of all observation data were calculated and published in overview tables at the end of each year.

From 1836 onwards, the observatory also took part in the meteorological observations, which were carried out on Sir John Herschel's suggestion during the times of equinoxes and solstices in 1840. Koller personally gave those fata to Herschel in 1838.

In 1840 and 1841, Koller did a very meritorious job by determining the hourly heat flow for Kremsmünster from the rich material of observations of temperatures measured from 1820 to 1839 according to Bessel's method for calculating periodic natural phenomena. Koller published it in the annual report of the "Francisco-Carolinum" museum in Linz in 1841. In an appendix to his article, he also gave a report "On the temperature of the spring water in Kremsmünster".

Koller completed a similar work in 1843 under the title "On the hourly changes in vapor pressure and humidity in the air". He used the data from ten-years of his psychrometric observations in Kremsmünster noted from 1833 to 1842. He published it again in the annual report of the "Francisco-Carolinum" Museum in Linz for the year 1843. That work was very deserving, since at that time only few labs continued such prolonged uninterrupted series of observations of haze, pressure, and relative humidity.

In 1840s, at the suggestion of the great scholars including Alexander von Humboldt and Carl Friedrich Gauss (Gauß), the investigations about the geomagnetism attracted the general attention of then natural scientists. Within a few years this branch of the natural research was developed into an independent science. Certainly, Koller did not want to stay behind. For more than a century, the Kremsmünster's scientific studies are operated with preference, in tune to the worldwide contributions to the research of the mysterious magnetic earth force. Koller founded his magnetic observatory in August 1839. At the time of its creation, it was the second in the Habsburg empire after Kreil's in Milan. Koller set up a Gaussian magnetometer in the large, beautiful observation room of the observatory with a four-pound rod needed to observe the variations in magnetic declination. Soon, in 1840, a bifilar apparatus with a 25-pound rod was used to observe the variations of the horizontal intensity.

Since 1815, determinations of magnetic declination with a declinatorium of Georg Friedrich Brander (* 1713 Regensburg; † 1783 Augsburg) have been made from time to time in Kremsmünster. Since 1832 they performed them regularly at the beginning or at the end of each month in the morning and afternoon, but that work did not provide the desired reliability due to the imperfect setup of the instruments. Koller accomplished his first observations with the Variations-Declinatorium in August 1839 and published his findings in "Gauss and Weber's results of the magnetic association" from the year 1839. That magazine entitled *Resultate aus den Beobachtungen des Magnetischen*

Vereins (*Results of the observations of Magnetic Association*) was edited by Carl Friedrich Gauss and Wilhelm Weber in 1836-1841. Starting from the October of the year 1840, Koller used both devices for observations and since then none of the data needed for Göttingen collections and the other one collected by the Londoner Royal Society has been neglected, unless there have been great obstacles.

Koller received the strongest support in his arduous work by the observatory staff, many professors, and younger Benedictines at the monastery. Their results were published in Gauss and Weber's journal and in "*Lamont's Annals for Meteorology and Earth Magnetism*".

In 1854 a complete compilation of all magnetic observations and determinations made in Kremsmünster from 1839 to 1850 (inclusive) appeared under the title: "About the magnetic observatory in Kremsmünster and the results derived from the observations from 1839 to 1850 by P. Augustin Reslhuber, director of the observatory." That work was published in Vienna by the imperial court and state printing house.

In June 1841, Koller built an iron-free station made of wood in the large garden of abbey, far from all buildings. There he permanently installed his Gaussian unifilar (single-file) magnetometer with a four-pound rod to determine the absolute declination and horizontal intensity. From time to time, absolute determinations of the magnetic quantities and comparisons of the data of this apparatus with the variation apparatuses (of probable Lamont's design) located in the observatory were carried out.

At the beginning of 1841, they began recording the data of the variation instruments three times per day at 8:00 a.m., 2:00 a.m. and 8:00 p.m. according to the local time of Göttingen. Those measurements are resumed to this day, so that Kremsmünster observatory has the longest continuous series of daily magnetic observations in the areas of former Habsburgian empire.

Before his appointment to Vienna, Koller also initiated the acquisition of Repsold's inclinatorium, which was only completed in 1848, after Koller's departure. With these instruments, Koller made multiple determinations of inclination in Vienna in 1848 and 1849 and sent them to the Kremsmünster observatory in June 1850. With gracious support from the abbot Thomas Mitterndorfer, Koller provided a portable passage instrument with 19' 5" lines of lens' aperture and a focal length of 21 inches for his iron-free magnetic observatory, including a golden pocket chronometer from H. Kessels' Altona manufacture, catalogued as No. 1304 at the Kremsmünster observatory.

Koller immensely enriched his Kremsmünster observatory, and all the scientific collections connected to it including the natural history collections, the physics cabinet, and the natural science library. He completely redesigned some branches of scientific research in his abbey. The magnetic observatory owes its new foundation to Koller, so that the era of Koller will always remain as one of the most brilliant in the annals of the college and the observatory.

With his departure to Vienna, Koller did not abandon his association with the Kremsmünster and the observatory that had become so dear to him, but remained the most devoted brother, sincere friend, counselor, and benefactor of the observatory. Every year, if professional business and health allowed, he visited his monastery and the observatory, always accompanied by his loyal friend professor S. Stampfer. A nice bunch of scientific news was presented and empowered at Kremsmünster whenever he was in the monastery.

In 1857 Koller proved to be a benefactor of the observatory by contributing a significant sum to the acquisition of a new refractor with 68 Parisian lines of lens opening, and the observatory's dialytic telescope, a type of achromatic telescope with a second correcting lens, from the Viennese workshop of Simon Plößl (Ploeßl, 1794 Vienna-1868 Vienna). It had 37 lines of lens opening and 40 inches focal length. In 1861 they followed the latest principles worked out by head of the astronomical-mechanical workshop of the polytechnic institute in Vienna Gustav Starke (1832 Vienna-1917), the son of Georg Christoph Starke (1794 Mühlhausen a. d. Unstrut in Thüringen-1865 Vienna). He made for Kremsmünster a passage instrument with a telescope. It had 24 lines of lens aperture and 24 inches of focal length. It had vertical circles with a diameter of ten (10) inches and horizontal circles with a diameter of eight (8) inches. With his gifts, Koller also enriched the library with numerous valuable works of scientific content.

Koller did not renounce science after his appointment to the civil service. Rather, through his long-term care, the research had become a necessity for him in his hours of rest, as several treatises published during this time and other treatises found in his manuscripts show clearly. In 1849 with his newly acquired PhD in philosophy Koller even appeared as a docent at the University of Vienna and read his lectures "on spherical astronomy", which were characterized by precision, clarity and beautiful language as his audience was crowded

(Klemenčič, 1981, 16). Koller also liked to attend the lectures of younger professors, partly for his own instructions, but even more to honor them in their active pursuits. So, Koller attended the lectures of professors Dr. Josef Petzval, Dr. Fr. Moth, Dr. Carl Hornstein, and Dr. Josef Stefan. In Koller's last summer semester he listened to the course of Dr. Theodor Oppolzer (1841 Prague-1886 Vienna), which he did not merely sketch out in Oppolzer's lecture rooms, but fully worked out at home and wrote down. In addition, he set up a small observatory for astronomy and meteorology in his Viennese apartment, insofar as space allowed, equipped with instruments from the best masters.

Ernst Wilhelm Ritter von Brücke (1819 Berlin-1892 Vienna) recommended J. Stefan to the counsellor in the ministry for Education and Religion (Worship) the Benedictine Marian Koller. Koller certainly already knew a lot about Stefan. So. Koller personally attended at least six semestrial courses of lectures in Stefan's classroom in 1862-1864, the last one also with the young L. Boltzmann at the same classroom. As a native Carniolan, Koller had great special sympathy for Stefan's Slovenian origin. In Klagenfurt, Stefan had graduated from the Benedictine grammar school when M. Koller's younger brother Andrej was a mayor there from 10 November 1850 to 13 April 1852, while Stefan was reading the textbooks of Koller's travel companion Kunzek. Stefan was a Klagenfurt student of Andrej's son-in-law, the Slavicist Anton Janežič, at his class of a Benedictine headmaster Karl Robida. Those Benedictine and family connections were especially pleasing to M. Koller. Upon Koller's recommendation, Stefan became the youngest professor of physics-mathematics at Viennese university and the co-director of the Physics Institute on 9 March 1863. Soon afterwards, Stefan was appointed a full member of Viennese academy of sciences as the first recipient of Ignaz Lieben Prize on 27 April 1865. Stefan took over the directorship of the Viennese Institute of Physics after Ettingshausen's retirement by imperial order on 1 October 1866, a dozen days after Koller's death. In 1875 Stefan became the secretary of the mathematical-Natural history class of Viennese Academy and from 1885 until his death he was the deputy president pf the Academy.[2]

At his ministerial office, Koller worked together with Johann Nepomuk Kleemann (Kleeman, 1808 Černovice (Tschernowitz) in Bohemian Brno-City District-1885). Kleemann used to be a professor of history and Latin philology

[2] Suess, Eduard. 1893. Die Feierliche Sitzung der Kaiserlichen Akademie der Wissenschaften, p. 256.

in Prague, Gorizia and later in Ljubljana, where he was also the headmaster of the Grammar School (k.k. Staatsgymnasium) until 1850. Kleemann replaced the deceased Koller in his office in 1867. Koller's collaborators were also Eduard Tomaschek (* 1810 Matzen in Lower Austria; † 11 December 1890 Vienna) who became a baron later, the minister's counsellor (Ministerialrat) Johann von Fontana and the cryptogamist Ludwig Samuel Joseph David Alexander baron Hohenbühel Heufler zu Rasen und Perdonegg (1817-1885). They worked under the command of Worship-Education Section chief the writer later baron Adolph knight Kriegs (1819 Vienna-1884 Vienna). Their minister of State was Richard count Beleredi (1823 Jimramov in Moravia-1902) in 1866. Beleredi replaced the liberal centrist politics of interior minister from 13 December 1860 to 26 June 1865 Anton Ritter von Schmerling (1805 Vienna-1893), but Beleredi's own conservative federation promoting the United States of Greater Austria collapsed by establishment of dualism ratified by the restored Diet of Hungary on 29 May 1867, eight months after Koller's death. As a Carniolan who knew all about the Slavic problems in dualism, Koller might not approve the Dualism, at least privately.

On 26 January 1848 Koller became a full member of Viennese Imperial Academy of Sciences (Čermelj, 1976 32; Vaniček, 1860, 3, 108; Šubic, 1902, 65; Stefan Archive Wien 106_I_3988 Documents of 9 February 1863 p. IV & 20 September 1866 pp. 11-12; Höflechner, 1994, 15). Koller was an active and rarely missing member of the Academy; only in his last two years he appeared less frequently. Therefore, many members of the Academy who were close to him mourned his death, including the professor mineralogist Zippe, Director Kreil, the president of academy baron Baumgartner and especially Koller's best friend his intimately connected professor Stampfer.

Doppler's Affair: how Koller, Stefan, Ludwig, Petzval, Ettingshausen, Mach, Exner took sides

On July 18, 1838 Ettingshausen, Marian Koller and August Kunzek von Lichton met the professor of mathematics at Polytechnic Christian Doppler (1803 Salzburg-7 March 1853 Venice) whom Petzval and Ettingshausen later forcefully expelled from Viennese institute of physics, certainly not without

Koller's administrative help. Koller personally attended and noted Petzval's lectures in 1850-1856, but not Doppler's Viennese courses. Koller also listened to quarrels between Petzval and Doppler in front of the Viennese academy, which finally ruined Doppler's health (Strnad, 2016. *Mala zgodovina Dopplerjevega pojava*. Ljubljana: DMFA; Gassauer, Theodor (* 1907), *Die wissenschaftliche Kontroverse zwischen Petzval und Doppler*, PhD 1950; Mach, 1860, 543-560; Mach, 1861 120-126). Petzval used Galilean imagination to suppose that nothing could be true if not following from his differential equations, while Doppler and his later fan E. Mach preferred more experimental approach. Eight years after their 1852-1853 dispute which killed Doppler and advanced Ettingshausen, in the Viennese lab of his doctoral advisor Ettingshausen the newly habilitated privatdocent Mach claimed that Petzval relied on theory while Doppler based at experiments, while Petzval stated that Doppler's equations cause negative and infinite musical tones as especially intriguing in Viennese worldwide center of music (Mach 1861, page 124). In that time in 1862 Petzval stopped lecturing on optic at Viennese university while he was regularly riding his black Arabian horse to his courses. During his Viennese jobs, Mach still used the concepts of atoms (Mach 1864, 69-70) before Mach took over his university chair in Graz in 1864-1867 as he was unable to get the Viennese chair of his ill supervisor Ettingshausen because Koller supported Stefan's candidature.

Of course, Koller and his travel companions avoided much contacts with the antimilitarist Bohemian mathematician Bernard Bolzano (1781–1848), who was dismissed from the new Prague university chair of philosophy of religion, even if Bolzano previously met Cauchy.

Ettingshausen's student Ernst Mach exceled with a device for checking the effect of Mach's model Doppler. Another supporter of J. Stefan, the physician-physiologist Brücke, came very harshly against Mach's design, because Brücke was a part of a complot conspiracy team which got rid of Doppler in the early years of the Viennese institute of physics. Stefan's antagonist S. Šubic supported Mach's apparatus later. In that first Mach's scientific contribution, his simple apparatus demonstrated that Doppler effect was real, at least for the sound which was so important for the Viennese musical scene with Stefan as a member of local musical organizations. Doppler's effect is oblivious today after the massive uses of trains and with all those pandemic covid-19 ambulances hurrying around; Tartini's third tone and Einstein's relativity also faced similar problems as they were on the limits of then experimental possibilities just like

Doppler's idea, which Doppler, unfortunately, tried to prove for much more complicated cosmic stars' light and not for the easy handy terrestrial sound. Mach's six-foot tube with a whistle at one end was mounted to rotate in a vertical plane. When the listener stood in the plane of the axis of rotation, no changes in pitch could be heard. But if the observer stood in the plane of rotation, fluctuations in pitch that corresponded to the speed of rotation could be heard. The application of this work to Doppler effects with light remained controversial for a while, but Mach is regarded as one of the first to realize the possibility of studying a spectrum of stars to understand their movements just like Doppler wished to do for double stars. In that way, the old conflict of the Swede Anders Jonas Ångström (1814-1874)-Petzval-Brücke-Ettingshausen against Doppler continued in Ettingshausen-Stefan-Boltzmann's antagonism against Mach. It was the old fight between new Viennese metropolis and old Prague metropolis which was vivid already in the Prague Jesuit J. Stepling's opposition to Bošković's fans who succeeded in other parts of Habsburgian monarchy except of the Slovak ex-Jesuit Maximillian Hell's (1720-1792) Viennese observatory soon taken over by Hell's helper Francis Treisnecker. Doppler's great work for the undulatory theory, which did not please his colleagues who used the somewhat different atomistics, was *Über den Einfluss der Bewegung des Fortpflanzungsmittels auf die Erscheinungen der Aether-, Luft- und Wasserwellen: Beitrag zur allgemeinen Wellenlehre.* It was published in Prague: Borrosch & André in 1847. The brothers Weber also published on that topics in 1825, while Grove (1843) and later the Benedictine Karl Robida supported the wave theory of heat which soon lost its momentum inside the leading Viennese and British mainstreams. Doppler noted the antagonism between Wilhelm Weber's ideas and Cauchy's formula:

$$F \propto r^{-2}$$

for spreading of the wave of the vibrant oxygen molecule or aether. Doppler noted that antagonism again in textbook interpretation of Johann Samuel Traugott Gehler (1751 Görlitz-1795 Leipzig) (Doppler 1847, 11) based on Poncelet's data, versus the opposite Gustav Theodor Fechner's (1801 Żarki Wielkie (Groß Särchen) in Polish Saxony-1887 Leipzig) experimental physiology (Doppler 1847, 13). Mach's clearly inherited Doppler's love for Poncelet and Fechner's bellowed optic illusions. Wilhelm Weber could be the other decisive guy even if Maxwell ideas soon surpassed Weber's schema even in J. Stefan's lectures attended by Koller in early 1860s, while Cauchy was much more Ettingshausen's model than Doppler-Mach's model, at least during

Cauchy's stay in Prague where he met Bolzano in 1834, and during Cauchy's later work in Gorizia which lasted for two years. Well, people get attached. Once you cut the umbilical cord they attached to other things: Doppler's effect is everyday reality at least for the ambulance. It is when you don't hear it, it's for you, together with a funny definition of a nice neighbourhood: a place you couldn't afford to live in (Charles Bukowski (1920-1994), *Pulp*, London: Virgin books, 2009, 95, 114, 131).

Doppler studied mathematics in Viennese polytechnic Institute in 1822-1825 and learned his philosophy in Salzburg. He mastered more of his higher mathematics, mechanics, and astronomy in Viennese University until 1829 when he became the assistant of the Viennese professor of higher mathematics and mechanics Adam baron Burg (* 1797; † 1882). Soon after Cauchy left Prague for Teplice (Teplitz) 90 km to the northwest in March 1835, in 1835 Doppler moved to the post of professor of mathematics in Prague Real School and a year later to the Prague Polytechnic. He stayed in Prague until 1847 where his research of colors of moving stars was published with the Royal Bohemian Society. Ernst Mach's father, Johann Mach, was just two years Doppler's younger and he knew all about Doppler's achievements.

Doppler researched the changing frequencies of radiated light from (double) stars because of their moving through vacuum in Prague in 1842, almost two and a half centuries after Kepler studied his snowflakes in Prague. Mach published a book on Doppler's theory in 1873. He was fascinated with Doppler's thinking and character,[3] as was the Bohemian mathematician Bolzano.

In 1852, at the Viennese Academy of Sciences, Ettingshausen assessed the discussion of the Prussian Grammar School (gymnasium) professor of mathematics Theodor Schönemann on the groups of Évariste Galois and on the balance made up of leverage. A few weeks later, in 1852, Ettingshausen rejected the theory of reflections and the refractions of light of his colleague and friend Petzval at the Viennese Academy while he also cited Exner's protegee Doppler

---

[3] Eden, Alec (1935 Essex-2011), *The Search for Christian Doppler*, Vienna: Springer, 1992; Schuster, 2011, 70; Peter Maria Schuster (* 1939 Vienna; December 26, 2019 Pöllauberg), *Moving the stars Christian Doppler, his life, his works and principle, and the world after*, Pöllauberg/Austria Hainault/UK Atascadero/USA Living Ed. 2005. Bettered Reprint: *Dem Christian Doppler zur Huldigung*. Pöllauberg/Österreich: Living Edition, 2017, Überarbeitete Auflage.

who took over his Viennese position in a competition against Ettingshausen, but Ettingshausen did not note Doppler as a friend. Exner and Doppler were in fact the leaders of hostile fractions which opposed Ettingshausen and other leaders at their mutual new Viennese academy, just like E. Mach used to be Stefan's antagonist later. There is something inherited into mutual competitions between the neighbouring metropolises of Vienna and Prague. Ettingshausen used the principle of maintaining fluctuation at the border of two media, as developed by Cauchy and his Russian student Ostrogradski. In a way, that idea of conservation resembled later Petzval's ad hoc postulate of conservation of frequency of a wave needed to refute Doppler's simple experimental formula. Petzval used to be an ordinary professor of mathematics in Buda (later part of Budapest) since 1835, and in 1837, in the race with the former Ljubljana professor Schulz, he became an ordinary professor of higher mathematics at the University of Vienna and retained his chair for four decades. As the old-fashioned Hungarian fencing and riding gentleman, he investigated photographic lenses and criticized the effect that his rival Doppler from the University of Vienna also researched. Doppler even tried to propose the academic prize for photography while Petzval was an expert for optical lenses and Ettingshausen used to be a local pioneer of photography. In 1856, Ettingshausen researched Cauchy's Gorizia equations of reflection and refraction of light in bodies with a large refraction index published in Paris in 1839. Those ideas were also used at the measurements of Ettingshausen's student, son-in-law, and predicted successor Grailich.

Ettingshausen was the closest associate of Cauchy in the Habsburg monarchy. Ettingshausen was brought to Vienna by his father and first studied Philosophy, Law and Artillery at Bombardierschule, where Jurij Vega previously lectured. Ettingshausen soon advanced his mathematical networks. In 1817 he became an associate professor of mathematics at the University of Innsbruck. From 1821 to 1835, he was a professor of higher mathematics at the University of Vienna, then he switched to the chair of physics, applied mathematics and mechanics. In 1826, together with Baumgartner, he began to publish a highly respected Viennese physics-mathematical magazine, but Koller's works were probably not widely printed there. On 4 March 1840, Ettingshausen first photographed through a microscope. In 1852 he became a professor of engineering at the Viennese Polytechnic Institute. In 1853 he took over and reorganized Doppler's Institute of Physics at the University of Vienna. Ettingshausen's and Petzval's Viennese group dismissed the Prague-initiated Doppler's effect as hoax and forced Christian Andreas Doppler (1803 Salzburg-

17 March 1853 Venice) to leave his Viennese institute of physics in three days which accelerated Doppler's death few months later despite of Prague Karl Kreil's and Bolzano's support, as they also agreed with Doppler that all bodies in cosmos move in 1843.[4] That movable cosmos could be annoying to Habsburgian catholic authorities as the expansion of Copernican doctrine, but E. Mach and the British and Netherland protestant support enabled full Doppler's posthumous rehabilitation.

Exner was unable to prevent Doppler's downfall as Koller's co-worker Franz Serafin Exner (1802 Vienna, † June 21, 1853 Padua) was already seriously sick with pneumonia and progressing lung disease. Exner went to Northern Italy in 1852 as ministerial commissioner for the Lombard-Venetian school system carrying out the study reform in those provinces which at that time were still part of the Habsburg Empire; he left Vienna together with his poor friend Doppler. Later, J. Stefan welcomed Franz Serafin Exner's son Franz Sigmund Exner as the new full member of Academy at the meeting of Academy on 8 October 1891 after Petzval died on 17 September 1891 (*Wien.Anz.* no. 19). Stefan also lectured on telephone and induction in his paper *O nekaterih poskusih z induktorjem v zemeljskem magnetnem polju* (*Über einige Versuche mit einem erdmagnetischen Inductor*), SAW 82, II, str. 1306–1313 (1880); *Wien.Anz.* 1880, pp. 262-263) and as the secretary of Academy read to the academician the second volume of journal *Monatshefte für Chemie und verwandten Theile andere Wissenschaften* on 4 March 1880 after the Academy received the journal on 1 March 1880 (*Wien.Anz.*, no. 6 p. 43). Earlier, Stefan also read there the first volume of the same journal.

(Exner-Baumgartner's) Bolzano-Doppler's Habsburg research group therefore lost their Habsburg duel with Ettingshausen-Petzval's group soon after Cauchy left Gorizia for Paris. Doppler published on magnetism, electricity, optics, and astronomy. He remains in the limelight of history of science mainly due to his discovery presented at the Royal Bohemian Society of Science on May 25, 1842 entitled *"On the colored light of the double stars and certain other stars of the heavens (Über das farbige Licht der Doppelsterne und einiger anderer Gestirne des Himmels)"*. Doppler described the shift of frequency which bears nowadays

---

[4] Strnad, Janez. 2016. Mala zgodovina Dopplerjevega pojava. Ljubljana: DMFA, pp. 15, 21, 116 This Strnad's almost singular original work of history of science was highly praised by his Student Alojz Kodre for Strnad's apparent academic excellence which was based on evil racist beliefs on the exclusivity of physicists.

his name. Doppler unfortunately applied his good theory to the wrong example of the light of the double starts which enabled the attack of Joseph Maximilian Petzval (1807 Spišská Belá in northern Slovakia-1891) who was much more versed in experimental optics and astronomy. Petzval relied on differential equation to oppose Doppler's experiments, which were based already on the Dane Ole Roemer's Parisian Observatory measurements of Jupiter's satellites in 1676.[5] The astronomer Koller certainly knew more about stars than Doppler.

On January 18, 1849, a Viennese academy's meteorological commission was set up, consisting of the full members baron Baumgartner, knight Ettingshausen, Koller's best friend S. Stampfer and Schrötter, as well as from the corresponding members Kunzek and Graz professor of applied math and physics Habsburgian Telegraph communication director named in 1850 Julius Wilhelm Gintl (1804 Prague-1883 Prague). On 15 March 1849 session of the new Viennese academy with Koller as its new member, Anton Schrötter von Kristelli became the commission's rapporteur and read his long report. Schrötter successfully proposed and established the more focussed Habsburgian meteorological-geomagnetical commission with Koller and Doppler as additional members (Schrötter 1863, 137-141). That was in fact a kind of family business: Ettingshausen's daughter Antonia Schrötter von Kristelli (1828–1916) married Ettingshausen's protegee-assistant Schrötter in 1848/1850. Baumgartner and Ettingshausen were brothers-in-law. Koller was Ettingshausen's best friend, while Doppler's mentor Exner belonged to their antagonist Viennese academic circle. The main task of that newly extended commission was to name a new director of the new Viennese meteorological-geomagnetical observatory and Koller was clever enough to push forward his old friend Kreil who happened to be also Doppler's best friend. Even if we do not now much about the direct relations between Koller and Doppler, Kreil was certainly their mutual friend.

As a reporter of commission, Schrötter's then submitted the following requests to the mathematical-naturalistic class of Academy for decision, which were subsequently approved:

---

[5] Alec Eden (* 1935). 1992, 2012. *The Search for Christian Doppler*. Wien: Springer-Verlag, 40; Strnad, Janez. 2016. Mala zgodovina Dopplerjevega pojava. Ljubljana: DMFA, pp. 35. 57; David Nolte (professor of physics and astronomy at Purdue University in West Lafayette, Indiana), The fall and rise of the Doppler effect, *Physics Today* 73/ 3 (1 March 2020): 30-35

1. To agree with the general terms of plan drawn up for the Central Station Vienna. 2. To decide that the Academy should turn to the Ministry of the Interior with the request for the Viennese own meteorologist, possibly under the title Director of the Meteorological Observatory - or rather "Institute", with an adjunct and a servant. Their duties are clear from Schrötter's draft, it only needs to be added that the meteorologists, as is the case with astronomers, would have to give regular lectures on their subject. Furthermore, the building of the necessary localities should be established as a part of this request since a meteorologist without an observatory would play a sad role. 3. The academy would provide the observatory with instruments and to maintain them for the future. Thanks to the great munificence of the Vice-President (Ettingshausen), the Academy is in the pleasant position of fulfilling this request; and instruments from various institutions would be gladly given to the new observatory, so that the burden imposed on the academy is little compared to the size and importance of the task.

All requests were unanimously approved. At the same time, the mathematical-naturalistic class of Academy named a special permanent commission for the management of all affairs relating to meteorological observations. They appointed Koller and Doppler as members of the Commission in addition to the previous ones.

In the meeting of the mathematic-naturalist Class dated October 9, 1851, they received the following letter from the then curator-guardian of the Academy and simultaneously the minster of interior in 1849-1859 baron Alexander Bach (1813 Loosdorf in Lower Austria-1893 Schöngrabern in Lower Austria): "With reference to the petition of 19 February 1850, number 214, I have the honor to inform the Imperial Academy of Sciences that His Majesty with the highest resolution of 23 July 1850 approved the establishment of a central institute in Vienna for meteorological and magnetic observations. They had determined that the staff of the institute should consist of a director with the salary of two thousand guilders, as well as one hundred and fifty guilders lodging contribution, an adjunct with the salary of eight hundred guilders and eighty gulden lodging contribution, two assistants with the salary of four hundred guilders and sixty guilders of lodging contribution; then a servant with the receipt of three hundred and sixty guilders annually. They have named the director of this institution the director of the Prague Observatory, Karl Kreil, with the rank and character of a full professor of physics at the University of Vienna and with the obligation to report on the results of his research.

On 15 March 1849 Schrötter proposed the tasks of the new observatory which already had a clear mark of Koller's expert hand:

In the interests of the history of the establishment of the currently existing k. k. Central - Institute for Meteorology and Geomagnetism in Vienna and the rectification of related facts, it is necessary to give this report at least in excerpts here by the Viennese academic Session published in Volume 2, p. 170. The subject of the report of the commission of the mathematical-scientific class of the imperial Academy is divided into two parts. The first concerns those in the railway stations and other important points of the great monarchy where the meteorological observations should be made. The second the more detailed regulations focus on the Central station in Vienna, the establishment of which the honored class has decided. Regarding the first point, the Commission has agreed to allocate the following observation sites with instruments in advance: The stations of Vienna, Olomouc, Brno (Brünn), Graz, (Koller's) Ljubljana railway station, Trieste maritime school, and the railway station of Gloggnitz (in the Neunkirchen district of Lower Austria surrounded by high mountains). Valuable observations were already made there, or at least men live there who are known to be inclined and suitable to devote themselves to this business with goal; it will only be necessary to combineand compare their instruments with the normal instruments in Vienna, to share the better instruments and to invite all gentlemen to join the draft instructions which was drawn up by director Kreil at the instigation of this honored class of Academy. These places are the following: Trient, Innsbruck, Salzburg, Troppau, Krakow, Lvov (Lemberg), Tarnow, Chernivtsi (now in Ukraine), Przemysl (now in in southeastern Poland), Melk (Mölk, with Puschl's Benedictine monastery of Lower Austria), (Koller's) Kremsmünster, Böckstein 10 km south of Salzburg, Prague, Senftenberg, Hradec Králové (Königgrätz), Deutschbrod, Leitmeritz, Pürglitz, Schlössl, Tetschen, Czaslau, Pilsen, Karlstein, Bleiberg, St. Paul, Gleichenberg, St. Lambrecht, Admont, Lienz, Stilfser Joch in South Tyrol, Klagenfurt, and Zadar (Zara). Hungary, Italy, and Croatia etc. have not been taken care off for the time being due to the current political situation (following the turmoil of Spring of nations). Only those places that are by the sea (like Zadar) must add the observations of the ebb and flow, the temperature of the sea, etc. All observers with whom the Academy will contact should use both the intended draft and with lined tables, in which the observations data are to be entered (in unified manner). The Commission has paid special attention to the second point, the establishment of the meteorological central station in Vienna, since it is of the opinion that the achievement of results which should directly promote science can only be achieved through complete centralization of all observations made in the whole scope of the monarchy. The Commission believed that it had to consider above all which observations should be made in the Central Station, whereby it was guided by the idea that the Academy should only come out with complete results corresponding to the current state of science, that it should do so where possible to go one step further, to a certain extent to better what we have neglected for so long. The (Koller's) advice given about this resulted in the following observations needed: 1. Air pressure. For this purpose, a standard barometer and a barometrograph must be set up in addition to a few ordinary portable barometers. 2. Temperature of the air. 3. Radiant heat (read on the actinometer invented by John Herschel in 1825 and by Claude Servais Mathias Pouillet). 4. Temperature of the soil at different depths. 5. Temperature of springs and the Danube. 6. The humidity of the air. 7. Rainfall. 8. Direction and strength of the winds. 9. Earthquakes. 10. Electricity of air (in atmosphere). 11. Complete magnetic observations, which so far in Habsburgian monarchy have only been made in (Kreil's) Milan, (Kreil's) Prague, (Koller's) Kremsmünster and Krakow. 12. Observations of clouds. 13. State of polarization and transparency of the atmosphere, blueness of the sky, dawn and evening redness, twilight. 14. Höfe, side suns and secondary moons etc., rainbows, fog, smoke seen above. 15. Northern lights. 16. Sunspots, zodiacal light. 17. Meteors, shooting stars. 18. Observations of vegetation (according to Quetelet's Instructions etc. with the additions from Kreil's Prague collaborator and later Keil's Viennese adjunct Karl Fritsch (* 1812 Prague; † 1879 Salzburg). 19. Observations on the periodic phenomena in the animal kingdom, such as the migration of fish, birds, metamorphoses of insects, etc., in accord with the observations made in (Quetelet's) Brussels. 20. Periodic phenomena in human social life, such as: dominant diseases, mortality etc. using the instructions given by the Pasteur's predecessor, forerunner of cell theory professor of anatomy at the University of Liege Theodor Schwann (* 1810 Neuss, † 1882 Köln), etc. 21. Temporary determination of the chemical conditions of the atmosphere. 22. Publication of all the observations made both in Vienna and in the other stations, in such a detail as is necessary for scientific use, and in general overviews that allow easy use of them in wider circles.

It could not have been the intention of the Commission to include in this report all the details about the instruments, the observations or methods of observation, the way in which the various data were obtained, compiled and published, etc., and it is therefore a long way off to believe that it gives more than the very outlines of what the imperial academy must accomplish if it is to be worthy of joining the ranks of similar institutions that have long existed in England, Belgium, Russia and other countries; but it will be entirely sufficient to show that the solution of the task which the academy has set itself is only possible if a specially appointed scholar devotes all his energies exclusively to it. Hitherto meteorology has for the most part been treated only as an appendix to physics and astronomy, and it is mostly the astronomers who cultivate it in practice, to which it owes its greatest extensions. There is probably no doubt that both sciences are closely related, roughly like physics and chemistry. The meteorologist particularly requires a lot of data from the astronomer and the methods of calculation and types of observation in both sciences are made according to the same principles, so that only a man with a thorough astronomical training can adequately fill his post as meteorologist in the sense just suggested. But it cannot be that both branches of human knowledge, which are so important and now so extensive, could be pursued by one individual, no more than is and can be the case with physics and chemistry. Experience also teaches that those astronomers, however talented and hardworking, who prefer to devote their attention to meteorology, are unable to serve astronomy in the same way. The rich material, which will surely soon flow into our central point in Vienna, would be dead capital, only suitable for increasing the mass of valuable data that is still unused in our offices, if the leaning spirit that connects them is missing. That makes science and industry accessible and useful. In meteorology isolated observations are of little value even more than in many other sciences; only when they are present in such a number and of such quality that one can derive laws from them does their usefulness and correctness appear even to the layman. It is therefore necessary to arrange the plan for such observations in such a way that it connects as much as possible with that of other countries, so that the results obtained in the Habsburgian monarchy can easily be incorporated into the meteorological network which has already spread across Europe, extended a great part of Asia and America, and even over individual points of Africa. The Commission believes that it must present the honored class of Academy with its views on the Viennese place which it considers most suitable for making meteorological observations. The choice could not be long in doubt, since the polytechnical institute is, according to its location, as well suited for this purpose as it is the only possible in such a large city as Vienna. This, however, requires that the facilities necessary for these observations should be established there. Fortunately, the construction of the building is such that it has the necessary terrace, etc. There are no difficulties, and the costs for rebuilding cannot be very considerable (Schrötter, *Wien.Ber.* 1849, 169-175; Schrötter, 1863).

Therefore, Koller and Doppler as the new member of academic commission together supported their mutual friend Kreil as the new director. The astronomer-meteorologist Karl Kreil, who worked in Prague together with his friend Doppler, mentioned in a letter written shortly after Doppler's death (27th April 1853) to the Secretary General of the Viennese Imperial Academy of Sciences Anton Schrötter: "Doppler's ill health began already in Prague, where numerous lectures before large number of listeners in overcrowded rooms began to take toll of his health." Karl Kreil supported the publication of his friend Doppler's lecture of May 25, 1842 on the colors of light of binary stars and some other celestial bodies before the Royal Society in Prague in the Astronomical and Meteorological Proceedings of 1844 even if Kreil failed to attend the talk in person. Kreil added a popular article on Doppler phenomenon. In 1845, Bernard Bolzano published an article in *Annalen der Physik* entitled Some remarks on the new theory of Mr. Professor Ch. Doppler. In the minutes of the Meeting of the Mathematical Section of the Royal Bohemian Society of Sciences held on 5th November 1843 the member František Palacký (Palacky) proposed Doppler as the associate member with support of Bolzano and several other members. On 31 December 1843 under the chairmanship of Bolzano, with Kreil and Exner also being among the other ten members present, Christian Doppler was elected as a full member of that Bohemian Society by nine votes with one against. In 1851 Kreil married Mathilde von Pflügl and later took care of Doppler's family after Doppler's death as his closest friend. Kreil become the brother-in-law of medical practitioner from Linz dr. Hermann von Pflugl (Pflügl 1814-1882), who as a widower married Doppler's eldest daughter Mathilde Doppler in 1862, just before Mathilde's caretaker and subsequently also a relative Kreil died on 21 December 1862.

In January 1852 in Venice, Doppler wrote by his own hand his last work on the possibility of determining of the number and the absolute distance of the atoms, as well as their reciprocal force of attraction. Doppler firstly illustrated his point with the cases of various simple solid bodies. There, Doppler proved that there was no real difference between his and Ettingshausen's atomism: the complot against Doppler-Exner's Viennese academical fraction had therefore purely personal goals. Some politics might be involved, but nobody except Bolzano was a pacifist or open critic of Catholicism (Alec Eden (* 1935). 1992, 2012. *The Search for Christian Doppler.* Wien: Springer-Verlag, 27, 32, 39, 67; Ewald Hiebl, Maurizio Musso (ed.), *Christian Doppler: Life and Work, Principle and Applications: Proceedings of the Commemorative Symposia in 2003, Salzburg, Prague, Vienna, Venice*, STARNA Ges.m.b.H., 2007, 39;

Bolzano, 1843, Ein Paar Bemerkungen tiber die neue Theorie in Herrn Professor Ch. Doppler's Schrift: "Uber das farbige Licht der Doppelsterne und einiger anderer Gestirne des Himmels." *Annal. der Physik und Chemie* 60: 83; Bolzano, 1847, Christian Doppler's neueste Leistungen auf dem Gebiet der physikalischen Apparatenlehre, Akustik, Optik und optischen Astronomie. *Annal. der Physik und Chemie*, 72: 530-555).

Kreil was Doppler's best friend and early follower of Doppler's effect, while Kreil was also the oldest friend and collaborator of Koller, who was a close friendly travel companion of Ettingshausen. Therefore, Koller was caught between two fires: since he did not publish his opinion about the quarrel by which Ettingshausen ousted Doppler, it is hard to determine whom Koller supported. Anyhow, Koller closely collaborated with Doppler to bring Kreil to Vienna. Exner-Doppler's quarrel against Ettingshausen-Petzval certainly mirrored the competition between Palacky's nationalistic semi-Protestant Prague supported by Bolzano and maybe even Exner, and the imperial Catholic Vienna with some Petzval's Budapest premises focusing the later Dual Monarchy of 1867 which diminished the anti-Habsburg Hungarian revolution of 1848–1849 and the unified imperial customs system established on 1 October 1851. The local imperial Catholic policy of Ettingshausen's friend Metternich, ousted in mid-March 1848, was involved after Metternich returned to Vienna in September 1851 just before Petzval publicly attacked Doppler. But there was no antagonists' difference in their mutual support of kinetic atomism which was later revoked only by Ernst Mach in Prague and by several less influential Habsburgians including Puschl and Stefan's teacher Robida.

The stage of that gravest Viennese scientific quarrel before Boltzmann ousted Mach and Nazis ousted S. Freud was prepared at a meeting of the Viennese academy on 22 January 1852. There Petzval read a paper criticizing Doppler's theory after Doppler defended his optics against the Jesuit Benedetto Sestini (Sestri, * 1816 Firenze; SJ 1836; † 1890 Frederick, Maryland). Andrea Caraffa's (1789 Torri; SJ; † 1845 Tivoli) student Sestini taught at Collegio Romano but flied from Spring of Nation to Georgetown's Woodstock College in 1848, just like Doppler escaped from the revolutionary Hungarians to Vienna. Sestini collaborated with Antonio Gross and his colleague Ignazio Cugnoni (1822 Roma-1903) worked as an observer and calculator in Collegio Romano in 1840-1845, but later excelled mostly as the photographer, architect, and engineer. As a professional astronomer Sestini argued that the physicist Doppler departed from the principles of Young, Fresnel and Arago regarding Baily's catalogue made with the help of the professional astronomer Koller. In his own way, Doppler challenged his foreign branches of sciences including astronomy and Petzval's mathematical optics, just like the violinist G. Tartini humiliated D'Alembert with his mathematical explanations of *terzo tuono* a century earlier. Tartini had to wait long for Helmholtz to prove his point, while Doppler was soon successful abroad, but Stefan was still somewhat afraid of Petzval and therefore avoided any direct mentions of Doppler's effect in Stefan's university lectures in early 1860s, a decade after Doppler passed away.

Few decades after Doppler, Einstein was more successful while challenging astronomers' anomalies in the orbit of Mercury because the astronomer Edington made use of post WW1 thirst for new strong belief in science (Benedetto Sestini, *Memoria sopra i colori delle stelle del Catalogo di Baily, osservati dal P.B. Sestini*, Roma: Marini, 1845: pp. IV, VI & *Memoria seconda*. 1847: pp. 12-14; Doppler, *Über das farbige Licht der Doppelsterne und einiger anderer Gestirne des Himmels*, Prague 1842; Anonymous, *L'Époque journal complet et universel Tous les journaux en un seul - on souscrit ici* (printed in October 1845-February 1847 under the ultra-conservative patronage of the minister of foreign affairs François Guizot), Paris, no. 133; Christian Andreas Doppler: Einige w*eitere Mittheilungen meine Theorie des farbigen Lichtes der Doppelsterne etc. betreffend*, 18 July 1850, *Wien.Ber.*, 6:. Reprint: 1850 *Annalen der Physik*, 3rd series, volume 81 (157): pp. 270-275; Christian Andreas Doppler: *Weitere Mittheilungen meine Theorie des farbigen Lichtes der Doppelsterne betreffend*, 26 July 1852, *Wien.Ber.*, 8/1: pp. 91–97, here page 94. Reprint: 1852 *Annalen der Physik*, Volume 161, Issue 3 pp. 371-378).

The focus on (Petzval's) optical-astronomical cases might have been Doppler's tactical error as defending Doppler's acoustic was much more promising with new railways onboard while the astronomical optics of Doppler's effects with enormous velocities looked like the science fiction. In the next paper of the same volume of *Wien.Ber.*, Mohs' student and academic successor the mineralogist Wilhelm Karl knight Haidinger (1795-1871 Vienna), published his *Ueber den Zusammenhang der Körperfarben, oder des farbig durchgelassenen, und der Oberflächenfarben, oder des farbig zurückgeworfenen Lichtes gewisser Körper*, pp. 97–133. He told the audience about Brewster's, Liebig's collaborator Wöhler's, John Herschel's and similar polarization or even fluorescent experiments to conclude with Cauchy's theory.

Enters the fencing expert Joseph Maximilian Petzval with his another attacking paper entitled: *Ueber ein allgemeines Princip der Undulationslehre: Gesetz der Erhaltung der Schwingungsdauer*, pp. 134–156. Ettingshausen was still lobbying in the back against Exner-Doppler tandem inside their never finished rivalry, which later brought additional Viennese physics institute to Exner's son, just like the eternal rivals IJS and nearby Physics faculty in Ljubljana started with the ousting of Anton Peterlin a century later. Petzval began by dividing the science to big one (his own) and the small one (Doppler's), just as there is a big and small war. Petzval ended with high towers needed to protect even the poet Schiller from harm. The great ocean of knowledge also has its dangerous spots, its cliffs, and shallows. If, therefore, one becomes aware that many of those who are keen on knowledge are shipwrecked on the cliff of a certain error, then it is time to raise a simple and great truth to the rank of a rule. Therefore, Petzval proposed against Doppler his and hoc coined: "principle of preservation of the oscillation duration (frequency)."

At a later meeting of Viennese academy on 21 May 1852 under the presidency of Andreas von Baumgartner, about sixty members and guests including Koller assembled to hear both sides of the argument, just two weeks before Koller's almost fatal accident on 3 June 1852 which might have been caused by Koller's absentminded distraction as he was thinking about the quarrels among his personal or at least academic friends and he consequently forgot to watch his step. Petzval began rhetorically by citing only Fresnel, Cauchy, and Poisson on the beginning (Petzval, 1852 *Wien.Ber.* page 568; Eden 1992, 58), just like Doppler quoted with caution J. Herschel's doubts about longitudinal waves in aether in Prague on 25 May 1842 (Doppler, 1842, On the Coloured Light of the Double Stars and Certain Other Stars of the Heavens, read at the *Meeting of the Section of Natural Sciences of Royal Bohemian Society for Sciences in Prague* on 25 May 1842, page 4 (# 2)). Petzval had some doubts in kinetic theory of mater while stating that the heat is either a substance and not a modification of the body, or the law of molecular action is only and exclusively Newton's, namely force as the inverse quadratic ratio of distance, for since I am only concerned with the matter (Petzval 1852 *Wien.Ber.* page 570), and not aether. Therefore, in the case of kinetic atomistic, Doppler was much closed to Ettingshausen than Petzval.

Petzval ended his attack by quoting Christoph Buys Ballot's Utrecht acoustic train experiment performed in June 1845 (*Wien.Ber.* page 586) but did not mention Doppler in that continuation of Petzval's 15 January 1852 attack published in February on page 134 of *Wien.Ber.* Doppler also answered rhetorically again by citing that same Christoph Buys Ballot's June 1845 Utrecht acoustic locomotive experiment (Doppler 1852 *Wien.Ber.* page 591). Christoph Buys Ballot might support Doppler's effect in acoustic, but not in optics where the available experimental velocities were far from the speed of light even in later Einstein's times. Ettingshausen finished their vigorous debate by his own two pages while citing Petzval as a friend and Doppler just as a colleague: Ettingshausen was obliviously more interested in in obtaining Doppler's position which was suddenly available because Doppler's and Exner's lungs were failing, while Ettingshausen might be sympathetic to Doppler's ideas anyway as he soon encouraged E. Mach to research them.

Ettingshausen and Petzval defeated Doppler, but Doppler was rehabilitated by Ernst Mach and many others while Ettingshausen and Petzval's fame and old age misfortunes never reached Doppler's Parnassus as Doppler's name is the most widely quoted today. Doppler kindly revenged on some spiritual ways to his once victorious opponents. Doppler in vain tried to get the teaching post even in Ljubljana or USA until he got his first profitable post as the teacher at Prague Polytechnic Institute with Bolzano's help only in 1835. Cauchy was in Prague in those times from August 1833 to March 1835. Later, Cauchy stayed with Bourbon court at 90 km north-western Bohemian Teplice (Teplitz) until October 1836 when they headed southwest to Gorizia.

After he returned to Paris, Cauchy was unable to provide much further direct help to Doppler while Bolzano died in 1848. In his preserved lectures of 1826-1839, it was too early for Koller's quotations of Doppler's novelty of 1842-1852, but even Stefan failed to do so in early 1860s, at least according to Koller's notes. E. Mach supported Doppler already in 1860s. Only in 1873 and 1878 Ettingshausen convinced E. Mach to experimentally prove Doppler's ideas which led to Mach's shock wave photography in Rijeka Torpedo factory later.

Reformer of Habsburgian Technical Education

It is not easy to determine Koller's role in final Doppler's ousting, although it must have been decisive as Koller's position in state ministry was similar to Exner's. Likewise, it is extremely difficult to fully depict Koller's effectiveness in his civil service because he avoided published reminiscences. Koller himself kept his official secret sacred even to his most trusted friends and did not talk about state affairs at all in any ordinary conversation. So much is certain that his appeal to the honorable post promptly received undivided applause: "That is the right man!" Was the general voice: "Koller is an excellent teacher who would have been an ornament for any university; as head of a study and educational institution, he knows the best all goods as well as much of the shortcomings of the previous teaching system. Koller as the man of exact sciences knows best what science needs."

Koller took over his lively part of the great and benevolent change that the education system in Habsburg monarchy had experienced since 1849, following the revolutionary Spring of Nations. In the reorganized Ministry of Education, he had to cultivate a field with a new Habsburgian seed: the foundation of technical the secondary schools and their upgradation at university level.

In 1851, Koller's six-class secondary real schools came into being. Their creation was almost exclusively the work of Koller, which he cultivated like his foster child until the end of his life and which he endeavored to bring to perfection. In the huge Habsburgian empire, partly from state funds, partly through contributions from corporations, Koller's new secondary schools quickly emerged as the first and surest basis for the prosperous upswing of Habsburgian industry.

Koller's utmost care was used to develop a proficient apprenticeship, to acquire teaching aids, and to create suitable localities for his real schools. He made it happen in the course of a decade, as he brought into good motion the educational system which was previously thought as hardly possible.

Even though calls for reform of these new schools were rightly made later, they were certainly a work that should not be underestimated at that time, when teaching was particularly poor in Habsburg monarchy, especially in that technical branch. Who knows the fruits that these schools have borne during their fifteen years of Koller's managing, who knows how many thousands of young, well-trained workers were brought into the various technical and industrial establishments by them? Even Nikola Tesla was one of those youngsters; certainly, he was just a twelve years old real school student in the time of Koller's death. Only posteriority will be able to appreciate the services which Koller rendered to Habsburgian prosperity. The fact that these merits were honored even in the highest places is demonstrated by the distinction with which the emperor Franz Joseph awarded the Knight's Cross of the Order of Leopold to Koller. In the lecture of then Minister of Culture and Education, the Count Leo Thun, the merits which Koller has "brilliantly earned for the rapid growth of secondary schools in Habsburgian monarchy" were expressly emphasized.

Koller conscientiously tried not to miss anything, but rather to move the started work quickly forward. He proved his devotions after he misfortunately broke his right upper arm by a fall from a moving train on 3 June 1852. He was unable to write for several months. In order not to fall behind with his work, which was becoming very urgent at that time, he kept at his own expense an official who wrote by Koller's dictate.

Koller's attention was enough for his first step, but the establishment of secondary schools was not enough. He understood very well that these schools are of great advantage and blessing for one of the most numerous parts of society, the middle class. Should industry and trade be thoroughly and universally raised, then those who are not suitable for a higher education, or who cannot obtain it because of lack of resources, must also be taken care of. He therefore arranged that secondary education. The practical courses for craftsmen and apprentices in the evenings, on Sundays and on festive days, so-called lower technical schools, were developed in many places, including the cities where Koller did not establish his new real schools.

After the secondary schools were organized and in good working order, Koller reorganized their upgrade, the Viennese Polytechnic.

The Viennese Polytechnic Institute was involved in the revolutionary events of 1848/49 with many of the students and professors organised themselves into their own Technical Corps in the "Academic Legion" of the militia. When teaching resumed in October 1849, the longstanding Director and founder Prechtl retired. His successor was the Professor of Mechanics and the Science of Machines, Adam knight Burg (1797-1882). In 1851 some of the students and assistants at the institute, including Burg's personal assistant Cäsar Bezard, caused a treasonous scandal. Bezard was convicted and executed in 1853 and Burg was relieved of his post in 1852. The Polytechnic Institute was now regarded as "politically unreliable". I was put under military leadership, initially in 1852/53 with Colonel Christian knight Platzer as Director, and then with the colonel Karl baron Smola from 1853 to 1858. In September 1858 a civilian scientist again became Director of the institution. He was the former Professor of Mineralogy at the Joanneum in Graz, Georg Haltmeyer (1803-1867). After much consultation and negotiation, Koller proposed a complete reorganisation of the institute, which was recorded in the new organisational statute approved by Emperor Franz Josef I on 17 October 1865. This new statute provided for the institute to be led not by a Director appointed by the government but by a Council of Professors, who would elect a Rector from among themselves. The Commercial Department was disbanded after at the Real Schools it had already been hived off in 1851. The Technical Department was restructured to form a General Department and four specialist schools: for roadbuilding and hydraulic engineering (Engineering), structural engineering (Construction), mechanical engineering (Mechanical Engineering) and technical chemistry (Chemical-Technical School). Furthermore, the Matura was made a compulsory requirement for studying at the institute and a final rigorous examination introduced for the first time, initially as voluntary. The new Koller's statute came into force for the 1866/67 academic year. Several months after Koller passed away, on 3 November 1866 the professor of higher geodesy and the spherical astronomy Joseph Philip Herr (1819 Vienna-1884) was elected as the first rector. Herr used to be an assistant and coauthor of Koller's best friend Simon Stampfer; therefore, Herr was obliviously Koller's (posthumous) choice, also because Herr was the editor of the influential *Magazine of the Austrian Association of Engineers and Architects (Zeitschrift des Österreichischen Ingenieur- und Architektenvereins)* from 1858 to 1867.

Koller transformed the old Prechtl's Viennese Polytechnic into a technical university functioning above the secondary schools for further training in the various subjects. Koller has been perfecting that task in his last years with tireless activity and prudence, great preference, and the happiest success. On November 28, 1860, the professorial college of the polytechnic handed over the reorganization statute, which had been drawn up to the ministry. The negotiations started, the statute was sanctioned by the emperor in the summer of 1865, and the college of professors had to submit their requests. Applications came to the ministry. Each one had to go through Koller's hands. In facts, nothing was too much for the power or his mind; he made his decisions with the usual sure tact and presented them to his minister. More than twenty professors had already been appointed, and it was already determined that the reorganized institute would come into being on 1 October 1866. Only a few appointments were missing, and the new building was ready and Koller's lively wish was fulfilled. However, he was deprived of the joy of celebrating the beautiful festive day of its opening.

Just as no new creation is perfect in all its parts from the very beginning, so it was with Koller's real schools; a work of that size and scope, which had so deep an impact on practical life, had to be put to the test at a time. The economic progress of the nations involved forms the rapid succession of wonderful discoveries and inventions; they demanded new approaches. Experience has shown that there are many things that need to be changed and improved in the system of secondary schools; Koller wanted to do that too. Voices could be heard saying that Koller was opposed to an additional reform of secondary schools; but this was not the case. Already at the beginning of 1865, a reorganization plan for the secondary schools was drawn up in the teaching council. After Koller had made the appropriate changes, he left them alone for the time being, since he wanted the Viennese polytechnic to be reorganized earlier. Koller said several times to his friends: "I would still like to solve these two tasks, the reorganization of the polytechnic and the reform of secondary real schools, then I'll retire to the calm of my dear Kremsmünster." He happily solved the first one. The second task was left to his successor for execution.

The minister entrusted Koller with several honorary orders and tasks. So, he went to Trieste to subject the local academy for nautical science and trade to an in-depth inspection in 1854; Koller put detailed elaborate on the necessary reorganization of this institution to the highest place. He not only received the fullest recognition but planed the reorganization which was carried out.

In 1857 Koller examined the completely organized Joseph's Polytechnic in Pest, and the upper secondary schools in Pest, Buda (Ofen) and Bratislava (Preßburg). From 31 October 1857 until the summer of 1861 Simon Šubic taught at the higher real school of Pest. In 1859/60 he was joined by the future Nikola Tesla's teacher Martin Sekulić. All of them benefited from Koller's advices.

In the same year 1857, at the request of the Minister of the Interior, Koller was chosen by the Ministry of Education as a representative for advising the establishment of secondary schools for agriculture and forestry. In those fields, his younger brother Jožef Koller proved to be a leading expert.

In 1859, Koller went to Graz on behalf of his minister, where he inspected the Joanneum established in 1811. The former professor of mathematics in Ljubljana Karl Hummel (1801-1879) taught physic at Graz University as a dean in 1859/60. Koller also carefully scrutinized the local Graz upper secondary school.

During such inspections, Koller was not satisfied with mere personal visits and inspection, but always provided precise and comprehensive reports on the results of the findings, which were always written by his hand.

These reports confirm the beneficial influence of not having a bureaucratic official at the head of the educational authority, but an excellent schoolman who knows how to appreciate the special and often very peculiar circumstances with his right scale.

Koller also played a key role in founding the k. k. Central Institute for Meteorology and Earth Magnetism in Vienna, and continually devoted the greatest attention to that most praiseworthy institute; the institution owes him the purchase of the library of its first director, Koller's friend Karl Kreil (1798–1862). Koller organized the introduction of daily telegraphic weather reports from several well-chosen observation stations in Habsburgian monarchy including Dežman's post in Ljubljana. Koller also enabled obtaining a grant to continued printing of the meteorological yearbooks, etc.

Koller's services to the state, especially to education and science, received the most general recognition, including the trust and distinction by the emperor, the Ministry of State and the flattering respect and veneration on the part of the teaching staff under his control. After Koller's death, the general mourning and the warm, truly touching obituaries filled all public papers.

Koller was a very noble, honest character. As a priest, as a teacher, a scholar, and a statesman, he always proved himself in his work as a person in the noblest sense of the word. His embodied humanity was the driving force behind his actions. Despite all his merits and awards, in his influential position, the simplest modesty adorned him. He was pushing nowhere, while his undemanding attitude forced everyone to respect. He was strict against himself while he was very mild in judging others; justice paired with equity was the guiding principle of his actions. Not servile to those above his official position, he was very condescending to everyone, friendly, and ready to serve. He was always a generous benefactor to his relatives, particularly to the poor students; hence he was generally respected. Marian Koller's name had the best sound anywhere, honorable in the history of the catholic Church, the State, the science and especially in the annals of his domestic monastery.

Cooperation with scholars related to his native Carniola

Slovene or even Slavic nationalism was a undesirable idea dogging any advances of Emperor Franz Joseph's officials after the Spring of Nations. Neither Koller nor J. Stefan liked to stand out in that direction. Of course, the support of Slovene and Carniolan scientists was within Koller's official possibilities, while Stefan was more than unhelping compatriot of Graz professor Simon Šubic.

Koller's collaboration with natural scientists in Carniola began with his patron Žiga Zois. Koller's Ljubljana teacher Johann Philip Neumann (* 1774; † 1849) used to be the first professor of physics at the Ljubljana Faculty of Philosophy who was never a monk. He later lectured as librarian at the Viennese Polytechnic, reorganized by Koller after Neumann retired in 1844. Neumann is today even more famous as the commissioner of music of Koller's few weeks younger peer Franz Schubert (* January 1797) who tried in vain to get a job in Ljubljana as a youngster.

In Vienna, Koller met again his former Ljubljana professor Jožef Jenko, as they shared a mutual friendship with Kopitar. Jenko took over his chair at Viennese university on 13 December 1819 to teach Leopold Schulz edler Strassnitzky (1803–1852) and maybe even to instruct Koller again. Janez Krstnik Kersnik and the Napoleonic Governor Duke of Dubrovnik Auguste-Frédéric-Louis Viesse de Marmont were the main Ljubljana scholars of the Illyrian provinces who frequented the baron Zois' household like Koller himself did. Kersnik's Ljubljana collaborator, Karl Hummel, married a girl from important noble family in Ljubljana eight years before he took over the prestigious Graz university chair of physics. He and E. Mach planned Simon Šubic as Hummel's replacement, but it never happened as Boltzmann used to be J. Stefan's candidate. Unfortunately, Simon Šubic did not receive Koller's special grace, although he was somewhat supported by Koller's friend Andreas von Baumgartner and Kreil's replacement Karl Jelinek (1822 Brno-1876 Vienna) who founded the *Österreichische Gesellschaft für Meteorologie* (Austrian Society of Meteorology) with Koller, Reslhuber and Littrow among its founding members and Dežman, Fellöcker, Reitlinger, and Stefan's collaborator the Viennese Theresianum teaching candidate Luca Žerjau residing at Wieden no. 17 among ordinary members on 16 November 1865.

Koller certainly supported his younger fellows the Benedictine physicists like Karl Robida (baptized Lucas, * 1804 Mala Vas on banks of Sava river by Ježica now north part of Ljubljana; OSB; † 1877 Klagenfurt) and Karl Puschl (Puschel, baptized Josef, * 1825 Wolfsbach in Lower Austria; OSB 1846 Seitenstetten; † 1912 Seitenstetten in Lower Austria 49 km east of Kremsmünster with Steyer city in between). Puschl taught mathematics, physics, and other subjects at monastic Benedictine gymnasiums in Melk and Seitenstetten between 1853-1871. Later, Puschl became gradually blind but continued to publish about aether mostly in *Wien.Ber.* (acts of Viennese academy) even if his ideas opposed the kinetical atomism of the secretary of that academy, J. Stefan. Robida and Puschl soon opposed mainstream physics of Ettingshausen and Stefan which might have echoed some hidden Benedictines' theological suspicion against then victorious Stefan's kinetical atomism. Koller probably did not go that far as ge used to be a born diplomat.

The extremely radical Catholic scholars such as Cauchy during his stay in Gorizia in 1836-1838, his protégé Franc knight Močnik and Kersnik's assistant the chemist Mihael Peternel (1808-1884) as first director of the Ljubljana Real high school in 1852-1860 founded by Koller (Klemenčič, 1981, 16) were especially close to Koller's heart because of his Catholic political orientation. Perhaps Koller met Cauchy in Prague or in Brader Bohemia in August 1833-1836 or in next months in Gorizia sometimes between 1836-1838. They certainly contacted one another in Paris in Autumn 1838. In fact, Cauchy mostly supported the Jesuits which were even more right wingers than the Benedictines. Cauchy's friend the former Jesuit Moigno published the translated articles of Koller's protegee Stefan in Paris.

More liberal intellectuals may have missed the true love of a Benedictine Koller, among them Ernst Mach's father under Gorjanci in Lower Carniola or professor of mathematics and astronomy in Ljubljana with Močnik and Peternel among his best students Leopold Schulz edler Strassnitzki as a member of parliament in Frankfurt during the Spring of Nations. Schulz and Koller were both members of Carniolan agricultural society and Schultz took over the chair of elementary mathematics at the Viennese Polytechnic on 29 January 1843, just before Koller began to reform it.

The liberal Germanophile meteorologists such as the principal of the Ljubljana grammar school and for some time also of real school Heinrich Mitteis, Karl Dežman or his sister Serafina Dežman were not Koller's best choices, although he collaborated with them as the initiator of the Viennese Central Institute of Meteorology and Earth Magnetism. Later, Koller's nephew became the leading local liberal German politician.

Ignac Klemenčič (1853-1901) might have been Boltzmann's successor in Graz if Koller had lived long enough; Koller's replacement Kleeman as an opportunist quarrelling with leading Carniolan politician Janez Bleiweis, of course, rejected even the recommendations of Klemenčič supported by Boltzmann and Fran Šuklje (1849 Ljubljana-1935 castle Kamen in Kandija by Novo Mesto). Kleeman stated that the Graz professorship was a political position to which Klemenčič, as a Slovenian nationalist, did not fit well enough. Kleeman certainly had a point as the nationalism gradually deteriorated the Habsburgian monarchy.

Boltzmann's brother-in-law the Styrian Slovene Anton Šantel (1845–1920) and Kranj-Ljubljana professor Michael Wurner (1829-1891) were also too young for Koller's beneficial influence. J. Stefan's Viennese collaborator Nicolò Vlacovich (* 1832; † 1890) lectured in Koper and after autumn 1863 in Trieste, where Koller had previously inspected the Academy of Nautical Science and Trade in 1854.

Koller's correspondent the count Francis Joseph Hanibal Hohenwart married the baroness Margaret Erberg, the younger sister of Joseph Kalasanc Erberg. Like his father's cousin Sigmund Hohenwart, Francis Jožef Hanibal Hohenwart often went to the climbing tours. He led a Carniolan provincial museum as president of the museum's curators in Ljubljana, and president of the Carniolan Agricultural and useful arts Society between 1827 and 1834; the councillor of that same Carniolan Agricultural and useful arts Society was already his father.[6] Koller was also its member. Because of his fooling around by his semi-noble behaviour, Francis Jožef Hanibal Hohenwart had to swallow many naughty jokes while he played Tulpenheim in the Linhart's theatre comedy "Mayor's

---

[6] SBL, 1:330-331; Smole, 1982, 620-621; Kovačič, 2002, 113; Bufon, 1971, 52.

daughter Micka". In fact, he was playing his very self in his cameo appearance. Francis Jožef Hanibal Hohenwart left his property in Carniola to his nephew Karl Sigmund Hohenwart, who was then finishing his studies of law in Vienna. Karl Sigmund Hohenwart became the famous politician and Viennese prime minister from February 1871 until October 1871. Thus, the high political functions intertwined with the scientific successes of Carniolan Hohenwart's family. The nephew of the owner of castle Kolovec Francis Jožef Hanibal, Karl Sigmund Hohenwart (Hoⵏhenwart, * 1824 Vienna; † 1899 Vienna) was repeatedly elected a deputy with the votes of the Slovenian voters in the Upper Carniola electoral district from 16 October 1872 to 22 January 1897. Although he did not represent exactly Slovenian rights, he defended the rights of the Land of Carniola with F. Šuklje's help. He organized his Slavic neighbours of the Old Slovenian camp into his political Viennese Hohenwart's Club which opposed the Habsburgian liberals in 1873/74. Earlier, Karl Sigmund Hohenwart's (* 1824) father the count Andrej Hohenwart (1794-1881) was an elected candidate for the Frankfurt parliament. As a new member of new parliament, he travelled to Frankfurt together with Anastasius Grün alias Anton Alexander Auersperg and others on 20 April 1848. The great grandfather of Francis Jožef Hanibal Hohenwart, (Leopold) Ludovik Hohenwart from Kolovec, was the uncle of John Louis Hohenwart (1644 Kolovec-1710), who sold the dominion of Brdo in 1670 and became a provincial's first deputy in 1689. Ludovik Hohenwart (Louis, Ludvik Leopold Hochenwarth, * 1693 Kolovec-1757) has published two astronomical works: Ways of observation of celestial phenomena, and The physical-mathematical discussion of comets. The younger Sigmund's brothers, Anton Hohenwart (* 1731 Koloveⵏ SJ 1748; † 1800)[7] and Janez Hohenwart (* 1732 Koloveⵏ SJ 1747; † 1771 Vienna), both taught philosophy; the first among them in Slavonska Požega in 1767-1771, the other in Passau in 1763 and at Theresianum in 1764-1771. On October 4, 1759, three brothers Anton, Janez and Sigmund Hohenwart donated their personal first masses at the Jesuitical church of St. Jakob in Ljubljana. Five years after Bernardin's solemn defence of his exam thesis in 1754, the Hohenwarts' family members were again reunited around that first-class social event: the personal first mass. Bernardin Hohenwart (Hohenwarth, * 1734 Ljubljana; SJ 1754 Vienna; † 1779 Ljubljana) helped his older brother Sigmund at the main altar. Their brother Jurij Jakob Hohenwart (* 1724; † 1808) has inherited Kolovec and later left it to his son the count Francis Jožef Hanibal Hohenwart (* 1771 Ljubljana; † 1844 Kolovec).

---

[7] Posthumously printed in Historia Annua, 439-461; Lukács, 1988, 2:570.

Koller and Francis Jožef Hanibal Hohenwart both mastered Slovenian language but preferred to correspond in German language as was the habit of their days. Although they were relatives, Hohenwart addressed Koller as "high Sir" because Koller was not yet a counsellor in those days, while Koller called Hohenwart "count". In the beginning of his letter to Koller, Francis Jožef Hanibal Hohenwart first mentioned the local director of normal lower school, the principal of Ljubljana's Imperial Royal Normal Main School appointed in 1825, Janez Nepomuk Schlakar (Schlackar, Šlakar, * 1791 Kamnik; † 1863 Ljubljana). Schlakar informed Hohenwart about the born Carniolan working in Kremsmünster who did not forget his fatherland: Koller. In 1803-1809 the future priest Schlakar (Šlakar) studied Grammar school with Kamnik priest Kristof Plankel's (Plankell) fellowship in Ljubljana together in the same grade with Ljubljana native Franc Koller one class before Wolfgang Marian Koller, Weber and the future judge Anton Čop (Ts□hopp, * 1786, Hraše by Les□e; † 1865 Graz). But Schlakar (Šlakar) escaped to Klagenfurt to avoid serving in Napoleonic armies. In Klagenfurt he finished his grammar school, some philosophical and theological studies before graduating in Ljubljana after Napoleon's downfall in 1814. Of course, avoiding service in French army was not uncommon, as the rector of the Ljubljana Lyceum, the later Archbishop Jožef Walland (Balant, Walant, 1763-1834), severely complained over the dropout of students who preferred the other safer side of Habsburgian border in fear of the French soldier's skirt (Bufon, 1970, 40).

Schlakar (Šlakar) certainly kept his connection with his fellow Upper Carniolan Koller forever. In 1821 Schlakar helped to establish a bank in Ljubljana. In 1850 Schlakar became provincial school superintendent certainly with a little Koller's help.

As a proud Carniolan the count Hohenwart therefore wished to be informed about Koller's merits and other news about Koller's whereabouts. On second (last) page of his letter Hohenwart also noted his own curatorship of Carniolan land museum and invited Koller to become a member of Museal society. In his next handwritten paragraph, he invited Koller to join the Carniolan Agricultural Society whose president Hohenwart used to be five years earlier between 1827 and 1834. He inherited that prestigious post from his father as the aging Žiga Zois declined to take it over. Thirteen days latter, Koller answered at full three pages. He received Hohenwart's letter after one week on 30 November 1839. Hohenwart's letter had no crossing-outs as it was probably dictated to the professional scribe, while Koller's letter had many reparations because he was just a poor monk writing on his own. On his first page, Koller mentioned his observation tower. Hohenwart demanded his biographical notes to make Koller a member of those Carniolan societies while suspecting that Koller might be his fellow (Upper) Carniolan and therefore also eligible to be listed in Jožef Kalasanc baron Erberg's (1771-1843) anthology of Carniolan literati (Versuch eines Entwurfes zu einer Literar-Geschichte für Crain) as predecessor of later lexibn of the knight Constant Wurzbach von Tannenberg (* 1818 Ljubljana; † 1893 Berchtesgaden). Despite of his brother-in-law Hohenwart's praise of Koller's merits, Erberg did not mention Koller in his anthology, probably because Koller was more than two decades his younger.

Koller told Hohenwart that he was indeed born in Upper Carniola in Bohinj and later also added his birthyear 1792. He stated: "My father was a loyal servant of an excellent gentleman, the well-known and revered baron Sigismund von Zois" (Koller, 1839, 1; Fellöcker, 1864, 247). Marian Koller studied in Ljubljana with professors Hladnik and Kersnik; he did not mention any of his other teachers. The baron Sigismund Zois wanted Koller to continue his studies in Banská Štiavnica (Schemnitz). On second page of his letter Koller noted his lecturing on natural history and physics, as well as his publications in H. Schumacher's *Astronomische Nachrichten*, and in 16th volume of *Annalen der k. k. Sternwarte zu Wien* where they printed his Observations of stars at Kremsmünster in 1839. The count Hohenwart might have heard about some of those magazines, at least by their names. On the last page of his letter, Koller listed his memberships in different scientific societies. For unclear reasons, he did not specifically mention his "scientific European tour" performed a year earlier in Autumn 1838. As a matter of fact, the rumours about that famous journey might be one of the reasons why Hohenwart was suddenly so much interested in Koller's work.

Hochgeborner,
Hochverehrter Herr Graf!

Ihre gütige Zuschrift vom 22. Nov. l. J. ...

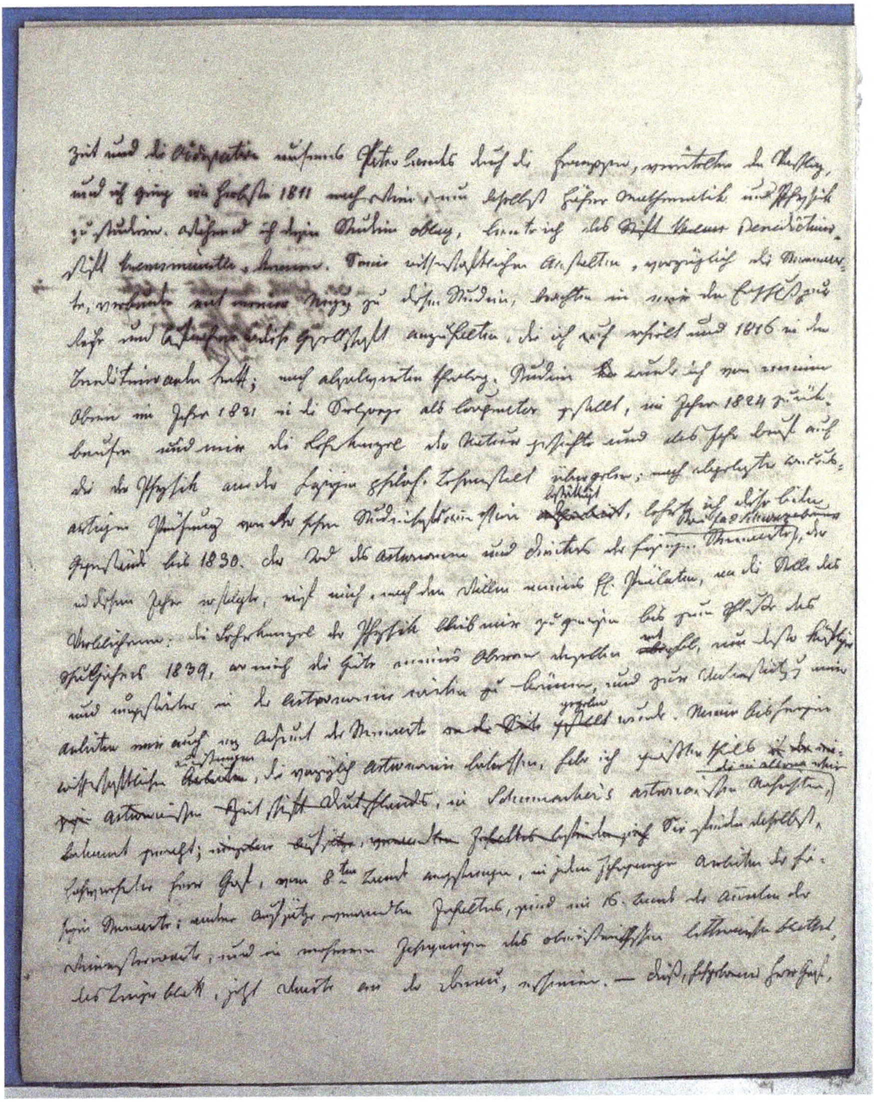

Figure 2: Koller's letter to the count Franz Joseph Hannibal Hohenwart signed in 1839, first page (Source: Directions-Archiv der Sternwarte Kremsmünster with the permission of the head Dr. Amand Kraml).

Certainly, as members of the local Inner Austrian elite, Koller family members had hard times to chose between German and Slovenian nationality as they used both languages. On 13 February 1896 Marian Koller's nephew Maximilian Josef Koller (* 6 December 1843 Klagenfurt) opposed the Slovene (-German) combined grammar school in Celje as the member of one of both German

Liberal parties, Styrian Land representative and the Rapporteur of the Styrian Provincial School Board, after that same Celje grammar school question caused the downfall of the government of the prime minister Alfred III Prince of Windisch-Grätz's (1851–1927) Federalist Party on 19 June 1895 (Cvirn, 1988, 43; Cvirn, 1997, 176; Stenographic minutes of the sessions of the Provincial Assembly of the Duchy of Styria 13 February 1896, pp. 440 and 441; Lampreht, 2018, 121). On the other hand, the top politicians of Hohenwarts' family publicly warned, that major part of Slovenian nation will find their elite redundant if the Slovenian nobles will continue with their liberal nationalistic Germanizing politics.

Koller for Habsburgian African Missions

Koller supported the pioneer of the missions in Sudan, his fellow Carniolan explorer of the White Nile Ignatius Knoblehar (Ignatius Knoblecher, Abuna Soliman, 1819 Škocijan near Mokronog-April 13, 1858 Naples). Knoblehar mostly invited his predominantly Slovene and Tyrolean colleagues to Sudan. Koller was one of the main pillars of the fraternity "Mary's Society for the Promotion of Catholic Missions in Central Africa" (Marien-Verein zur Beförderung der katholischen Mission in Central-Africa, Marienvereins-Comité zu Wien), founded in Vienna in 1850 and introduced by the bishops to all Habsburgian dioceses for fundraising. In Viennese yearly of that society Knoblehar's reports were printed, including the travelogue of his sailing on the ship Stella Matutina (Morning Star). Those yearly notes were regularly concluded by Fries' reports on the collection of contributions in the first seven volumes of the *Jahresberichte des Marienvereins zur Beförderung der katholischen Mission in Central-Afrika* from 1852 to 1858 with emperor himself among the donators (Klemenčič, 1981, 16). Along with the extremely active member Koller, the president of the fraternity Mary's Society for the Promotion of Catholic Missions in Central Africa was Koller's fellow Carniolan Andreas Meschutar (Andrej Mešutar, 1791 Selo near Ljubljana-1865 Baden near Vienna). Meschutar was a nominal bishop in Sardika (Serdica, Сердика in central Sofia) since 1853, court counsellor, member of Carniolan agricultural society, honorary member of Academy of Sciences in Vienna, and a section head of the Ministry of Culture. Among the members of Mary's Society for the

Promotion of Catholic Missions in Central Africa were also the court counsellor J.B. Altmann, son of the former boss of Koller's employer Moritz II Count Fries von Friesenberg (1804-1887) as treasurer of the fraternity, and Karl von Hammer-Purgstall (1817 Vienna-1879 Trieste). Karl von Hammer-Purgstall was the son of the first president of Viennese academy the orientalist baron Joseph Hammer-Purgstall (1774 Graz-1856 Vienna) of Carniolan origins. Karl von Hammer-Purgstall worked as the Viennese parliament deputy for Styrian areas west of Graz including Hartberg, Feldbach, Weiz and Fürstenfeld in 1870s. He on centralist liberal political side like Koller's nephew Andreas Koller. The other members of Fraternity included the PhD of jurisprudence ministerial counsellor of ministry of trade (Handel) the Catholic of Jewish origin Carl Ferdinand knight Hock (baron in 1859, * 1808 Prague; † 1869 Vienna), priest of Viennese St. Peter church the Jesuitical consistorial counsellor Maximilian Leopold Horny (Horni, * 1787; † 1857), and former Swiss Protestant Friedrich Emanuel von Hurter-Amman (* 1787 Schaffhausen, Switzerland; † 1865 Graz). Hurter-Amman Emanuel achieved his PhD in theology to become a state historiographer as Joseph Hammer-Purgstall's collaborator at the ministry for foreign affairs. He was also a court counsellor. The other members of Fraternity included the chamberlain of *Central-Severinus-Vereines (Katholiken-Verein,* Catholic Society) Heinrich count O'Donnel. Hurter-Amman Emanuel was also a member of that *Central-Severinus-Vereines KatholikenVerein.* O'Donnel was a relative of Irish officer in Habsburgian army Maximilian Karl Lamoral count O'Donnell von Tyrconnell (Donel, * 1812; † 1895) who saved emperor's life in 1853 (Mešutar, *Das Comité* 1851, 8; Mešutar, *Die Mission* 1851, 8; Kaiserseder, 2013, 66; *Hof- und Staats-Handbuch* 1853: pp. 40, 52, 123, 152, 179, 345, 347).

That international combination of former Jews, former Protestants, Irishmen, freemasons, German nationalists, and trade experts made the support for Habsburg African imperialism extremely effective, except for the lack of widespread use of quinine. One of supporters of Habsburgian African imperialism was a deputy president of the Viennese Geographical Society, the statistician-freemason Friedrich Wilhelm Otto Ludwig baron Reden (pseudonym Friedrich Wiemund, * 1804 castle Weidlingshausen of Dörentrup in Lippe 40 km southwest of Hannover (not Klosterneuburg 15 km north of Vienna); † 1857 Vienna). Reden was a former liberal leftist nationalist deputy of Frankfurt parliament forced to live Hanover for Vienna in 1854 where he helped organizing the Viennese statistical congress shortly before his death in 1857.

The Habsburgian efforts in the Sudan benefitted from the missionaries' activities. Habsburgian ambitions were restricted to the development of regions' intensive trade, which would be controlled by the Habsburg Empire. The missionaries explored the unknown terrain, gathered resources, initiated the barter trade with the population along the White Nile and established the first forms of colonial education. Thus, they fulfilled the first stages of colonization as precursors of a Habsburgian imperialism in Sudan (Kaiserseder, 2013, 145). Their advances were resembled those used by the missionaries in China. Consequently, the former Habsburgians of Venetia and Lombardy later colonized the Sudanese neighbour Ethiopia under Italian flag in 1936-1941 to propel the deadly Eritrean and Tigray conflicts in 2020/21.

Mešutar, Koller, Knoblehar, Dovjak and many other Carniolans made the Habsburg Sudanese missions partially Slovene, although the somewhat elitist Knoblehar preferred to see Tyrolean missionaries in his camp.

Knoblehar used to send his and his companions' scientific observations to Koller. He measured pressure with Heinrich Kappeller's barometer (Kapeller the younger inherited his firm from his father who established it in 1830s at Wien-Margareten, Franzensgasse 13 later also at V Bezirk Kettenbrückengasse 9 and in Budapest outpost to produce even the clockworks, thermometers, and weather houses in 1850s-1913 by his Institut für physikalische und meteorologische Instrumente), temperature, wind direction, cloudiness, and Blue Nile water level at its junction with the White Nile in Khartoum. The water level was in fact measured by Martin Dovjak (Dovyak, * 30 January 1821 Šentjernej in Lower Carniola; † January 1854 Sudan) in Khartoum on the banks of Nile from 14 June 1852 to 14 November 1852 just before that poor guy died there on the spot because almost no necessary quinine was available in those regions. The measurements were then edited by Koller for publication in the *Bulletin of the Viennese Central Station for Meteorology and Earth magnetism* established by Koller's initiative, and in the Viennese Academic Gazettes (Kreil, Resultate 1858, 37-68; Kreil, Über zwei Reihen 1857, 476-488; Reden, 1857, 150–160; Lukas, Koller, Kreil and others, 1859, 499-527; Knoblehar, Koller, 1859, 528; Knoblehar, Koller, 1859, 529-533; Lukas, 1859, 534–536; Klun, 1851).

Kreil wrote on his opening 37th page of his paper in 1859: "I owe my data to Ministerial Councilor Koller, which were sent to him by Pro-vicar Knoblehar." On page 66, Kreil also summarized Knoblehar's notes on earthquakes. Later in 1865, Theodor Kotschy (Teodor Koczy, * 1813 Ustroń in Polish Silesia; † 1866

Vienna) wrote about Knoblehar's plants stored in the herbariums of the Imperial Palace of Vienna after Kotschy's own travels along the Nile River. Kotschy began his Middle East travels as an assistant to the geologist Joseph Ritter von Russegger (1802–1863).

Despite the idealized image of the Habsburgian Sudanese imperial politics with its final collapse, a couple of years later in 1864-1869 followed the Mexican imperialist interventions of the Habsburgs with new emperor Maximilian sailing on Tegetthoff's Novara boat. The criticism also resonated, as it was in many ways an expensive adventuristic attempt. Despite being known in Peru from ancient times, the large-scale European use of quinine against malaria started only around 1850, to late for Konoblehar whose efforts collapsed with his death. Without the quinine most of central and north Europeans could not withstand the equatorial malaria. Therefore, Knoblehar promoted impossible attempts by the backhand colonialism of a dying Habsburgian empire. Without its traditional Turkish enemies onboard, the Habsburgians were desperately trying to hide domestic national problems and ethnic conflicts. Among the skeptics joining Knoblehar was mainly the Ljubljana gunsmith-master Jakob Šašel (Schaschel, 1832 Kappel (Kapla pri Borovljah)-1902 Karlovac) who worked as the famous gunsmith in Karlovac after his disappointed return from Sudanese Africa in 1857. Nikola Tesla knew him very well, as in Karlovac Tesla frequently met his son the writer-priest Jakob Šašel (1862 Karlovac-1911 Mahićno suburb village of Karlovac).

Farewell

Koller enjoyed solid and lasting health, which only suffered three major disorders during his 74 years. In his youth he survived his smallpox disease. In 1828 on the way from his Carniolan birthplace to the monastery he contracted a nervous fever due to cooling, from which he completely recovered after a few weeks. On June 3, 1852, he had the misfortune of falling on a rail journey from Vienna to Brunn am Gebirge because he stepped off the wagon of his train that had not yet fully stopped. He broke his brachium; had a railroad keeper not helped Koller quickly enough, the greatest misfortune could have happened. After a few weeks, Koller had recovered from this accident to such an extent that he was able to perform his duties, but the weakness and sensitivity of the injured point to the effects of the weather persisted for a long time. Koller was finally cured after his baths in Baden near Vienna, in today's second greatest Czech spa Teplice (Teplitz), and at the biggest spa town in modern Slovakia Piešťany (Pischtian). Although Koller's physical strength slowly decreased with age, his mind was fresh, his judgment was safe and sharp, his memory very faithful, his prudence clear and unclouded, his zeal in office tireless. With youthful energy he was still attached to sciences in his old age. On the last evening before his fatal illness, he was still answering the questions about astronomy to a younger scholar.

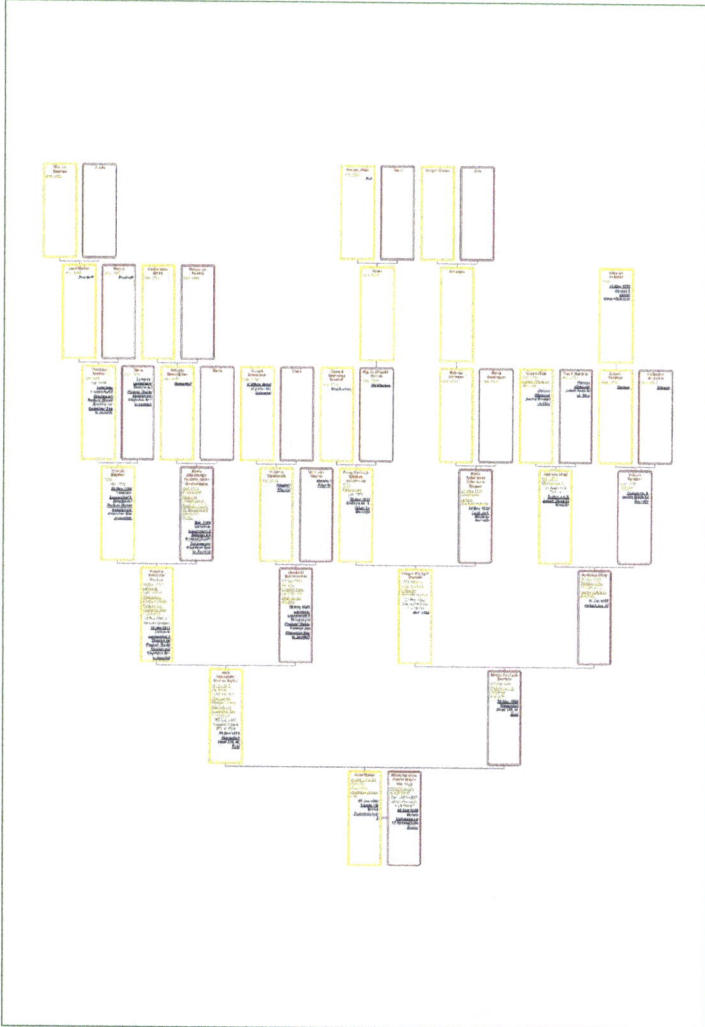

Figure 3: Stefan's genealogy

On September 17, 1866 in the afternoon, Koller paid last respect to the deceased Viennese professor of natural history Friese, who died of cholera. Johann Nepomuk Friese (1792 Chomutov in Bohemia-15 September 1866 Vienna) used to be a professor of Natural Sciences at the University of Innsbruck in 1819-1847, and then the professor of General Natural Sciences at the University of Vienna in 1847-1866.

With several friends and acquaintances of the deceased, Koller was standing in the church close behind Johann Nepomuk Friese's coffin. The bad smell of corpse disgusted him. He felt a chill, his face went pale; After noticing that trouble, his companion asked: "Sir counsellor, are you missing something?" Answer: "The smell is really disgusting! I will go outside."

Koller then went for a long walk to the Prater. He came home soaked in sweat and changed his clothes; the following night was restless. On September 18, Koller worked late into the night when cholera broke out violently; the hurriedly called doctor recognized the greatest danger. At ten o'clock in the morning of September 19, the patient received sacraments of death while in full consciousness.

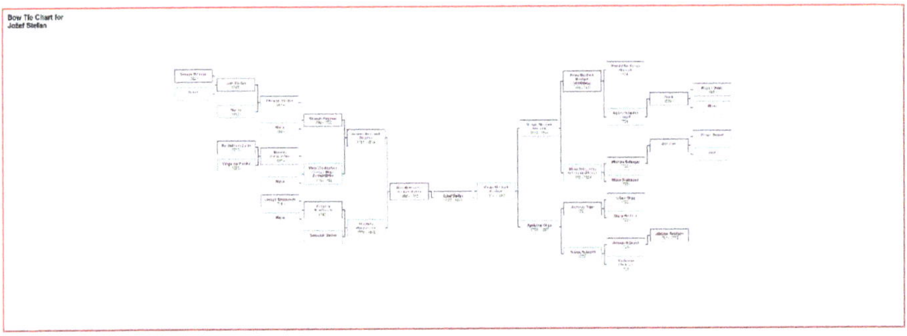

Figure 4: Stefan's ancestors

At 11:00 am, Koller asked for paper and pencil as he wanted to write down his last will. With great effort and reluctance to be helpful, Koller wrote his last lines, which are difficult to decipher:

"When moving from the general to the specialist knowledge… (the following words illegible) "that you find." "Every piece has a scientific and an administrative side, so it takes more time than anywhere else." Without a doubt, these thoughts refer to the reorganization of the polytechnic, which Koller's reformed recently and planned more advancements.

At noon Koller received no more medical help. He died in the evening at 6 a.m. On September 21 in the afternoon at the Augustinian court parish church the abbot Othmar Helferstorfer (* 1810 Baden; † 1880 Vienna) blessed Koller's remains in front of many participants. In the Matzleinsdorf cemetery Koller was buried for his eternal rest.

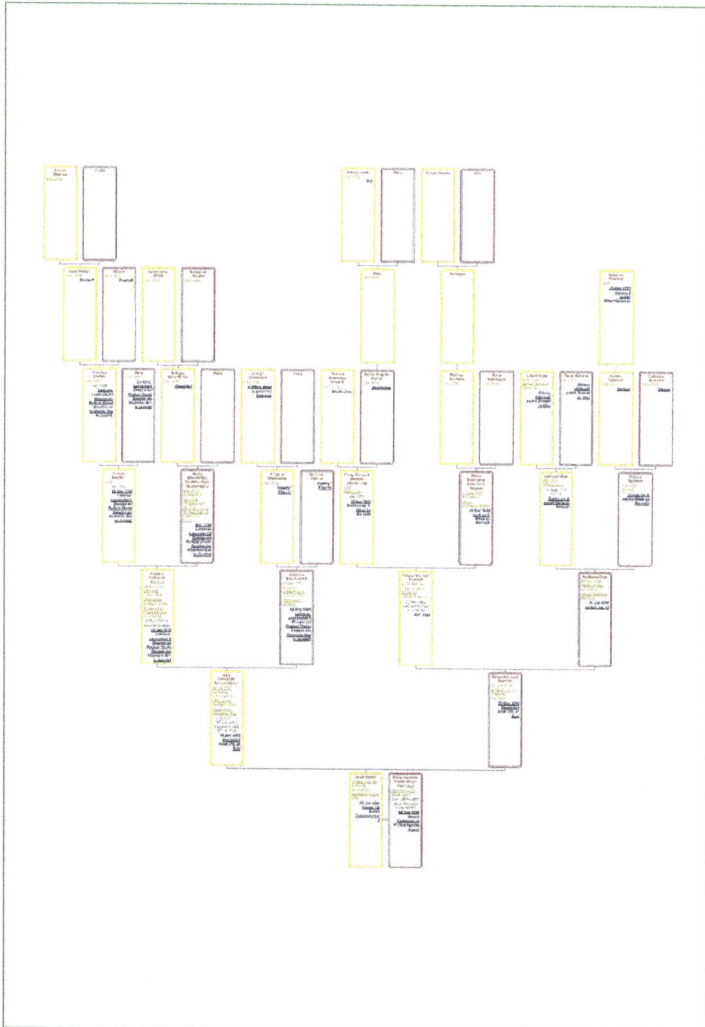

Figure 5: Stefan's ancestors by all branches

Koller had scheduled his departure from Vienna for a visit to Kremsmünster on Monday, September 23. The church had one of the most conscientious priests in Koller, the state one of its most loyal and zealous servants, science a tireless sponsor. The Kremsmünster Abbey has lost its greatest ornament.

The Carniolan writer Josip Stritar, then a Viennese student, wrote to his fellow Fran Levstik that the Cholera had picked up: "The ministerial councilor for education Koller, a well-known honest man born in Carniola: the man bore death in his own name (Bufon, 1970, 40). It was a kind of play of words in Slovenian language which uses a word "Kolera" for Cholera.

In 1854 the Englishman John Snow (1813 York–1858 London) found a link between cholera and contaminated drinking water, but the dangers of dead contaminated bodies were not clear yet. The fourth cholera pandemic which killed Koller lasted from 1863 to 1875. It spread from India to Naples and Spain. In 1883, Robert Koch (1843–1910) used his microscope to identify V. cholerae as the bacillus causing the disease. The first effective human vaccine was developed in 1885 by Catalonian physician Jaume Ferran i Clua (1851 Corbera d'Ebre in Tarragona of Catalonia-1929 Barcelona). He was the first to immunize humans against a bacterial disease by the method of Louis Pasteur (1822-1895). However, his vaccines and inoculations were rather controversial; therefore his peers and several investigation commissions rejected his proposals. The Russian-Ukrainian-Jewish bacteriologist Waldemar Mordecai Haffkine (1860 Odessa-1930) successfully developed the first human cholera vaccine at the new Parisian Pasteur's institute in July 1892, too late for Koller.

Conclusive Notes on Koller's Merits

Koller's astronomy has more than just historical values. His predecessor in Kremsmünster observed John Dalton minimum during the period of low number of sunspots in times of the reduced solar activity from about 1790 to 1830, corresponding to the period of solar cycles 4 (6) to 7 (Hayakawa et al., 2020). Koller took over the management of the observatory in 1830; of course, he had been involved in measurements for a decade and a half before, and probably edited the notes of his predecessors, just like Hallerstein did in Beijing a century earlier.

Like the prior Maunder minimum and the assumed even earlier Spörer minimum, Dalton minimum coincided with a period of lowered Earth's globally averaged surface temperatures, reduced by about 1°C in then German-speaking countries. Probably the parallel rise of volcanism is largely responsible for then cooling trend, as our ancestors survived the year 1816 without summer mainly due to the explosion of Mount Tambora in Indonesia in previous year as one of the two largest eruptions during the last two millennia. Those increased volcanic eruptions were probably triggered by a simultaneous decrease in the incoming solar energy.

Koller was certainly still interested in sunspots after he left Kremsmünster for Vienna. On 18 July 1850 on his four pages, Koller revalued and commented Böhm's paper offered to Academy on 17 January 1850 and revised in Innsbruck on 2 February 1850. The professor of mathematics at University Innsbruck who replaced Kreil in Prague observatory after that same publication in 1852, Josef Georg Böhm (* 1807 Rožďalovice in Bohemia; † 1868 Prague), proposed for the Viennese Academic publication his observations of sunspots used for the determination of the solar rotation (Böhm, Beobachtungen von Sonnenflecken und Bestimmung der Rotations-Elemente der Sonne). Next after Koller's valuation in the same volume, Koller's acquaintance Doppler defended the priority of his discovery of law now named by Doppler. After Koller's positive evaluation, Böhm's analysis of observations completed in 1832-1836 was published in *Denkschriften* of Academy of Sciences in Vienna at volume 3 of 1852, followed by Bohm's other longer paper entitled: Observations of sunspots in the years 1833, 1834, 1835 and 1836 from the Viennese university observatory. Böhm observed sunspots as the assistant of J.J. Littrow from 2 May 1833 until 26 July 1836 in frequent collaborations with nearby Koller's observatory. Therefore, Böhm observed after Dalton minimum which lasted from 1790 to 1830. Böhm noticed 88 sunspots during his 149 observations. He edited them by the analytical geometry and Gauss' method of least-squares. He used a cartography of the sunspots available after the efforts of Galileo's Jesuitical enemy Scheiner, followed by Cassini and Philippe de La Hire (1640-1718). Their beginnings were continued by Lalande and the director of Parisian observatory Jean Baptiste Joseph chevalier Delambre (1749-1822) (Koller, 1850, 151, 152). Koller was still unaware of the earliest now known report on planned sunspot observation of 364 BC preserved at comments to the Chinese star catalogue compiled by Gan De (Lord Gan, Gan Gong, 甘德, born in the State of Qi, flourished in 4th century BC) or by or by Shi Shen (石申, Shih

Shen, Shi Shenfu, fl. 4th century BC, born in the State of We). Böhm concluded that the sunspots are equally frequent on both solar hemispheres. According to Böhm, the sunspots are infrequent around the solar equator and at the solar latitudes above 35 degrees where they appear only rarely, while they occur most frequently in the intermediate zone (Koller, 1850, 153). Seventh solar cycle at the beginning of Böhm's observations lasted 10.5 years, beginning in May 1823 and ending in November 1833, thus overlapping the Dalton Minimum. The eighth solar cycle lasted 9.7 years, beginning in November 1833 and ending in July 1843; its beginnings covered the majority of Böhm's observations. Indeed, early in their approximately 11-year long solar cycles discovered by Samuel Heinrich Schwabe (1789 Dessau in Saxony-1875 Dessau) in 1843 and promoted by Wolf, sunspots appear in the higher solar latitudes and then move towards the equator as the cycle approaches maximum, according to Spörer's law. Böhm did not notice that but tried to find the average latitudes, which does not make much sense after the different ideas explaining sunspots soon prevailed in his era. A decade after Koller's comments of Böhm's work, Spörer's law predicted the variation of sunspot latitudes during a solar cycle. It was discovered by the English amateur astronomer Richard Christopher Carrington (1826 London areas-1875) around the year 1861. Carrington's work was refined by the German astronomer from Encke's Berliner observatory, Gustav Spörer (1822-1895). Both were just unknown teenagers during Koller's visits in their areas in 1838. Koller and Böhm provided no quoted references except for Böhm's introductory notes on earlier J.J. Littrow's paper, and Böhm avoided any formal thanks to Koller for his positive valuation of Böhm's paper which would be common today.

Koller's Lectures about Physics-Chemistry in Kremsmünster

Marian Koller taught his courses "Vorlesungen aus der gesamten Physik in der 8ten Classe des Lyzeum" an der philosophischen Lehranstalt zu Kremsmünster for more than a dozen years in 1826-1839. He prepared them in two bulky booklets: 15a (I. Abteilung: Chemie, Mechanik, Akustik on 694 pages together with later added title page) and 15b (II. Abteilung: Optik, Magnetismus, Elektrizität, on 672 pages together with later added title page).

Koller noted the beginnings of his Kremsmünster courses in his first notebook (15a Abteilung) entitled *Chemie, Mechanik, Akustik*, on 694 pages together with later added title page. He never published anything about chemistry in print: in Jahresberichten des Museums „Francisco-Carolinum" in Linz he published about the local waters only by measuring their temperature in his Untersuchung über die Temperatur des Quellenwassers zu Kremsmünster in 1841 by commenting on early data of J. Herschel (p. 3), B. Franklin (p. 4), Kurfürst Karl Philipp Theodor's (Carl, * 1724 ⎡astle Drogenbus⎡h by Brussels; † 1799 Munich residence) Societas meteorologica palatina established in 1780 as 3rd class of Mannheimer Akademie der Wissenschaften (p. 4), Lamont, and Dove (p. 6). Koller was not involved in publications about hydromechanics either. In his treatises: On the calculation of periodic natural phenomena (Über die Berechnung periodischer Naturerscheinungen) read in Vienna on 13 April 1848 and in his Contribution to the theory of the tubular level (Beitrag zur Theorie der Röhrenlibelle) read in Brno on 11 January 1865, Koller treated his topics by the purely mathematical standpoints, which was his usual habit.

Koller began his course of physics-chemistry with comments on his friend Baumgartner's new textbook about Physics (Baumgartner, *Die Naturlehre*, Vienna: Heubner, January 1826, # 5). For the good observer, time and space are extensive. The measure of extension is needed which Koller expressed in the form of simple equations. He praised the Nonius as a system for taking finer measurements on circular instruments and micrometer-screw (Koller 15a: page 2; Baumgartner, 1826, 5 # 19, 10, # 21) Koller discussed the impenetrability. He believed in the atomistic system according to the corpuscular theory of Leucippus, Democritus, Epicurus, Gassendi, Descartes, Le Sage, Prevost, and Berzelius. The forces in matter are repulsion and attraction. Koller interpreted the impenetrability and porosity by the ideas of Santorio from Koper, Dodart, Seguard, Lavoisier, and Haüy. The mercury thermometer (Koller 15a: pages 3, 5-8; Baumgartner # 23) followed from the invention of thermometer by Santorio, Galileo, Drebbel and the great Newton by his observations in 1709. The divisibility of bodies and their gravity were determined by Wollaston's experiments with platinum while Baumgartner in his textbook mostly noted Boyle. Better thermometers were designed by De L'Isle in Petersburg in 1733, and by Celsius in Uppsala in 1742. The gravity, cohesion, and aggregational states were decisive for their production (Koller 15a: pages 11-16; Baumgartner

32, # 33, # 42). The inner constitution of matter was Koller's cause of the light, electricity, and magnetism. Koller described an imponderable aether stuff, known as the media of radiating substances, which Baumgartner did not mention on that occasion in his textbook. All stuff is divided into solids and liquids (Baumgartner, 1826, 24). Considering gases and vapours, Koller noted Faraday's research like Baumgartner did, but Koller additionally acknowledged the translator of Faraday's work and expert on electrolysis, Antoine Alexandre Brutus Bussy (1794 Marseille-1882 Paris) (Koller 15a: 17; Baumgartner # 42, # 43, 1826, 23-24), even if Koller did not report meeting Bussy in 1838. Faraday's important research was accomplished in 1823. In 1823 Faraday took over several liquids in a drip-able state to transform them into permanent elastic body (a gas). He hermetically sealed and encapsulated them by large pressure to transform them into gases, including ammonia and many other liquids. To Baumgartner's paragraph # 43 Koller added the newer merits of the baron Charles Cagniard de la Tour, who experimented with aether, alcohol, water and similar substances for three years (Koller 15a: 17-18).

Koller (15a: 19; Baumgartner's # 42 and # 43) listed four kinds of imponderables: electricity-magnetism, and light-warmth stuff, as it was usual in his 19th century. He additionally noted 58 kinds of ponderable stuff, while 11 kinds among them were not yet brought into fluid state, with some non-metallic stuffs included: the oxygen, hydrogen, iodine. Koller listed 41 kinds of metals including aluminium and mercury.

Koller provided updated notes on eight Baumgartner's paragraphs (Baumgartner # 44 to # 51; Koller 15a: 19). He began by elasticity (Baumgartner # 44) and the substances not yet brought into fluid states: oxygen, nitrogen, hydrogen, chlorine, iodine, selenium, phosphorus, fluor, borine, CO_2, and some metals (Baumgartner # 50 page 29; # 51). The chemical affinity was important for the chemical compounds while Koller summed masses of compounds calculated by his mathematical aids. Koller's fractional distillation was no more noted in that paragraph of Baumgartner's textbook. As a devoted priest, Koller believed in (almost divine) great force needed for ordering, following the data of Claude-Louis Berthollet (1748-1822) (Koller 15a: 20, 22, 23 at bottom, 25). Berthollet was the first prominent chemist to reject the phlogiston theory and accept the oxygen theory of combustion on Lavoisier's premises in Berthollet's *Essai de Statique Chimique* published in Paris in 1803. The cohesion and expansion were discussed by Berthollet as the path to Berthollet's controversy versus Proust's Law of Definite Proportions published in 1797, when Koller was still a kid. While Proust believed that chemical compounds are composed of a fixed ratio of their constituent elements irrespective of the methods of their production, Berthollet believed that this ratio can change according to the ratio of the reactants initially taken. Although Proust proved his theory by accurate measurements, his theory was not immediately accepted partially due to Berthollet's authority, just like Petzval's criticism of Doppler's effect in Vienna. Proust's law was finally accepted when Berzelius confirmed it in 1811, but it was found later that Berthollet was not completely wrong because there exists a class of compounds that do not obey the law of definite proportions. These non-stoichiometric compounds are today named berthollides in his honor. Berzelius and Gay-Lussac were the most influential Koller's authorities in chemistry (Koller 15a: 26, 27, 29, 32). Berzelius isolated the metal thorium in 1824. Koller also noted Berthollet on several occasions (Koller 15a: 29, 33, 35; Baumgartner 1826 # 51, # 51; Schmidt, p. 199). Etienne François Geoffroy (1672-1731) published the earliest affinity table in 1718. Koller connected his cohesion force with the Vital Force as a kind of energy in Koller's somewhat Catholic Pantheon of Vitalism (Koller 15a: 36, 39). At the second quarter of 19th century, Koller's fellow Carniolan priest in Šentvid (Podnanos) and Vipava Matija Vertovec (* 1784 Šmarje on Kras; † 1851 Šentvid (later renamed to Podnanos) took over his vitalism views on organic chemistry from the leading chemist Berzelius. According to Berzelius' theory, the organic matter cannot be obtained outside the living bodies. However, in 1828, Liebig's friend Friedrich Wöhler synthesized the urine, which was not yet a defeat of the Vitalism theory, as it triggered many experiments, which, together with the Law on conservation of energy, put vitality of vitalism views outside the mainstream at least inside the organic chemistry, despite of some Pasteur's claims. Liebig's position from the years 1838-1846 was getting closer than Vertovec's own ideas, as Liebig claimed that it will soon be possible to synthesize all genres of organic molecules, but that life will never be created in lab, which finally also became obsolete after the genetic experiments a century later despite of the oppositions of the churches. The theory of vitalism of all organic matter only remained useful in physiology,[8] emphasised mostly that the chemical processes controlled the organic matter only after the organic matter is no longer alive. This statement is to be traced by Robida in 1849, similarly as in Erjavec's translation of Schoedler in 1870 as well as in the files of Vertovec himself.[9] In their notes, with the only exception

---

[8] Senčar-Čupovič, 1983, 117.
[9] Vertovec, 1847, 45 and 146.

of Robida, they explicitly performed the concept of Life force (the power of life in Vertovec's data and life-reproducing (Oživljajoča) power at Erjavec's notes), which operates in organic, but not in inorganic substances.

Koller (15a: 40, # 53) remembered Carl Friedrich Wenzel's (1740 Dresden–1793) *Lehre Von Der Verwandtschaft Der Korper* (1777) where Wenzel determined the reaction rates of various chemicals, establishing that the amount of metal that dissolves in an acid is proportional to the concentration of acid in the solution, based on equivalent weights of acids and bases. Wenzel's work was bettered by Jeremias Benjamin Richter (1762 Jelenia Góra in Polish Silesia–1807) who used to be fascinated with Lavoisier's role of mathematics in chemistry and atomistic in 1796-1798, promoted by Richter's work focused on the newly discovered "semimetal" uranium *Über die neueren Gegenstände in der Chemie* published in Wroclaw in 1792–1802. Koller certainly approved Richter's Galilean fascination praising mathematics. Martin Heinrich Klaproth announced the discovery of a new metal uranium while he was working in his experimental laboratory in Berlin in 1789. Berthollet-Proust's controversy was still important part of Koller's Francophonic mind until Stockholm based Berzelius finally decided for Proust after the Swedes with Gustav II Adolph and Linnaeus reshaped European politics and sciences. Berzelius's stuff was positively and negatively influenced by the dimension of atoms (Koller 15a: 41-42, 43 bottom, 44, 45 bottom of left and right columns, # 54). On the other hand, Berzelius blocked the acceptance of Avogadro's ideas as Berzelius supported old atomic numbers with the water wrongly noted as HO (Koller 15a: 61, 63; Baumgartner 1st edition 1824, 2nd edition 1826; 3rd edition 1829 # 56; 5th edition 1836, 7th edition in 1842; Berzelius's *Chemie* p. 141).

After that foreword full of chemistry, Koller still relied on the data of chemists in his Introduction to Physics in broader sense (Koller 15a: 65 #1, #2). Baumgartner did not have a separate numbering of paragraphs # in his physics part of a textbook, while Koller strictly followed 3rd Baumgartner's edition just for the data of Koller's chemistry up to Koller's page 63.

The physics in broader sense of word relied on observation and experimentation as anticipated by Francis Bacon Verulam in 1560 (Koller 15a: 68-69 # 4, # 5). As the priest, Koller loved the generalizations like the Basic Forces of Nature which also covered his bellowed astronomy including the terms like equator, satellites of planets and all those millions of bodies (Koller 15a: 70, 74-75 # 8 & # 9, 80, 84). In his first part, Koller noted the General facts of bodies including again the atoms of Democritus, Epicurus, Pierre Prevost (1751 Geneva-1839 Geneva), Dalton, and Berzelius (Koller 15a: 74-75 # 15 & # 16, 97). The other important concepts included the porosity, inertia, heating, and cooling measured by the thermometers of Santorius, Galileo, Drebbel, and Fredrik Rudberg's (1800 Norrköping-14 June 1839 Uppsala) descriptions (Rudberg, Ueber die Construction des Thermometers, *Poggendorff Annalen* 40/116/1 (1837): 39-62; Koller 15a: 100 # 22, 104 # 23, 107-108, 109).

Carl Wilhelm Scheele (1742–1786) researched chloride in 1774. Davy provided other breakthroughs while chlor-gas was explored by Reaumur and Faraday liquified it. Antoine Jérôme Balard (1802–1876 Paris) discovered bromine in 1826. Iodine was used in manufacture near Paris in 1821 (Koller 15a: 113-115 # 63, 123 # 66). Philippe le Bon (Lebon D'Humbersin, 1767-1804 Paris) patented apparatus called thermo-lamp in 1799, while Johann Joseph von Prechtl developed gas lighting in the Habsburgian monarchy. The distillation was used for better lighting and gradually they even applied some electrical lights which won the game much later (Koller 15a: 136-139, #77 & #78). In Montpellier Jérôme Balard (1802 Montpellier-1876 Paris) discovered bromine in seawater as chemical assistant at the faculty of sciences of his native town in 1826. Koller added his own annotations about Justus baron Liebig's (1803-1873) saltpeter and Georges-Simon Serullas (1774 Poncin-1832 Paris). Serullas was a professor of pharmacy notable for his first published work on iodoform, an early antiseptic and disinfectant (Koller 15a: 144, 146).

Hydrogen was also researched by Scheele while the chemical harmonica became a special case of the singing flame demonstration apparatus. Faraday proposed that the flame was extinguished or rekindled by the hot burner at the same frequency as the singing sound. Wheatstone and Tyndall used a rotating mirror to show that Faraday's hypothesis was, indeed, correct. Koller also explained to his students the facts about tellurium, selenium and kalium, as well as the fermentation (Koller 15a: 149-150, 153, 161; Baumgartner 1826, # 98 pp. 51-52 at the concluding chapter of chemistry before statics).

The research of carbon belonged already to Newton's semi-alchemistic traditions of 1694 even if Newton's research of alchemy was still a carefully guarded secret in Koller's times (Koller 15a: 165). The carbonic acid was researched by Nicholas-Théodore de Saussure, a Swiss plant physiologist and plant chemist. In 1797 he published three articles about carbonic acid's formation in plant tissues in the *Annales de chimie ("Annals of Chemistry")*. In

1813, H. Davy also published on carbonic acid and his work was noted again in 1837 (Koller 15a: 166).

Koller draw two pictures of the distillation machines. The important question of binding-proportion as a relationship involving forces was discussed by Berthollet in his disputes against Proust. Koller was able to offer a table of elements with their symbols beginning with 1st Oxygen, 2nd Chlorine, 3rd Iodine and ending with 51st Kalium. Those numbering seems to be quite arbitrary, although Koller made some calculation and summation of atomic weights of the elements (Koller 15a: 169-171 # 50-# 51). Mendeleev's periodic table appeared only after Koller's death, in 1869.

Koller's Physics-Mechanics

After his basic chemistry, Koller switched to mechanics by his analytical determining of the location of a point on the plane (Koller 15a: 175, 180). In his statics, Koller discussed the balance-equilibrium of forces acting between bodies (Koller 15a: 187 with figure 13; Baumgartner 3rd edition of 1829 # 90 II 1st paragraph of page 68).

Next, Koller discussed the states of solid bodies with annotation about Haüy's crystals formed as prisms, and the isomorphism described by Eilhard Mitscherlich (1794-1863) in 1819 (Koller, 15a: 203-204; Baumgartner 1826 # 154 p. 81, # 158, # 160).

Koller connected his equilibrium of forces in bodies, dynamics, and temperature with the old authorities including Aristoteles and Descartes. Koller's ideas about the centre of gravity relied on the textbook of Piarist Remigius Döttler (1741-1812) which Koller used during his studies in Linz, while Baumgartner did not mention it anymore (Koller, 15a: 212, 220, 224-225, 232-233 Baumgartner # 54, # 78).

Koller discussed the centre of gravity, levers, and pendulums of his Carniolan compatriot the professor Jurij Vega (Vega, page 305; Koller, 15a: 233-234, 237, 241-242; Baumgartner # 93, 101). Baumgartner did not mention Vega anymore, while Koller praised Vega several times in his lectures, maybe even because of

Koller's Carniolan proudness. Koller illustrated gear wheels and the wheel of Johann Andreas von Segner (Ján Andrej, János András, 1704 Bratislava-1777 Halle) as a first waterjet which resembled one type of modern lawn sprinkler (Koller, 15a: 242-243 Fig. 58; Baumgartner 1829 3: # 107, # 108, 112 # 147). Koller certainly commented just mathematically interesting paragraphs of Baumgartner's textbook including elasticity. He tried to upgrade and modernize Baumgartner's textbook. His friend Baumgartner probably used Koller's suggestions in his later textbook editions (Koller, 15a: 245, 248; Baumgartner # 113, # 122).

Lavoisier-Laplace's pyrometer for measurement of expansion of bodies by temperature was described by Biot (Koller, 15a: 248; Baumgartner 1829 # 143 in a chapter belonging to hydrostatic; Baumgartner 1826 2: page 89 # 165). Dulong and Petit determined the expansion of solid bodies with their thermometer, while Baumgartner also used Johann Holzmann's Viennese thermometer (Despretz, pages 80f; Koller, 151: 249 # 144; Baumgartner 1826 2: page 90 # 169; Baumgartner 1829 3: pages 137 & 138-139 by # 177 & # 179). César-Mansuète Despretz (1791 Lessines in today Belgium-1863 Paris) published his *Traitè élémentaire de physique... Quatrième édition, revue, augmentée, et accompagnée de dix-sept planches* in Paris in 1836. Breguet's spiral bimetallic thermometer was used for measuring, sometimes involving three metallic plates of platinum, gold, silver (Koller, 15a: 250 with illustration no. 42; Baumgartner 1826 2: 90 # 169). Abraham-Louis Breguet (1747 Neuchâtel-1823) was also praised by Koller's Ljubljana professor Gunz and Gunz's later successor Schulz von Strassnitzky. Breguet and Johann Holzmann were both watchmakers, but their branch of craft of those days might be today divided into several subbranches. Koller also noted the clocks with pendulum but did not mention the Viennese Holzmann. Joseph Perkins' apparat for the determination of compressibility of water was described in Gilbert's *Annalen* and in *Phil.Trans.* in 1820 (Koller, 251 # 148; Baumgartner 1826 2: 91 # 170). Koller discussed the Viennese-Carinthian professor of British origin Herbert's apparatus for determining of compressibility of water at Koller's illustration no. 18 (also by fig. II) but failed to mention the merits of Herbert's Jesuitical student Anton Ambschell, who translated Herbert's Latin language work into German language as a professor in Ljubljana. The higher salt-trade inspector in Braunschweig Rudolf Adam Abich (* 1738 Osterode; † 1809 Schöningen) designed another tool for the measurement of compressibility of water which Gauss's professor of mathematics in Braunschweig Eberhardt August Wilhelm von Zimmermann (1743 Uelzen-1815 Braunschweig) described in his book.

Abich's soon married the daughter of the mineralogist Martin Heinrich Klaproth. The expert of thermoelectricity professor of medicine at Kiel Christoph Heinrich Pfaff (1773 Stuttgart-1852 Kiel in Holstein) also researched the compressibility of water: by water's expansion while heated, it was oblivious that water could be compressed, but it took a time to develop the necessary early experimental tools with Idrija Hg involved. Koller's cited Haüy's textbook and Laplace's experiments with the mercury and illustrated Torricellian vacuum barometer by Baumgartner's data (Baumgartner 1829 Neue Auflage 3: 131 # 172; Koller, 15a: 251-152 # 148, # 160, # 162; Zimmermann, *Ueber die Elasticität des Wassers, theoretisch und historisch entworfen ... Zugleich eine Ankündigung einer neuen hiehergehörgien Maschine*, Leipzig: Weygand, 1779; Beobachtete Entstehung einer Blitzröhre durch den Blitz; Magnetisirung durch den Blitz; Christoph Heinrich Pfaff, Versuche über die Zusammendrückbarkeit des Wassers; und Controverse über das Seewasser, *Annalen der Physik*, 72/9 (1822): 111-112).

Baumgartner's 3rd edition appeared in Vienna in 1829. It was dedicated to Baumgartner's brother-in-law Ettingshausen with an introduction dated in Vienna in January 1829, while 4th bettered edition appeared in Vienna in 1832 again dedicated to Baumgartner's brother-in-law Ettingshausen with an introduction dated in Vienna in January 1832. Therefore, Koller wrote those notes in late 1829, in 1830 or in 1831.

Koller discussed the aerometers and aerostatics of Herbert, Perkins and Abich, Zimmermann, Altsted and Pfaff, Sturm and Calladon (Koller, 15a: 252-254 # 162; Baumgartner 3: # 162, # 170 page 130). The hydrostatic paradox and buoyance were also important including Bramah's hydrostatic press of 1796 (Koller, 15a: # 178, # 185 pp. 256, 261, 269-270, 273). In fact, the locksmith Joseph Bramah (1748 Stainborough Lane Farm, Barnsley Yorkshire-1814 London) patented his invention of hydraulic press in London in 1795. Koller also dealt with the capillarity, temperature, thermometers, fluids, and the determination of temperatures of the solids and fluids (Koller, 15a: 288 # 201, 294, 296). Dulong and Petit's tools were compared to Lavoisier and Laplace's barometer filled by mercury or alcohol. Jean-André Deluc (De Luc), a housekeeper and clerk to the Royal Society George Gilpin (* 1754 Kirkgate in Leeds, Yorkshire; † 1810), and Charles Brian Blagden FRS (1748-1820) contributed a lot, as well as Louis Lefèvre-Gineau (1751 Authe (Ardennes)-1829) who researched the structure of water. In 1772, the teenager George Gilpin joined Cook's Second Voyage as a servant to astronomer William Wales whose brother-in-law he became later. At RS, Gilpin collaborated a lot with H. Cavendish (Koller, 15a: 297-300; not mentioned at Baumgartner's 3rd edition). Gustaf Gabriel Hällström (1775 Finnland-1844) contributed to the establishment of an astronomical observatory in Turku to initiate the earliest systematic meteorological observations in Finland. Gmelin in Petersburg, Biot and the Swiss meteorologist Jean-André Deluc (de Luc, 1727-1817) researched the adhesion of fluids as did Hällström, Gilespie (Gillespie), and Dalton (Koller, 15a: 302, 305-308).

Koller's determination of the specific weights of fluids and solids was followed by determination of the specific weights of drippable fluids. Hällström researched the specific weights of solids while the aerometer with a scale was used by René Antoine Ferchault de Réaumur besides Baumé's barometer and an aerometer with a weight (Koller, 15a: 309, 313, 319-321 # 186; not in Baumgartner's 3rd edition but in Baumgartner 4: 1832 # 186). Koller mentioned the statics of expansive bodies, therefore of gases (Koller, 15a: 323; Baumgartner 2: 1826 # 217).

Koller illustrated Cavendish's bottles as well as the nicely illustrated bottle and apparatus of Peter Woulfe (1727–1803) needed to demonstrate Mariotte's law (Koller, 328-331 # 220, # 223). Koller discussed the hydraulic presses, Heron's fountain, and air rifle which Koller acquired for his Kremsmünster lab in 1833. Koller later crossed out his own comments on Pierre Louis Dulong (* 1785; † 19 July 1838) and Alexis Thérése Petit's (* 1791; † 1820) experiments. Petit was a brother-in-law of Koller's later acquaintance Arago, but Petit and Dulong both died before Koller visited Paris in autumn 1838 (Koller, 15a: 339-342, 352 # 238). Cavallo experimented with a parallelepiped, but Avignon based Montgolfier's aerostat and other devices proved to be decisive until Blanchard's aeronautic successes (Koller, 15a: 362-363).

The aeronauts flying above Adriatic included the Roman Gaetano Grassetti, Pasquale Andreoli (1774 Falconara Marittima-1837 Gela, Caltanissetta) of Ancona, Francesco Zambeccari (1752 Bologna-1812). They researched the elasticity of gases. The same did Blanchard, Rozier, Charles, Robert, Dalton, Gay-Lussac, and Humboldt, while Koller did not mention his Carniolan compatriot Gregor Kraškovič. The affinity and atomism were Dalton's best ideas, but Bertholet, Gay-Lussac, Humboldt, and Thomson published against Dalton, also because the private tutor Dalton (1766-1844) in Manchester was not quite a member of their own exclusive academic milieu. The specific weight of water was determined by Davy's and Faraday's experiments. Koller added his own updated note about Johann Samuel Traugott Gehler's (1751-1795) posthumous edition of *Physikalisches Wörterbuch, oder, Versuch einer Erklärung der vornehmsten Begriffe und Künstwörter der Naturlehre* which appeared in Leipzig in 1825, and about Claude-Servais-Mathias Pouillet's *Élémens de physique expérimentale et de météorologie* published in Paris in 1829. As a Francophile student of Haüy's textbooks in Ljubljana, Koller cited Haüy's work in French language and praised Biot's textbooks (Biot, *Précis élémentaire de physique expérimentale ouvrage destiné a l'enseignement public Tome 1* printed in Paris in 1824; Biot, *Traité de physique expérimentale et mathématique, par J.B Biot... Tome premier* of Paris 1816). Dalton's (partial) pressure of gas was still a novelty in those Koller's times, as well as the behaviours of liquids in free air detected already by Musschenbroek (Koller 15a: 368, 372 # 256, 374, 381, 387-388, 391-392, 398-401 # 263, 403-404; Francesco Zambeccari, Gaetano Grassetti, Pasquale Andreoli, *Relazione del viaggio aereo intrapreso dal nob. sig. co: Francesco Zambeccari, dal signor dottor Grassetti di Roma, e dal signor Pasquale Andreoli di Ancona*).

The equilibrium of gases in air involved vacuum as designed in the apparatus of Johann Joseph von Prechtl (1778 Bischofsheim, now Lower Franconia-1854 Vienna) who was the director of the Viennese Polytechnic until 1849 and among the founding members of the Vienness academy. Therefore, Prechtl retired soon after the rebellious students ruined the reputation of his Polytechnic in imperial eyes, and soon after Koller became a minister's official in charge for Prechtl's Polytechnic. Prechtl's tool was needed for the measurements of the elasticity of vapour in atmosphere as it was changing by the temperature od air (Koller 15a: 410 # 199; Baumgartner-Ettingshausen's *Zeitschrift für Physik und Mathematik's* ten volumes published in Vienna in 1826-1832, here volume 1: pp. 383-390).

Like Stefan after him, Koller discussed the properties of vapors and laws of evaporation including the Glasgow apparatus of Dr. Andrew Ure FRS (1778-1857) which Koller adorned with a nice illustration (Koller 15a: 414 # 250). Ure was a Scottish physician, a founder of Andersonian Institution, which became the University of Strathclyde. He used to be foremost the consulting chemist, scriptural geologist, and early business theorist. Koller discussed elasticity by the example of deadly accident in manufacture in November 1815, probably in Habsburgian monarchy. Deluc measured the pressure, as did Gilbert and Wollaston. As a lecturer interested in history of technology, Koller described and illustrated the event of 1663 when Samuel Sorbière visited Edward's workshop at Vauxhall of south London to see and describe the hydraulic machine which Edward Somerset 2nd Marquess of Worcester invented as one of the earliest steam engines. Koller accompanied it's description by his own nice illustration. The idea was bettered by Savery, Cowley, Newcomen, as well as by Watt's condenser in 1769 and 1774 and by the patent of J. Perkins issued in London in 1823 (Koller 15a: 417, 420 # 253, 422, 423, 424, 427-428, 437, 442-443). Koller described Saussure's hygrometer and Gay-Lussac's merits. Koller added Dalton's table which Koller later crossed out, and the Habsburgian local designs of Mathäseus, the preacher at Joachimsthal in Bohemia (Jáchymov) who died in 1568. Savary in 1698, Papin, Newcomen, Robison, and Black made additional contributions as narrated by Davy's and Faraday's pupil John Millington (1779 Hammersmith, Middlesex-1868 USA) (John Millington, *An epitome of the elementary principles of natural and experimental philosophy* translated as *Grundriß der theoretischen und Experimental-Physik / 1, Enthaltend die Eigenschaften, die der Materie im Allgemeinen zukommen; die Mechanik; Pneumatik; Akustik; Hydrostatik und Hydraulik: nebst einem Bericht über die Entstehung, Vervollkommnung und den gegenwärtigen Stand der Dampfmaschine; Mit 14 lithographirten Tafeln. / nach dem Englischen Originale von John Millington,* published in Weimar: Landes-Industrie-Comptoir, 1825; Koller, 15a: 442 # 260, 451 # 261, 452-455).

Koller also offered a sketch of a turbine of Joseph Perkins designed in London in 1823 as an echo of Watt's ideas advanced in Britain in 1822 (Koller 15a: 463, 467, 471). Leslie's hygrometer and August's Berliner psychrometer were already described by Baumgartner (Baumgartner's *Zeitschrift* volume 4; Koller 15a: 474 # 264, 477). Under his special subtitle "Psychrometer", Koller listed the contributions of Daniel, as well as the mechanician of Jena university employed after 1819 Friedrich Körner (* 1778 Weimar; † 1847 Jena). Körner is especially favoured as an early boss of Carl Zeiss (1816 Weimar-1888 Jena). Koller mentioned Saussure while including long French citation from Gay-Lussac's textbook, which was published in Paris without permission of the angry Gay-Lussac (Gay-Lussac. *Leçons de physique* volume 1 Paris, by notes of Grosselin, 1828; Koller 15a: 482).

Koller's Dynamics (of Gases, Fluids and Solids)

Koller noted his dynamic as a part of his hydrodynamics which he explained again to his students in his much later notes of lectures of his same course. Koller described fall-apparatus of George Atwood, FRS (1745 Westminster, London-1807 London). Atwood's machine was commonly used in classrooms for the demonstrations of Newton's laws. As Atwood was also a renowned chess player like Koller himself, which made his tools very welcomed in Koller's courses. Koller praised the professor Jakob Philipp Kulik (* 1793 Lemberg (now Lviv in Ukraine). In 1816, Kulik became professor of physics at the Lyceum in Graz. The position and swinging of a pendulum were researched for ballistic purposes in the military base of Woolwich by the major general sir William Reid FRS (1791-1858) who used to be a Scottish military engineer, administrator, and meteorologist (Koller 15a: 494, 499, 503, 505-506; Baumgartner 3: # 282).

Besides Baumgartner's textbooks, Koller mostly relied on the textbook of Georg Gottlieb Schmidt (1768-1837) who used to be the physicist and mathematician in Gießen (Koller 15a: 512; Georg Gottlieb Schmidt *Handbuch der Naturlehre*, Giessen, 1801-1803; Georg Gottlieb Schmidt & Melos, *Naturlehre für Bürger- und Volksschulen, sowie die untern Klassen der Gymnasien,* 1836). In 1785 in Göttingen Schmidt studied with Abraham Gotthelf Kästner and Georg Christoph Lichtenberg. He dealt primarily with the determination of mechanical and thermal properties of gases, which is why his students jokingly nicknamed him "Luftschmidt".

Koller discussed the straight central collisions between elastic bodies, also measured by pendulum for military needs. Construction of the pendulum was decisive for its ballistic uses as found already by the Habsburgian artillery officer Jurij Vega (Koller 15a: 522-523, 525, 527-528, 533; Vega, *Mechanic of Solids*, Wien: Joseph Tendler, 1809, part III Lemmas II and VII). Koller also discussed and illustrated the absolute friction as a problem of an arm of level. After more that two decades of its printout, the second edition of Vega's textbook was still used as best source for dynamic by Vega's Carniolan compatriot Koller! Koller might use that Vega's textbook already as a student in Ljubljana or later in Linz and Kremsmünster. To explain friction, Koller also utilized the works of Coulomb and the data of baron Gaspard Clair François

Marie Riche de Prony (1755-1839). Prony was a French mathematician and engineer, who researched hydraulics. The German translation of his book was certainly in Koller's possession. Felix Fontana of North Italy was also among Koller's favourites, as were the experiments with one arm of a balance immersed in water of John Theophilus Desaguliers FRS (1683-1744) who researched Newtonian laws in 1717. As a leading freemason, Desaguliers might not have been Koller's favourite model (Vega 1809 III: 333f; Gaspard Prony, *Neue Architektura hydraulica,* Frankfurt am Main: Andreäischen Buchhandlung, 1794-1801; Koller 15a: 537, 541, 544-547).

The motion of expansive bodies (gases) was a part of Koller's pneumatic based on Coulomb's and Prony's textbook about density and pressure of gases, which later deeply interested J. Stefan (Koller 15a: 548-549, 561; Baumgartner 3: # 272, # 276).

Koller's Acoustics with Oscillations

Acoustic was a part of higher analysis of oscillations inside Koller's course, as Koller used to be mostly interested in mathematical part of physics (Koller 15a: p. 570 # 377, p. 573; Baumgartner 3rd edition # 307 page 229). Koller discussed the experiments of both brothers Weber about the part of water on vessel's surface experiencing the wave motions. Wilhelm Eduard Weber and Ernst Heinrich Weber published their book about wave theory based on experiments in Leipzig in 1825 (Koller 15a: 580-581 # 287, # 295 at the end, 584; Baumgartner 3rd edition: page 222 # 295, page 219 at end of # 291; in Baumgartner's 4th edition Wien: J.G. Heubner, 1831, acoustic had a separate numeration of paragraphs). Vacuum and the hydrodynamics of vibrations were researched by Schmidt, the member of Napoleonic Egyptian expedition Prony's collaborator Pierre-Simon Girard (1765 Caen-1836), and the baron Charles Cagniard de la Tour. Koller added praise of Schmidt's and Anbidsson's data. Most of all, Koller inserted Hachette's long French report about the works of Gay-Lussac and also about the measurements of Charles-Bernard Desormes's (1777-1862) and his son-in-law Nicolas Clément (Clément Desormes after 1820, 1779-1841) (Pouillet 1: 234; Koller 15a: 586-588; Hachette, De l'écoulement des fluides aériformes dans l'air atmosphérique et de l'action combinée du choc de l'air et de la pression atmosphérique, *Annales de chimie et de physique*, 1827, 35: 34-53; Nicolas Clément Desormes, *Annales de chimie et de physique*, 1827, 36: 69-80). Koller had in mind Hachette and not the guy with similar surname of Desiré-Raoul Rochette (1790-1854), a French archaeologist and numismatics.

Koller added his special paragraph "E" about the longitudinal vibrations in self-flowing gaseous elastic bodies, monochords, and other musical instruments like the clarinet or fagot. Koller discussed the spreading of sounds effected by the data measured by thermometer and barometer to determine Laplace's elasticity. In 1787, Ernst Florens Friedrich Chladni (1756-1827 Wroclaw) as well as a father and son Soemmering researched the sounds of tuning fork as an acoustic resonator. Samuel Thomas Soemmering (Sömmerring, 1755 Torun (Toruń) in north-central Poland-1830 Frankfurt) was the son of the physician Johann Thomas Sömmerring. He wrote about impacts of sound on the human body (Samuel Thomas Soemmering, *Über das Organ der Seele* 1796; Koller 15a: 600 # 333 & # 334, 603, 605 (not in Baumgartner's 1st editions of 1824), 611, 615 # 384, 616-617, 628).

Koller discussed Charles Wheatstone's Mikroskon as stereoscopic-style microscope, Berzelius publication of 1829, Savart's measurements of elasticity and the chemical harmonica (Koller 15a: 629-630 # 396, 636-638 # 405).

In 1670 Samuel Moreland designed a vibration tube and in 1671 he claimed credit for inventing the speaking trumpet, an early form of megaphone, probably inspired by Athanasius Kircher's Jesuitical Roman design. Koller also discussed combinations tone of his neighbour of Piran and Padua, the violinist Giuseppe Tartini (1692-1770) (Koller 15a: 644, 646, 649 # 371 4th part, 650, 654, 665).

Koller Marian signed himself with his full name and surname while he discussed echo as a state of reflected sound. He began his data with a formal title resembling a kind of published paper, which eventually never appeared in print (Koller 15a: 670-674).

During his later preparations, Koller added some acoustic into his second notebook of lectures. There, Koller discussed case of velocity of sound in elastic metal as proportional to elasticity which became main point of later Lamé-Clapeyron-J. Stefan's research in early 1860s (Koller 15b: 147-148).

Koller's pendulum and pair of forces

Koller discussed dynamic at the slope again in connection with a horizon and a pendulum (Koller 15a: 666 # 252 and # 253, 667). On that occasion, Koller did not mention the military ballistic measurements of pendulums of Jurij Vega and the officers in Woolwich.

Later, Koller added some of his notes about the mechanics of pendulum into his second notebook of preparations for his courses of lectures. There, in the middle of his lecturing notes about electricity, Koller added a short note about the reversive pendulum enabling the regulation of the clocks and the correlation of amplitude. By changing the length of the pendulum measuring seconds, we had to consider the resistance-friction of air for the right correction of clocks even if Koller avoided to connect it with any kind of atomistic view (Koller 15b: 123). In 1817, the reversible free swinging pendulum invented by British physicist and army captain Henry Kater (1777 Bristol-1835) as a gravimeter instrument measuring the local acceleration of gravity.

In those times, the researchers expanded an important theory of pairs of forces, probably also as the explanation of transversal magnetic forces (Koller 15a: 678f with 15 pages of text of paragraphs filled by somewhat mixed topics). Koller discussed the antiparallel and equal forces, levers, and momentum of force presented as the diagonals of parallelepiped. Almost half of a century later, the pairs of forces became a favourite research topics of Tesla's teacher Martin Sekulić (Koller 15a: 679-680, 683-684, 690).

Koller's Optics, Magnetism, and Electricity (with Heat)

Overview

Koller's "Vorlesungen aus der gesamten Physik in der 8ten Classe des Lyzeum" an der philosophischen Lehranstalt zu Kremsmünster were updated by Koller during more than a dozen years in 1826-1839. Koller's 15b II. Abteilung: Optik,

Magnetismus, Elektrizität numbered 672 pages together with later added title page. There, Koller lectured about the heat were discussed after page 15b: 289 inside his course of electricity as the heat was not considered so much important title like the other kinds of then "imponderables", even if Baumgartner in 1826 in his 2nd edition discussed the heat in independent paragraph number 2 of his 2nd part, after optics and before electricity, while magnetism was his last 4th paragraph. 1st part was mechanics for both, Koller and Baumgartner. Baumgartner's 3rd part was: 1. Physical Astronomy – 2. physical Geography – 3. Physical meteorology. Koller apparently did not discuss any of those except some meteorological height of atmosphere at the end of his 15b preparations for lectures.

That Koller's second booklet 15b has four A3 leaves (Halbbogen) of small format, and fifteen A3 leaves (Halbbogen) of great format of Koller's lectures about Oersted and Faraday's discoveries about the electrical current, at least so the later not very exact librarian noted in the headline of Koller's booklet. In fact, electrical current was just a part of that Koller's course.

Koller's optics

As distinguished astronomer, Koller paid special attention to the double refraction (birefringence). Koller discussed the border between two mediums, birefringence in calcite Iceland spar (Iceland crystal) and the refraction of rays (Koller 15b: 82-83). Koller's figures 1 and 2 presented path of ordinary and extraordinary rays. To better and update the data of his friend Baumgartner's textbook, Koller noted in his right column reserved for comments the experiments of Becquerel's protegee, the Parisian engineer-optician Jean-Baptiste-François Soleil (1798-1878). After 1819, Soleil's eponymous company made optical instruments for notable researchers including Fresnel, Arago, Foucault and Jacques Babinet (1794-1872). In autumn 1838, Koller met Arago and Babinet. Koller discussed the birefringence in crystals with two optical axes connected to their elasticity (Koller 15b: 84-87). He explained birefringence as well as the polarisation by refraction, intensity of polarized light, and the angle of total polarization by simple Fresnel's undulation theory (Koller 15b: 89-94). Koller noted the polarisation by absorption, interference of polarized light, and polarized vibration, colours of crystal sheets in polarized light measured along its axes as Newton's colour rings in two-axed crystals and tourmaline plates. Koller's opinion about the polarized light followed Fresnel's hypothesis, including circularly polarized light as Fresnel's multiplication of light in quartz (Koller 15b: 96-105). After Fresnel's breakthrough, the elliptic polarization as another modification was discovered by David Brewster (1781–1868) in his *Treatise on Optics* in 1831. There Brewster coned his useful term elliptic polarization (Koller 15b: 106-108).

To discuss diffraction of light, Koller used the development into Fourier series by sines (Koller 15b: 143-144; Baumgartner # 78 page 49). Fourier was initially also a favourite J. Stefan's author of new French school which silently abandoned the restrictions of Laplace's calcul propre, which might have been still favoured by the differential equations of Stefan's teacher Petzval.

Koller continued his course focused on the science of light called optics in connection with his bellowed astronomy by discussing the velocity of light as determined by Galileo and by the Danish astronomer Römer's observations of Jupiter's satellites in 1675 (Koller 15b: 179, 182, repeated by Koller 15b: 223). Koller noted the measurements of huge force of light and brightness, initially without mentioning Lambert. He probably had in mind Johannes Kepler's concept of the radiation pressure, used to explain a tail of a comet which is always pointing away from the Sun in 1619. The idea of electromagnetic radiation's momentum was supported by James Clerk Maxwell in 1862; it was experimentally proven by the Russian Pyotr Lebedev in 1900 and by Ernest Fox Nichols and Gordon Ferrie Hull's radiometer in 1901. Evidently, Koller was thinking in that direction much earlier.

Koller put special attention on Wollaston's reflective goniometer (1809) bettered by Biot (1811) and Baumgartner for the measuring of angles in crystals (Baumgartner, Ein neues Reflexions-Goniometer, *Annalen der Physik*, 1822, 71/5; 1-6). Koller added a long Latin language citation from the textbook of Abraham Gotthelf Kästner (Kaestner, * 1719 Leipzig; † 1800 Göttingen) whom G. Gruber considered as his teacher. Koller used Kästner's academic anthology (Koller 15b: 189-190, 195, 198-199; Kästner, *Dissertationes mathematicae et physicae quas Societati regiae scientiarum Gottingensi a. 1756-1766 exhibuit Abraham Gotthelf Kaestner*, Altenburgi: Richter, 1771). Koller also noted Polemoscope as a type of periscope constructed by von Havel (Jan Hevelius) in 1637 and published decade later (Jan Hevelius, *Selenographia: sive Lunae Descriptio (Selenography, or A Description of The Moon)* Gdansk 1647; Koller 15b: 199). Koller was fascinated by the magic perspective, by the catoptric, and by the chemist and pharmacist Nicolas Lemery and César-Mansuète Despretz's waves of light described in 1836. As fan of the French, Koller praised the balloonist Jacques Alexandre César Charles' (1746-1823) megascop as a solar microscope used to obtain a projection of images enlarged with objects illuminated by a set of mirrors. Informed by his north-Italian connections, Koller praised the professor of mathematics in Modena (1815-1825) and then chief astronomer to the Grand Duke of Tuscany, Giovanni Battista Amici (1786-1863). In 1833 Amici designed his cato-dioptric microscope as a compound "periscope" instrument, which moved microscopic viewing from the horizontal to vertical position. Koller endorsed César-Mansuète Despretz's explanation according to the undulatory hypothesis as the third part of all rays of light went through the medium of transparent bodies (Koller 15b: 208, 218-220). Thomas Young described those vibrations by the spread of light in media of uniform density, while on the border between air and glass they produced oscillation needed for interference. Koller also respected somewhat obsolete Siméon Denis Poisson's (1781-1840) opinion about the corpuscular light behind obstacle in its dark spots, which enabled Fresnel's Parisian academic award despite of opposite ideas of Poisson (Koller 15b: 221). Koller noted Newtonian rays based on the theory of "fits" passing by access to easy refraction as a fourth part of all light waves. Koller discussed velocity of light also with the data of Bouguer and Lambert's refraction of light in air by citing the textbook of Georg Gottlieb Schmidt (1768-1837) (Koller 15b: 223). Koller praised the photometer as its three types of it were in common use by 1861: Benjamin Thomson Rumford's photometer, Ritchie's photometer, and photometers using the extinction of shadows, which was the most precise in those times (Koller 15b: 230).

For Koller, the achromatic prisms of Bošković's times were still a curious refutation of Newtonian optics (Koller 15b: 243-251). The achromatism followed from the works of Euler in 1747, Samuel Klingensterna (1698-1765) in Sweden in 1752 and 1765. The decisive step was Dollond's industry using several types of glass. Few years after those notes, in autumn 1838 Koller met Dollond's descendants in London. That experience certainly made him extremely proud.

Koller's birefringence of Bartholin's Copenhagen calcite as Iceland spar was the extension of the conflicting early ideas of Huygens and Newton. As an astronomer, Koller praised the micrometer (Koller 15b: 257-262, 272). He obliviously wrote about optics after 1824 and before 2nd edition of

Baumgartner's textbook appearing in January 1826, where Baumgartner on pages 380-381 cited Newton and Robert Smith's *Complete system of Opticks* translated by Abraham Gotthelf Kästner (* 1719 Leipzig; † 1800 Göttingen) as *Vollständiger Lehrbegriff der Optik nach Robert Smith's Englischen mit Änderungen und Zusätzen ausgearbeitet von Abr. Gotth. Kästner* in Altenburg/Leipzig in 1755. Baumgartner also quoted Euler, Bošković's work of 1785, Leopoldo Nobili's research in Milano in 1820, J. Priestey and Henry Coddington's book *An elementary treatise on optics*, Cambridge: J. Deighton, 1823. Henry Coddington (1798/99 Oldbridge, County Meath, Ireland-1845 Rome) was an Anglo-Irish fellow-tutor of Trinity College. On 23 October 1823 Coddington cited the undulatory fan T. Young and the corpuscular fan Biot but not yet Fresnel while on his initial page no. 1 he preferred agnostic view about the nature of light. Baumgarner still retained some agnostic points regarding the nature of light (Baumgarner 1826 2: 256-259). Baumgarner cited anti-Newtonian Goethe's *Farbenlehre* on p. 312, Young on pp. 331, 334, and Arago-Fresnel on pp. 333-334, 355. Baumgartner quoted Brewster, Wollaston, Malus, Biot, Eilhard Mitscherlich, Fresnel, Poisson, Brewster, and J. Herschel's (1820) research of birefringence on pp. 358, 364-377, and Seebeck-Brewster work on polarization at pp. 378-380. In his textbook of 1824, cited Seebeck, Samuel Clarke's Newtonians, Kästner's student professor in Halle Georg Simon Klügel's (1739-1812) *Analytische Dioptrik* of 1778, Newton, Euler, Bošković, R. Smith, Priestley, but not yet Leopoldo Nobili or Henry Coddington (Baumgartner 1824 1 (15b): 148). For Baumgartner and Koller of those times, the aether was to light the same as air was to their sound: the media needed for propagation (Baumgartner 1824 1 (15b): p. 5-6, 107). Fresnel published his pure-transversal hypothesis of polarization as *Calcul des teintes..."* (*"calculation of the hues..."*) in Arago's *Annales* in June 1821. It was too early for Baumgartner and Koller to mention it in 1824, as that transversality troubled astronomers because planets seemingly travelled without friction of any involved aether. Even in Baumgartner's second edition of 1826, the interference was explained by undulations but not by particles, and quite the opposite (vice versa) went for polarization with no transversal waves mentioned (Baumgartner in 1826 2: pp. 2: 357-358). Baumgartner seemingly endorsed both opposite ideas, particles and waves, in the way quantum mechanic did a century later. Even in his edition of 1836 Baumgartner did not solve the problem on his pages 294-296, with Encke's comet included, as Baumgartner filled cosmos with the troubling undulations of light, polarizations (Baumgartner 1836 p. 375) and interferences (Baumgartner 1836 p. 396). He noted transversal vibrations just for sound

(Baumgartner 1836 pp. 269, 272) and mentioned Lamé. That was why in 1860s Koller named Stefan's lectures "Fresnel's theory of optics" as he was still educated in doubts of true nature of light while Stefan absolutely grew up with Fresnel's transversal waves as the only alternative. Therefore, Carnot was also not included in Koller's and Baumgartner's output of 1820s and 1830s before Clapeyron's praised Carnot in 1834.

Koller's electrodynamics

In the first of his three parts of lectures about the electricity designed as "A) Electromagnetic phenomena", Koller admired Faraday's discoveries promoting electrodynamics (Koller 15b: 4-5). Ampère put forward those ideas by his tube of papier or wood curved into a cylinder and spirally overlapped by the copper wire. Ampère designed a coil to form a magnet whose poles followed the directions of the current.

Oersted's discovery stimulated all physicists and Faraday among them with his multiplicator. Ampère designed flat spindle, Nobili and Antinori followed with their discoveries. Koller underlined with his red ink all surnames noted in that part of his course (Koller 15b: 6). Leopoldo Nobili (1784 Trassilico in Toscana-1835 Florence) collaborated with Vincenzo Antinori (1792 Florence–1865 Florence) who founded the Italian Meteorological Archive and researched the electromagnetic induction. The priest Salvatore del Negro (* 1768 Veniɔe; † 31 January 1839 Padua) became the professor of physics in university of Padua in 1831. He experimented with electromagnets and from 1832 he researched the locomotion of electrical battery. He built an electric motor that could lift 180 grams by the height of one meter in one minute. It was probably the first electric engine with a quantifiable power output. Hippolyte Pixii (1808-1835) also constructed his own alternator. In 1809, Dal Negro invented the oligochronometer, an instrument needed for the precise measurements of the smallest fractions of time, which was the secret wish of Galileo two centuries earlier. Since 1794, Dal Negro was a member of the Academy of Padua, in 1835 he was admitted to the Accademia dei XL, then he became the academician in Modena at today's Accademia Nazionale delle Scienze. Besides other learned societies, he was also a member of the Accademia Virgiliana in Mantua. In 1838 the emperor Ferdinand I awarded him the Order of the Iron Crown (Salvatore del Negro, *Nuove esperienze ed osservazioni elettro-magnetiche*, Padua, 1831). Koller probably noted so many now almost forgotten (north) Italians because as a Benedictine Koller had many connections there as he was also a member of Paduan learned society. Benedictine monasticism is fundamentally different from other Western religious orders insofar as its individual communities are not part of a religious order with "Generalates" and "Superiors General". Each Benedictine house is independent and governed by an abbot while the Benedictines swear to reside in their domestic monastery. In Koller's times, two Benedictines ruled as Popes. The Pope Pius VII (1742 Cesena in Papal states-1823) reigned in 1800-1823, and the Pope Gregory XVI (1765 Belluno in Venetian republic-1846 Rome) reigned in 1831-1846.

Koller's electromagnets figured as magnet-models with physiological, optical, and chemical effects. Koller's magnet produced light and sparks by Parisian designs of Hippolyte Pixii (1808-1835). As an instrument maker Pixii built an early form of alternating current electrical generator in 1832 (Koller 15b: 7, 9-11).

Clarke, Saxton and von Ettingshausen later advanced the apparatus with additional condenser and electroscope (Koller 15b: 12, 490). In his 1845 edition of textbook, Baumgartner noted Pixii's work of 1832 bettered by Saxon, Clarke, von Ettingshausen, but also by dr. Franz Petrina (František Adam Petřina, * 1799 Semily, Bohemia; † 1855 Prague). Ettingshausen's collaborator-student Petrina taught as the imperial-royal professor in Linz in 1837-1844 while he published on kaleidoscope in *Ann.Phys.* in 1840 and about the voltaic experiments in the *Wiener Zeitschrift für Physik, Chemie und Mineralogie* in the same year 1840. In 1847 he wrote about electroscope in Prague and in *Nuovo Cimento*. Carl Wilhelm Hermann Brandes (1814 Wroclaw-1843), W. Julius, H. Michaelis in 1844, and Pouillet in his textbook translated-bettered by Müller in 1845 additionally noted Emil Stöhrer's work in Leipzig in 1843. Irishman Edward M. Clarke had started out as an optician, first in Dublin. By the 1830s, he had moved to London with his firm "Clarke optician and magnetician, inventor & manufacturer, 11 Lowther Arcade, Strand, London". In 1835, Clarke brought out a 'magnetic electrical machine' for which he was accused of plagiarism by Joseph Saxton, an American instrument maker employed at the Gallery of Practical Science known as Adelaide Gallery in London. Saxton himself was allegedly plagiarising Hyppolyte Pixii's machine in 1832, but to this day the undecided priority is still out on that issue. In fact, one plagiary accused another of his own crime, just like good old Publius Clodius Pulcher.

Ampère's hypothesis about the nature of magnet was Koller's favourite besides Faraday's proposed magneto-electric phenomena. Arago put magnet on motive disc to enable Faraday's rotation current. Arago used multiplicator's wires of copper or lead and put his collector on the disc (Koller 15b: 12-14).

The electric agents and Ampère's magnetic rotation-electrodynamic spiral make so-called rotational-magnetism. That invention was declared by Arago with a current working on the periphery. Power of a big magnet moves the magnetic needle to produce chemical and thermal effects as the cause of electricity in the center of electrical voltage-tension. The electromagnetic appearances included the polarity as researched by Franklin, Beccaria, and Wilson. Next advancements were produced by Cavallo's Leyden jar's magnetization and in Copenhagen by Oersted's influence of electricity on the magnetic needle as declared by Ampère (Koller 15b: 16-18, 20-23). Some used even 1200 pairs of plates for their galvanic element. Next in his course, Koller discussed geomagnetism, exceptionally with almost no illustrations even if precisely that might have been his own favorite topics (Koller 15b: 25).
Each molecule of a wire becomes magnetically polarized while the electrical current passes through. That effect of a magnet on a movable polar wire was discussed by Biot, Savart, Schmidt, Felix Savary, and Peter Barlow (Koller 15b: 27-28). The milestone was Ampère's (1775-1836) article (Ampère, Théorie des phénomènes électro-dynamiques, uniquement déduite de l'expérience, *Memoirs de l'Academie* 6, 1823, with further editions appearing in 1825, 1826, 1827; Koller 15b: 29).

Johann Salomo Christof Schweigger's (1779–1857) galvanometer was developed in Erlangen in 1820, while he corresponded with Goethe. Koller illustrated his achievements and added the data of A. Volta and P. Configliachi's student in Pavia Stefano Giovanni Marianini (1790 Zeme by Mortara, Novara-1866 Modena) (Koller 15b: 31-32). Marianini became a docent at Pavia University in 1817. He taught in Venezia in 1821-1835, and in Modena from 1836. He researched the electromotors in 1823-1840, and especially in 1825, including his measurements of the forces of geomagnetism. Nobili's galvanometers designed by Ruhmkorff were popular, as was the multiplier of Swiss physicist Jean Daniel Colladon (1802 Geneva-1893 Geneva) who worked in the French laboratories of Ampère and Fourier. Colladon discovered light fibre guided by total internal reflection. In 1831, he measured the polarity of torpedo fish in French coast but published about it only in 1836. Colladon nearly missed the discovery of Faraday's induction. Colladon studied one of Koller's favourite topics, the atmospheric electricity. In Paris, Colladon connected his galvanometer to the guiding wires of kites which were then flown at great heights as the repetition of probably imagined B. Franklin's experiments (Koller 15b: 33-34). Heinrich Daniel Ruhmkorff (Rühmkorff, 1803 Hannover-1877 Paris) produced his great generator while Leopoldo Nobili (1784 Trassilico in Toscana-1835 Florence) developed the astatic galvanometer in 1825.

By Oersted's example, Ampère proved action to be equal to reaction while the directing of the axes of magnet by the electrical current was examined by the Swiss physicist Auguste-Arthur de La Rive (1801 Geneva-1873 Marseille), who was one of the founders of the electrochemical theory of batteries (Koller 15b: 34).

Davy and Arago designed their friction experiments (Koller 15b: 35). Felix Savary researched the force of magnet. With his magnet, he applied a transversal force on a current-carrying wire, which was first observed by Wollaston and Faraday (Koller 15b: 38-39). The rotation phenomena was first demonstrated by Davy (Koller 15b: 41). Certainly, Koller's numeration of paragraphs did not always follow Baumgartner's first three editions of textbook where the paragraphs about electricity continued those of mechanic and consequently had much higher numbers. In Baumgartner's 4th edition of a textbook appearing in 1832 each branch of physics already had separate numbering of paragraphs, but after mechanics there was optics and magnetism before electricity. Moreover, Baumgartner mentioned just Colladon's acoustic experiments.

Koller's magnetism of earth's hemispheres followed Faraday's explanations (Koller 15b: 43-44). Koller added informal annotation about stability deduced from his own measurements of magnetism of the Earth after he influenced the geomagnetic force by rotatory motion (Koller 15b: 45); he apparently rotated his conductor in geomagnetic field. Koller illustrated the effects by arrows, the hemispheres, and conductors to prove his transversally working force by Ampère's theory. Ampère examined cross sections of his working current, as did Johann Wilhelm Ritter and Oersted. Ampère's small magnets were called electromagnetic solenoids by a Greek word for a cylindrical coil of wire acting as a magnet when carrying electric current (Koller 15b: 47, 48-50, 51-52).

The second part of Koller's notes about electricity was designed as "B) Thermoelectric phenomena". The main inventors were Oersted and Thomas Johann Seebeck (1770 Tallinn, Estonia-1831 Berlin) in Berlin in 1822. The tourmaline and the changing temperature of natural crystals was called pyroelectricity by David Brewster in 1824, while the stick of antimony was also examined. Seebeck's chains defined eastern and western orientation of tellurium to rearrange the metals into a scale by their thermoelectricity, similar like previous arrangements of metals by Voltaic abilities. So-called Darcet's alloying (Legierung) was discussed by Jean Darcet (d'Arcet, 1724-1801 Paris) in 1789, at the beginning of French revolution (Koller 15b: 53-54, 55, 56, 556, 652). Darcet was a chemist and the director of the porcelain works at Sèvres who introduced low-fusing metal alloy, which was also used in dentistry somewhat later. Koller explained to his students a theory of Seebeck's effect (Koller 15b: 57), as well as Oersted's and Fourier's polygon as a thermoelectric battery which was adorned by Koller's illustration (Koller 15b: 58-59). Oersted and Fourier with their thermoelectrical battery, as well as Becquerel in February 1827, also participated in research as Koller commented in his right column (Koller 15b: 61). Antoine César Becquerel (1788-1878) became a French pioneer in the study of electric phenomena. Koller again described the polarity of Earth by Ampère's hypothesis (Koller 15b: 63).

The third part of Koller's description of electrical phenomena was called "C) Electrochemical Appearances". They were not explained by contact (of Volta's theory), but by the changing of chemical constitution of metals (Koller 15b: 64). Wollaston used both plates outside the fluid like in Zamboni's dry battery. Plates with zinc were covered by copper. Koller met Wollaston few years later in London in 1838. Davy examined seawater for the Royal Society of London (Koller 15b: 65, 461, 463, 568-570, 610). The negative electric state was proved by conduction through the leaves of plants by Davy, and by Dumas' conduction of water in Paris. Jean Baptiste André Dumas's (1800-1884) ideas on organic analysis and synthesis challenged Berzelius's electrochemical dualism of all molecules where salts are composed of basic and acidic oxides. Dumas experimented with water at porcelain works at Sèvres. The mutual relations between chemistry and electricity were noted by three letters of H. Davy translated as: *Sur les Relations qui existent entre les actions chimiques et les actions électriques* (Koller 15b: 67-69). That was Koller's somewhat inexact citation of Davy's paper *Sur les relations qui existent entre les actions électriques et les actions chimiques,* printed in *Annales de Chemie* in Paris in

1826 in Volume 3: Tome XXXIII. It was translated from Davy's paper read in London on 6 June 1826, three years before Davy's death.

Koller praised Avogadro, Oersted, Becquerel and other physicists including Davy's experiments. They also used wires of platinum (Koller 15b: 69, 71, 72). Albert van Beek (Beck, 1787 Utrecht -1856 Utrecht) researched the positive effects of zinc as the protective coating for corrosion. He corresponded with Ampère in 1821 and reported about his success in Utrecht in 1831 as he later mentioned in an article about the protection of ships. It contained Albert van Beek's comments on an article of Sir Davy concerning the alleged property of tin to protect iron from oxidation in sea water (Albert van Beek, Bemerkungen über einen Aufsatz von Hrn. Davy, die vorgebliche Eigenschaft des Zinns, das Eisen vor Oxydation im Meerwasser zu schützen, betreffend, *Leibig's Annals of Pharmacy* in 1838, as translated from the *Annals de chemie et physiqie*; Koller 15b: 74).

The galvanic multiplicators of platinum were bettered by Schweigger's multiplicator while the Francophonic Koller especially praised Becquerel (Koller 15b: 76).

Koller's Chemistry - electrolysis

Koller was interested in chemical decompositions by electricity in binary compounds of dual atoms of ammoniac or chlorides as explained by Faraday's quantity of electricity that will cause a chemical change of one equivalent weight unit, and by atomistic view of electrodes (pole wires, Koller 15b: 115-116, 118). Faraday's measurements convinced Koller about the chemical equivalent's theory of affinity with both kinds of electricity participating in compounds of bodies as they were supposedly needed to discharge a conductor (Koller 15b: 120-122).

Koller's chemical affinity involved the collisions between the smallest particles of bodies of similar kind to propel the affinity by (electrical) current. They produce the chemical moment as the function of mass of the colliding smallest bodies; it happens in saltpeter or kalium by the greatest distribution. That was pretty much Koller's kinetic atomist view, which was two decades later professed by Clausius, Maxwell and Stefan! Koller certainly expressed there his atomistic viewpoint to his students. The diminished cohesion of the smallest particles makes bodies liquid at their changed temperature (Koller 15b: 125-126). The fluidity of oxygen inside its compounds connects the liquids' constitution to their chemical affinity, also in saltpeter or turpentine. Koller discussed the atomic weights of hydrogen and ammoniac in his right column reserved for modernized Baumgartner's data (Koller 15b: 128-130). Such mathematical treatment of the atomic weights was extremely popular scientific entertainment of those times and ultimately produced Mendeleev's periodic table in 1869, just five years after the Englishman John A.R. Newlands put forward his Law of Octaves.

Thermoelectricity

After his lectures on chemistry or at least electrolysis, Koller discussed the electricity again with his thoughts concentrated on the multipliers including the intensities of currents and their conductions in secondary wires, sources of electricity, and their influences of magnetic needle (Koller 15b: 131, 134). He applied those theoretic laws to the Seebeck's thermopiles arranged by Leopoldo Nobili and Melloni in 1831. In France and Italy Macedonio Melloni (1798-1854) dealt with thermal radiation to develop the thermopile for quantitative measurements of the intensity of the radiation with fifty-six thermoelements of antimony and bismuth. In his multiplier Pouillet used a compound of copper, named the royal of copper (Koller 15b: 136-137). Koller researched the intensity of current caused by multiplicator to research the effect of current on magnetic needle, which is proportional to the distance and intensity (Koller 15b: 141).

Electromagnetism

Koller admired Ampère's collaborator Felix Savary (1797 Paris-1841 Estagel near Perpignan) who worked with Ampère when electromagnetism was just at its beginning. Savary issued his famous *Mémoire sur l'application du calcul aux phenomènes électro-dynamique* in 1823 (Koller 15b: 155). In 1827, Savary published about rotation of magnets. In 1827 Savary studied the intensity of magnetism produced by an electrical discharge and applied the laws of gravity to determine the orbits of double stars circulating close around each other. Of course, Savary's astronomy was especially appealing to Koller. Koller analysed the effect of linear infinite current on the pole of a magnet in the approach close to Stefan's lectures a quarter of century later as Stefan was similarly preoccupied with the mathematics related to geometrical shapes of his models (Koller 15b: 156). Koller checked Biot and Ampère's theories of solenoids in his anotations about Koller's own experiments with the magnetic needle performed in his Kremsmünster laboratory (Koller 15b: 177). There Koller proved the formation of attraction and repulsion of electrical current as the position of its equilibrium increases by the small elements and decreases with the small elements; that approach of infinitely small steps was useful for the needs of Koller's favourite mathematical calculus. Koller soon used the electrochemical cell invented by a British chemist and meteorologist John Frederic Daniell (1790-1845) in 1836. Koller certainly got Daniell's cell very quickly, at least during his Londoner visits in autumn 1838, although Koller did not mention Daniell in his travelogues. Koller's cations and anions did not travel with the same speed which supported the opinions of Faraday's Londoner friend Daniell. Koller rotated his magnet around the small polar conductor wrapped around the cylinder's axis. Koller used Barlow's wheel with greatest elements as an early demonstration of a homopolar motor, designed and built by English mathematician and physicist Peter Barlow (1776-1862 London) in 1822. Barlow collaborated with Dollond on achromatic lens and criticized the ideas of electromagnetic telegraph in 1824, which Koller certainly disliked because Koller supported the telegraph of John Herschel's Scottish protegee William Ritchie (abt. 1790-1837). Barlow's criticism somehow dogged the early development of telegraphs in Britain. Koller also experimented with the reaction force of the magnetism of Earth (Koller 15b: 178) in his contributions to Gauss-Weber's research projects. Koller examined the inertia of Ampère's solenoid and checked the rotation of poles with Faraday's stative made from platinum conductors.

Koller promoted his own future studies of geomagnetism published after the year 1839 in the magazine *Resultate aus den Beobachtungen des Magnetischen Vereins*, edited by Carl Friedrich Gauss and Wilhelm Weber (Koller 15b: 345-346; Baumgartner, 1826 # 291 p. 401, # 293 p. 433). As usual in his courses, Koller began with some history including the linen-draper William Henley's (Henly, † 1779 London) electrostatic experiments with quadrant and electrometer performed around 1772-1774. Koller noted an English arctic explorer William Scoresby (1789-1857) who criticized Koller's later friend Airy (Koller 15b: 351; Baumgartner # 399 (537)). Koller illustrated a more complicated Coulomb's torsion wire balance in a suspended magnetostatic experiment (Koller 15b: 352-353; Baumgartner 1826, # 340 (470)). Those data formed the introduction to Koller's beloved geomagnetism (15b: 356-360; Baumgartner 1826 # 375, 378 & # 379, pp. 483, 493, 512). Koller praised the measurements of Biot, the determinations of meridians of Paris, Paraguay, and Cayenne. Koller also noted measurements in Lapland, probably those accomplished by Maximilian Hell. Koller also praised Halley's measurement of geomagnetism extended from 52 degrees north to 52 degrees south performed on a pink *Paramour* from September 1699 until 6 September 1700, two years before Halley visited Carniola of then still childish Koller. The Lutheran chemist-physicist Adolph Theodor Kupffer (Adol'f Iakovlevich Kupfer, 1799 Jelgava-1865 St Petersburg) studied mineralogy with Christian Samuel Weiss in Berlin. Kupffer researched the terrestrial magnetism in Kazan after 1824 and founded meteorological observatory about 1829 in connection with arctic expedition at Tobolsk and Irkutsk of the Russian admiral Pyotr Fyodorovich Anjou (Пётр Фёдорович Анжу, 1796-1869) and a Baltic German explorer-seaman of Russian navy Ferdinand Friedrich Georg Ludwig baron Wrangel (Фердина́нд Петро́вич Вра́нгель, 1797-1870). Kupffer researched black prismatic crystals at the granite of mountains of Verkhoturye (Verchoturie, Верхоту́рье) in the middle Ural Mountains together with Georg Adolf Erman (Georgius Adolphus, * 1806 Berlin; † 1877 Berlin), a physiꞏist and geoscientist, who travelled from Petersburg to Tobolsk together with the Norwegian astronomer Christopher Hansteen (* 1784 Christiania (Oslo); † 1873 Oslo) to accomplish their geographical-geomagnetical measurements in Tobolsk, Irkutsk, and Jakutsk in July 1828. Koller's interests in Arctic ices soon prompted Stefan's use of the artic expeditions' data for Stefan's movable thermodynamic border. Koller's knowledge about artic expeditions certainly opened doors to Stefan's research of movable phase border in melting ice four decades later. Arctic explorations soon became the basis of Stefan's problem of movable barrier between the ice and water.

Koller discussed Cassini and Arago's magnetic deviation as the error induced to compass by local magnetic fields, which must be considered along with the magnetic declination if accurate data are to be calculated (Koller 15b: 361; Baumgartner 1826 p. 484 # 376). More loosely, "magnetic deviation" means the same as "magnetic declination" connected with Sunspots' maximum which belonged to the most fruitful research of Koller's observatory.

Koller noted Peter Barlow's research of magnetic needle. Barlow experimented together with John Bonnycastle (1751-1821), a self-educated British master of mathematic. Bonnycastle eventually became an instructor of mathematics at the Royal Military Academy in Woolwich where Barlow's artillery school was also located (Koller 15b: 367-369, 371, 376; Baumgartner 1826 page 477 # 364 (500)). Koller mostly quoted those English data by Schmidt's textbook printed in Giessen.

Koller discussed electric current, Arago-Barlow's rotation of magnet and Arago's disc including its advancements by Seebeck, Colladon, Prevost, Nobili, Liberato Baccelli (1772 Lucca–1835 Lucca), Poisson, and by the research of violet rays of Lady Sommerville in the summer of 1825 (Koller 15b: 387, 389, 392-394, # 347; Baumgartner 1826 p. 481). William Wallace's and after 1827 Lord Brougham's correspondent Mary Fairfax Somerville (* 1780-1872 Naples) began her scientific investigations in the summer of 1825, when she experimented with magnetism. In 1826 she presented to the Royal Society her paper entitled "The Magnetic Properties of the Violet Rays of the Solar Spectrum". In 1835 she and Caroline Herschel were elected to the Royal Astronomical Society, as the first women to receive such an honor. The other advancements included the work of Phillip Jolly (1809-1884) who was the friend of Frenchman Jacques Frédéric Saigey (1797-1871).

Electrochemical appearances

Koller again repeated his introduction to electricity by explaining the meaning of electron borrowed from the Greek language (Koller 15b: 402, 405). Stephen Gray detected the electricity transferred by contacts using conductors in 1729. Jacques Charles François Sturm (1803 Geneve-1855 Paris) and Sturm's friend Colladon did not confirm Grotthuss's ideas as the conductivity of their water did not change between the pressures of 1 to 30 atmospheres, which Baumgartner did not know in 1826. Christian Johann Dietrich Theodor von Grotthuss' (1785 Leipzig-1822 Lithuania) charge was not transported by the movement of particles but by breaking and reshaping of their bonds. Grotthuss published there the first basically correct concept for the charge transport in electrolytes designed during his stay in Italy in 1806. Koller discussed Coulomb's balance as electrometer with small sphere made of rubber lacquer and put at the end of Coulomb's balance (Koller 15b: 407-408; Baumgartner 1826 # 274 pages 423-424). Koller supposed that the velocity of electricity seems to be infinitely great according to the measurements of his days; at least people were unable to measure it during his era. Three decades later in 1857, Gustav Kirchhoff researched the speed of electric signal along a non-leaking wire and concluded that the velocity of propagation of an electric wave is very nearly equal to speed of light in a vacuum. That velocity was later measured by Koller's academical great-grandson, the Carniolan Ignac Klemenčič, who used to be a student of Stefan's student Boltzmann.

Koller provided a table of abilities to produce a frictional electricity by different substances, arranged between the fur of cat's negative electricity and positive electricity produced by the rubbing of a wool rag. Koller used Biot's table of influences of changed temperature as Koller's updated Baumgartner's data of 1826.

Koller provided a table of repulsive forces (Koller 15b: 409; Baumgartner 1826 # 275 page 424). They were the function of dimensions of both spheres at the ends of Coulomb's torsion balance and their mutual distances, as proved by Koller's own measurements.

Koller noted the loss of body's electricity because of the humidity of air, which seem to be a problem resembling Koller's and later Stefan's loosing of heat by conductivity (Koller 15b: 411 # 278). Koller also considered the timing because the repulsive force of his electrified sphere of diameter 5-6 lines depended on time. In shortest time, the (electrified) state of air became proportional to the electrical force, but in those times Koller had no proper tools to measure that time exactly. Even if function of time might be beyond the European experimental capacities of 1820s, the space dependency was not, as it became clear that the free electricity (resembling electrons half of a century later) flows just through the surface and not inside the body and therefore does not cause the chemical effects inside the body (Koller 15b: 413; Baumgartner, 1826 # 280 page 427). That phenomenon was later called a skin effect for AC, first described by Horace Lamb in *Phil.Trans.* for the case of spherical conductors in January 1883. It was generalised to conductors of any shape by Oliver Heaviside in 1885. For DC, the skin effect does not occur as the entire cross section of the conductor carries the DC current. Obliviously, Baumgartner (and Koller) understood skin effect in 1820s, even if they did not relate it to then still unresearched AC current. The first alternator producing the alternating current (AC) was a dynamo electric generator based on Michael Faraday's principles constructed by the French instrument maker Hippolyte Pixii in 1832, therefore few years after those Koller's lectures. Pixii later added a commutator to his device to produce then more commonly used direct current, which angered the Graz student Nikola Tesla by its sparking. The earliest recorded practical application of alternating current (AC) belongs only to Guillaume Duchenne in 1855, therefore a quarter of a century after those Koller's writings. Koller never met the Parisian instrument maker Pixii who prematurely died three and a half years before Koller's Parisian visit.

As most of professors of his times, Koller loved previous century historical roots of his modern explanations including Coulomb or Tiberius Cavallo's collector (also called Nicholson's duplicator), described by the Leipzig based Johann Samuel Traugott Gehler (1751-1795). Koller also praised the zinc-plate doubler as an application invented by Abraham Bennet (1749 Derbyshire-1799) in 1787-1789 (Koller 15b: 415-416; Baumgartner 1826 # 302 page 437). It was a cleverly designed device for the continuous doubling of an initial small charge through a sequence of operations with three plates.

Koller discussed the force of cohesion of the curator of the mathematical and physics cabinet of the Royal Bavarian Academy of Sciences in Munich Julius Conrad knight Yelin (1771 Wassertrüdingen in Bayern-1826 Edinburg) and Becquerel to explain the changes of electricity by pressure and temperature in the spirit of J. Fourier and later Lamé's and Stefan's research (Koller 15b: 419, 657, 667; Baumgartner 1826 # 308 page 440, page 399 of # 397-398). The main example of Koller was a tourmaline, initially brought in larger quantities to Europe from Sri Lanka by the Dutch East India Company and popularized by David Brewster's term pyroelectricity in 1824 (Koller 15b: 423; Baumgartner 1826 # 309 page 441). The schorl (black tourmaline) was classified by Haüy whose textbooks Koller used as the student in Napoleonic Ljubljana. Nicolas Lemery (Lémery, 1645 Rouen-1715) analyzed that "Pierre de Ceylan" in 1717 while Torbern Olof Bergman made his first attempted chemical analyzes needed for mineral determination in 1779. Koller added to Baumgartner's edition of 1826 the consequences of Becquerel's experiments (Becquerel, Parisian *Annales de chimie et de physique* 1828 volume 37: pp. 5-15). Becquerel's mysterious hypothesis involved atoms as a source of electrical forces like those involved in the heat's generating the electricity in tourmaline. There were some speculations about the manner how the atoms produce electricity as Becquerel published at his concluding page no. 14. Comparatively exotic tourmaline was a rare object for the early researchers of electricity as its importance resembled the similarly exotic torpedo fish (electric rays) known to ancient Greeks as a fairy tale monster while its scientific examinations were not widely known before Cavallo's research in 1795.

Koller still echoed the early Volta's debates about the source of electricity including the electricity by contact in Volta's metals (Koller 15b: 424; Baumgartner 1826 # 311 page 442). Koller explained the Habsburgian North Italian Stefano Giovanni Marianini's (1790 Zeme by Mortara, Novara-1866 Modena) data and Davy's dependence of electric force on temperature in metals and fluids (Koller 15b: 426-427; Baumgartner 1826 # 312 page 443). Koller was certainly especially interested in achievements of Habsburgians like Friedrich count Stadion (of Thannhausen line, * 1774 Bamberg, † 1821 Naples), a chemist and landowner in Bohemia employed as a canon in Bamberg. In 1815, independently of earlier Humphry Davy's work in 1811, Stadion discovered chlorine dioxide and in 1816 he researched the perchloric acid and announced his results in *Ann.Phys.* In 1817 he published on the improvement of galvanic cells. His father was Joseph Georg Johann Nepomuk count Stadion (1749 Mainz-1814 Trhanov, Domažlice District in Plzeň Region of Czechia) of Thannhausen-Philippin Line who used to be a count of Mainz and the higher silver courtier at Viennese court. He was married to Sophie baroness Wambolt von Umstadt. Koller and Baumgartner were also aware of the new American contributions of Robert Hare (1781-1858), a chemist and experimental producer of chemical apparatus including Hare's calorimeter, the deflagrator arranged for the partial condensation of a multicomponent vapor stream on zinc plates, copper, and the oxyhydrogen blowtorch. Hare served as a professor of chemistry at the University of Pennsylvania for nearly three decades. Toward the end of his life Hare was an outspoken supporter of spiritualism, just like Crookes used to be later, which was probably not to Koller's and J. Stefan's liking (Koller 15b: 429, 443; Baumgartner 1826 # 315 page 447, # 321-322 page 454).

Koller discussed the electrolysis of mercury put on the thin surface of other liquid, for example sulfuric acid (Koller 15b: 431; Baumgartner 1826 no. 6 of # 319 page 449). It was charged by inserted platinum wire in experiments promoted by Davy and J. Herschel in *Annales de Chimie et de Physique* at volume 28 in 1825. Friedlieb Ferdinand Runge (1794 Hamburg-1867 near Berlin), as an analytical chemist teaching in Wroclaw until 1831, identified the mydriatic (pupil dilating) effects of belladonna. In 1819 he presented it to the fascinated Goethe who gave him some coffee from which Runge soon extracted caffeine. Koller commented the effect of heat on fluid between the plates of battery and discussed the effect of friction in electrolysis of water dissolved into oxygen and hydrogen as described by William Nicholson and Anthony Carlisle in 1800. Their achievements were reported by Benjamin Scholz (* 1786 Rosswald (Slezké Rudolti□e in Bohemia); † 1833 Heiligenstadt in Vienna). Scholz replaced Prechtl as professor of chemistry at the Viennese Polytechnic. In 1827, like Baumgartner later, Scholz became the director of porcelain manufacture in Vienna-Rossau and a head of another manufacture of mirrors. Scholz published one of the main Koller's textbook sources *Anfangsgründe der Physik als Vorbereitung zum Studium der Chemie* in Wien by J. G. Heubner in 1821, and the *Lehrbuch der Chemie: In zwey Bänden*, printed in Wien by S.F. Beck in 1829-1831 (Koller 15b: 433; Baumgartner 1826 nos. 7 and 9 of # 319 page 450). Koller followed Davy's example of 1806 as Koller obliviously performed his own chemical experiments with deoxidation to examine the constitutional particles of oxygen, chlorine, and iodine, therefore the molecules. Koller also quoted the research of water and oxygen presented by Koller's Ljubljana professor of mathematics Jožef Jenko (1776 Kranj-1858 Vienna) in Jenko's effort to determine the electric state of metals in battery (Koller 15b: 434-435).

Koller reported on charging of a battery (Ladungsäule) of Johann Wilhelm Ritter, later disputed by Jean-Baptiste Van Mons (1765 Brussels-1842 Leuven) in Brussels and by Volta's friend Luigi Valentino Brugnatelli in Pavia in 1805 (Koller 15b: 439-440; Baumgartner 1826 # 333 page 459). Koller additionally clarified the phenomena of Ritter's charging of battery with its included perishable substances. Koller denoted the chemical elements in his formula by rectangles, differently from the decades later Loschmidt's and Kekulé's circles. Koller probably wrote those lines before he heard about the premature death of H. Davy (1778-29 May 1829).

Koller certainly praised the professor of imperial Habsburgian Lyceum in Verona, Koller's colleague priest Giuseppe Zamboni (1776 Arbizzano in province of Verona-1846 Verona). In 1812 Zamboni invented an early electric dry battery resembling the earlier voltaic pile. In 1819, its advanced version was announced by the medical councilor physician Karl Christoph Friedrich Jäger (1773 Tübingen, † 1828 Stuttgart), or less probably by his brother Georg Friedrich Jäger (1785 Stuttgart-1866 Stuttgart) (Jäger, Ueber die electrischen Wirkungen so viel wie möglich ausgetrockneter Papiersäulen, published in *Annalen der Physik und der physikalischen Chemie*, 62/7 (1819): 227-246; Koller 15b: 441, 443, 449; Baumgartner 1826 # 320, 321, # 322 on page 454, # 330 page 457). Two small Zamboni's batteries were used by Becquerel in his design equipped with the electroscope of Johann Gottlieb Friedrich von Bohnenberger (1765 Simmozheim, Württemberg-1831).

Koller's increased heat changed the length and thickness of a wire with its nature included, as proved by great Davy's battery consisting of 50 pairs of platina immersed in oil and water to present the influences on nature of a wire (Koller 15b: 445; Baumgartner 1826 # 326 page 455). Koller approved Biot's ideas focused on electrical conduction as a function of temperature, which certainly also pleased J. Stefan (Koller 15b: 447; Baumgartner 1826 # 326 page 455 no. 2).

Koller recalled how Marianini examined the conductivity of fluids which quickly increased with temperature even in less conductive fluids; in the solution of salts, the conductivity is proportional to their concentration. Koller performed his own experiments while changing the temperature of his platinum wires. He endorsed Davy's conduction and electromagnetic effects in fluids including the electromotors themselves advanced by Becquerel's experiments (Koller 15b: 449, 451-453; Baumgartner 1826 # 330 page 457, # 332 page 458, # 334 page 459).

Unlike later J. Stefan, Koller was still in doubts about the theory of electricity in 1820s. Therefore, he described both, the unitarist ideas of Aepinus and Franklin, and the dualist suppositions of Robert Symmer (1707–1763). Symmer was a Scottish philosopher and physicist, known principally for the now defunct dual fluid theory of electricity. Koller followed the description of the astronomer Johann Tobias Mayer (1752 Göttingen-1830 Göttingen) who used to be Tobias Mayer's († 1762 Göttingen) son and Kästner's student. Johann Tobias Mayer published his *Naturlehre* in Göttingen in 1823. Koller compared the measurements of temperature to check both opinions about the electrical fluid(s) (Koller 15b: 455, 459; Baumgartner 1826 # 338 page 463). He even discussed three opinions of his contemporaries, which were soon resolved by Faraday's victory in J. Stefan's era: 1) Volta and many others had a contact as the cause of electric phenomena, 2) Davy upgraded Volta's opinion to endorse electrical action and intensity, 3) Wollaston, Felice Fontana's collaborator Giovanni Valentino Mattia Fabbroni (1752 Firenze-1822 Pisa), De La Rive, and others promoted their theory of chemical action which was first advanced and later developed by Oersted, Becquerel, Ritchie, Pouillet, and Christian Friedrich Schönbein (1799 Metzingen in Swabia-29 August 1868 Basel). Schönbein was a professor in Tubingen and in Basel after 1827. In 1838, he invented the fuel cell independently of Grove, and soon afterwards detected ozone as well as an explosive artillery smokeless mixture of nitric acid and sulfuric acid. Even if Franklin was in fact already victorious, Koller noted that Franklin's hypothesis was not without trouble, while Symmer also had a point. Koller detailly discussed the phenomena of repulsion and contraction proposed by Franklin. Symmer's dualism also explains the repulsion, atoms, and their influence in bodies, at least by Koller's opinion in those times (Koller 15b: 457, Baumgartner 1826 # 339 page 463).

Koller discussed the smallest particles and praised Davy's use of copper for protecting ships from seawater, published in 1825 in French translation as "*Relations…*" It was extremely important for emerging Habsburgian Adriatic navy and later Ignatius Knoblehar's expeditions. Koller's and Baumgartner's chief references were Franklin's Letters about Electricity published in Leipzig in 1758, and George John Singer's (1786–1817 London) Londoner lectures with Faraday attending as published in Singer's bok (Singer, *Elements of Electricity and Electro-chemistry*, London, 1814; Koller 15b: 461, 463, 609-613 as a clear copy; Baumgartner 1826 # 340 page 464 2nd part). Singer's work was translated in Wroclaw and Milano in 1819. As the local-patriots, Baumgartner and Koller loved to praise the Habsburgian Balthasar Stanislaus Leschan (Lešan * Petrinja in Military Krajina). Balthasar Leschan was probably born in Petrinja as a son of Styrian official sent to Military Krajina. Leschan obtained his PhD in philosophy to become the member of faculty of Medicine of Viennese university. Leschan published his Viennese medicinal dissertation *Electrologiae pura e principia generalia* in August 1825 and printed it next year in translation as *Grundzuege der reinen Elektricitaetslehre* in Vienna by Carl Gerold. He used to be Jacquin's Viennese assistant for chemistry and resided at the Viennese Bauernmarkte no. 288 in 1830 (Leschan, 1826, p. VIII; *Hoff- und Staat Wien* 1826: 117, 1830: 112, 1831: 110). By 1831 he already left Vienna, but he was still a member of Viennese medical faculty. Probably, he was on his way to lower Styria as he was not noted in the schematism of *Hoff- und Staat* in 1825 or 1832. Koller certainly knew Leschan very well. Koller's best friend Simon Stamfer taught mathematics in Salzburg at those times while the Benedictine Dominic Erlacher taught mathematics in Kremsmünster already in 1818, when Koller lectured as a provisional professor of Natural History in Kremsmünster (*Hoff- und Staat* 1818: page 159, 1832, 1835). In 1826, Leschan's *Grundzuege der reinen Elektricitaetslehre* was dedicated to the count Thomas Keglević of Buzin (Tamas, 1768-1850 Lobor in Croatian Zagorje) who worked as the actuary of higher schools in Zagreb district in 1802. According to Leschan's introduction dated on 1 July 1825, the lightning is not the only natural electricity. Leschan borrowed his data from Davy, Berzelius, Biot, and Singer. Leschan was also grateful to Jacquin. Electricity accompanies the changing of aggregational state as noted by Seebeck, Grotthuss, and by the unitary one-fluid theory of Paul Traugott Meißner (Meissner, * 1778 Medias by Sibu in Transylvania; † 1864 Vienna-Hernals) (Leschan, 1826, final page 158). Leschan almost followed the explanations published in Baumgartner's textbook of 1824. In 1893 his probable son the physician Lešan (Lešun) from Ljutomer participated in the exhumation of the remains of count Janko Drašković who died during his spa treatment in Radgona. The count's bones were transported to his final burial site in Croatia. Lešan family still lives in the areas of Ljutomer particularly in settlement Lešane in municipality of Apače by Gornja Radgona 30 km northwest of Ljutomer. The Croatian members of the family Lesan are rare, residing mostly in Krapina Županija and Slavonski Brod and by the Bosnian border.

Koller reported about the effects of magnet on electricity as researched by Biot, Savary, and Schmidt who found that the deflection force on magnet is inversely proportional to distance, while good or bad conductivity is irrelevant (Koller 15b: 465, Baumgartner 1826 # 382 no. 6 at page 488). Koller updated

Baumgartner's notes by Marianini's research. Marianini was a professor of physics in Venetia who found deflection force of magnetic needle caused by elements of zinc and copper as proportional to the surface, therefore, not only to the distance. The size of (magnetic) deflection is not function of the material of plates. The intensity of voltage also effects deflection of magnetic needle which is proportional to the number of pairs of plates.

Koller provided illustrations resembling Faraday's lines of force (Koller 15b: 470 as clear-copied at 15b: 482, 493-494). Koller discussed the magneto-electric effects of Ampère (1831) and Faraday's research on 24 November 1831, including Faraday's letter to Hachette communicated to the Parisian Academy. There, Faraday reported about the tube of paper or wood wrapped by spiral of copper-wire to make electromagnetic cylinder which relocates the magnetic needle by its oscillations. Koller wrote his footnotes like a book prepared for publication which never occurred in print (Koller 15b: 475-476; resembling Baumgartner 1826 page 487 # 381).

Koller discussed Ampère's theory of magnets and Faraday's discoveries in Koller's original footnote about L. Nobili and the knight Vincenzo Antinori (1792–1865). From 1829 to 1859 Antinori was a director of the Regal Museum of Physics and Natural History in Florence where he researched the electromagnetic induction with Leopoldo Nobili. He had originally attracted Nobili to teach physics in Florence, as he had invited Giovanni Battista Amici to teach astronomy. Koller quoted newest Antinori's Florentine publications as translated for *Ann.Phys.* in 1833. In 1824 in Modena, Nobili researched magnetic curves, later called by Faraday's name, while Ampère's theory had filled great holes in sciences to make Koller his fan forever (Koller 15b: 478).

Besides a magnet of Faraday, Koller outlined the *Ann.Phys.* publication of Georg Friedrich Pohl (* 1788 Stettin; † 1849 Wro□law (Breslau)). Pohl attended Hegel's lectures in Berlin, supported Naturphilosophie, described the electromagnet as chemical phenomena, refused validity of Newtonian gravitation for electromagnetic laws, and designed the first electromagnet in Germany in 1828 by using the spiral of copper (Koller 15b: 484). Pohl died of cholera like Koller did later. Koller praised Pohl's galvanometer with needle inclining to south or north. Koller also noted the priest Salvatore del Negro as professor of physics in Padua with his New experiments on effects of magnetism on electromagnetic spiral and a new battery (Koller 15b: 486; Salvatore del Negro, Baumgartner's *Zeitschrift* in 1832). There, Baumgartner printed the letter received from Negro. Baumgartner translated it into German language, as was usual in those times. Koller later added note in his commentary column about E. Lenz's law of magnetic effect on spiral, republished in *Ann.Phys.* in 1835 from the *Annals of Petersburg Academy* of 7 November 1832 (Koller 15b: 487). Emil Lenz (Эмилий Христианович Ленц, 1804 Tartu (Dorpat)-10 February 1865 Rome) was a Russian physicist of Baltic German descent who formulated Lenz's law of electrodynamics in 1834 and electroplated a lot together with Jacobi in 1839. Koller certainly did not dare to be any kind of a Russophile, but he quoted the Russians' achievements anyway.

Koller also noted that the magnets emit light as was discovered by Faraday and Watkins in London for their publication in *Ann.Phys.* in 1833. Again, Koller did not quote the original English language print, which may indicate lack of English sources or insufficient knowledge of English language, or both. Francis Watkins (1800-1847) of the firm Watkins and Hill resided at Charing Cross in 1834. He was a manufacturer of scientific optical instruments who became a member of the Institution of Civil Engineers as a subscriber to Faraday's lectures in Londoners' extremely productive mixture of aristocracy and businessfolks of those days (Koller 15b: 488).

Koller used Nobili-Antinori's data published in Baumgartner's *Zeitschrift*. Nobili, Antinori, Faraday, and Friedrich Strehlke (1797 Funkenmühle in West Prussia-1886 Gdansk (Danzig)) at Real-Gymnasiums of St. Peter and Paul in Danzig used a permanent horseshoe magnet with a winding armature described at *Ann.Phys.* of 1832 by Poggendorff's report, signed only with a letter P. Faraday's and Pixii's work on decomposition of water apparatus was reported in Parisian *Annales de Chimie et de Physique* by Jean Nicolas Pierre Hachette (1769 Mézières-1834 Paris) who used to be Monge's and Napoleon's fan as a French mathematician-geometer. After Napoleonic downfall, Hachette certainly had problems, but still retained some influence in Parisian academy (Koller 15b: 489).

In his note-column Koller especially commented the achievements of mechanic Sexton in London. In fact, Koller had in mind the Washington based Joseph Saxton (*1799 Huntingdon, Pennsylvania; † 1873 Washington, D.C.). Saxton only resided in London in 1828-1837, mostly employed by the Adelaide Gallery of Practical Science. There he modified Pixii's generator to design his own model (Joseph Saxton, Description of a Revolving Keeper Magnet, for Producing Electrical Currents, *Journal of the Franklin Institute* 1834, vol. 13, pp. 155-156; Koller 15b: 490). Koller described Coulomb's magnetic torsion-rotary balance as published by Baumgartner's *Naturlehre* (5th volume 1836 p. 522f; there Koller probably confused 5th with 4th edition (1832) of his friend Baumgartner's textbook).

Koller repeated his history of electromagnetism beginning with Beccaria, including the achievements of Avogadro, Oersted and Becquerel (Koller 15b: 495-505, 507f, 572, 659f). Those Koller's notes were written in much smaller letters as a clear copy with most calculations omitted. They were probably the transcript made by Koller's assistant. He stated that the effects of electricity, together with its phenomena of heat, were caused by electrical voltage. He mentioned the mutual influences between electricity, magnetism, heat, and chemical transformations. They are all caused by the effects of the same force whose agent was not known in those times, as Koller admitted (Koller 15b: 507, 580). As the admirer of order, he described electrical data in three sequences: A) Electromagnetic phenomena; B) Thermoelectric appearances (thermoelectricity); and C) Electrochemical appearances.

A) Electromagnetic phenomena included the lightning. It could change the polarity of a magnet, which already Franklin and Gabriel Gruber knew. The inventive Franklin, Beccaria, Wilson, and Cavallo tried to effect their magnetic needles by discharged Leyden Jars, but they were left without decisive results until Oersted succeeded in Copenhagen (Koller 15b: 508).

Koller offered to his students no less than five cases of circumstances of deflection of the magnetic needle as a function of the distance between the needle and the polar wire (electrodes) (Koller 15b: 511). As primary Francophonic educated teen, Koller was aware of then French-British competition which enabled H. Davy and Arago to promote their two rival explanations of electromagnetism and optics. One of them was a concept of action-at-distance, supported by the Newtonians, Oersted, Weber, and maybe Koller himself at last before the Spring of Nations. It was later advanced by Helmholtz's protegee Hertz in his tractate (H. Hertz, *On the Relations Between Maxwell's Fundamental Electromagnetic Equations and The Fundamental Equations of its antagonistic Electromagnetics* (*Ueber die Beziehungen zwischen den Maxwell'schen electrodynamischen Grundgleichungen und den Grundgleichungen der gegnerischen Electrodynamik*) Leipzig 1884; Koller 15b: 512). They refused the later victorious supporters of the electromagnetic field including Faraday, Maxwell, and J. Stefan (Fricke (Frické), Haworth.

1982. *Two rival programs in 19th century classical electrodynamics: action-at-a-distance versus field theories*. PhD thesis, The London School of Economics).

Koller accepted ideas of his model Ampère by noting every molecule of polar-wire as magnetized (Koller 15b: 514, 523, repeated as clear copy on pp. 584-585). Koller knew that Becquerel's conduction is not proportional to the surface of a body, but to the mass of a body as the magnetism inside the electrical machines follows several rules including small Ampère's models in some extremely short distances from electrical current, caused by frictions in conductivity sparks. That was an idea of magnetism of the marchese Cosimo Ridolfi (1794 Firenze-1865), later prime minister of Firenze in 1847-1848. The agronomist Ridolfi corresponded with the early Ampère's follower the Piarist Liberato Baccelli (1772 Lucca-1835) who used to be a professor of physics at University of Bologna and a member of Società Italiana (Koller 15b: 530, 535). Koller knew that the Leyden Jar has a great magnetic force as proved by the magnetic pole of a needle in Felix Savary's spiral (Koller 15b: 536-537). In third and last Koller's idea about the magnetism inside the electrical machines, he endorsed Savary's strong magnets to speculate about the causes of those effects. Koller supported the fundamental novelty about the nature of any force known up to date to describe magnet's effects by the transversal-force of the polar wire. The rotation of that magnet was examined by Wollaston and Faraday in a glass-vessel full of mercury. T. Young's idea of transversality was borrowed from hydrodynamics just like D. Bernoulli's earlier kinetic atomic theory, but only Fresnel had the courage to propose the predominant (indeed almost unique as was nearly proved later) transversal waves in luminiferous aether. That Fresnel's extreme stance discouraged even initially supportive Arago and probably even Koller's early astronomical reasoning, where the friction of cosmic earther was not welcomed. It took Faraday's mathematically untrained mind to propose until then unimaginative transversal force of magnetism. Ampère's Newtonian action-at-distance requested the classical longitudinal force but Hermann Grassmann's (1809 Szczecin-1877 Szczecin) paper Neue Theorie der Elektrodynamik (*Annalen der Physik und Chemie* 1845) used the derivation from Biot-Savart's law and Lorentz force to endorse only the transversal magnetic forces for several generations to come until after WW2. Anyway, many modern researchers believe in additional longitudinal components again.

Koller explained the rotation phenomena of Davy with an illustration resembling electrolysis where the mercury at its ending extremity had a conical shape with a pike (Koller 15b: 540-541). That conus responds to repulsion by the centrifugal force of particles of mercury. By Koller, the transversal force of polar-wire and the reaction of a magnet should be explained by north-pole, south-pole, equatorial, and geomagnetism with the use of Faraday's approach (Koller 15b: 542-543). But Koller also praised Ampère's electrical current where the magnetism of molecules themselves will be rounded by such current cross-sections. In the non-magnetic bodies Ampère found those electric currents in all directions (Koller 15b: 549). He used then popular wise analogy of Ampère's electromagnetic spiral called the electromagnetic solenoid (Koller 15b: 552).

Next, Koller discussed his "B) Thermoelectric appearances (thermoelectricity)" by his more Germanized data provided by Oersted and Seebeck in Berlin. Their metal also got its magnetic polarizations, which was already known for the tourmaline-shaped minerals after Haüy and Brewster researched changes of crystal temperatures under the influences of magnets. That initially noticed smallest effects became big like the butterfly's flapping wings effects, as Koller told his doubting laughing students about the bright future of electromagnetism. Similarly, some now parapsychological effects including the longitudinal electromagnetic forces-waves might became important in the future; those small things can have non-linear (exponential) impacts on a complex system, like the virulent pandemics or a chain reaction. More precise measuring by the sophisticated detecting instruments enable our consciousness of small previously neglected effects. Koller loved the interrelation between heat, magnetism and electricity as proposed by Seebeck in his Goethean polarity of the Nature (*Naturphilosophie*) as opposed to their antagonists' Newtonian concepts of force of Ampère and several of his fellow Frenchmen (Koller 15b: 553, 601f as clear copy, 654f as repeating). The optics as Fresnel's French Ampère-related product was initially not included in that high level Naturphilosophie, although Seebeck also worked on the anti-Newtonian Theory of Colours with Goethe. Optics became relevant partner of electromagnetism only later by the efforts of Maxwellians, which also affected the marketing of electromagnetic and optical inventions as the main motor of new industrial revolution up to the new era of Web.

Besides Seebeck, Koller also praised Cambridge based research of the Scottish chemist Stokes's Professor James John Cumming (1777 London-1861) who galvanized his students and electrocuted a poor cat (Koller 15b: 56, 556, 652; Baumgartner 1826 page 397). Without knowing about Seebeck's accomplishments, Cumming made his experiments for his Second Report on thermo-electricity in 1832. Koller noted the scale-series of thermo-electric abilities of metallic alloys of Jean Pierre Joseph d'Arcet (1777 Paris-1844 Paris). A son of the director of porcelain works at Sèvres, d'Arcet, produced bronze and bronze-like alloys. As Stefan later, Koller was also fascinated by the thermal effects and endorsed the influence of the temperature on electricity of metals including their conductivity (Koller 15b: 557). Koller cited and illustrated the research of force of electrical current designed by Oersted and Fourier's joint experiments with polygons made of bismuth and antimony (Koller 15b: 558). Koller rightfully predicted to his astonished and even ridiculously laughing students that the thermoelectricity of contact-electricity will make a revolution in chemistry by their thermochemical batteries, which Volta's batteries began. Koller's prediction of great future of thermoelectric piles was ridiculed for a while, as his Kremsmünster students were laughing. The thermoelectrical batteries' voltages soon became important: in middle of 19th century the thermo-electric battery was often used in place of galvanic batteries with their numerous copper and bismuth wires. The researchers connected them in series as R. Guelcher's (Gülcher) thermoelectric pile (thermopile) useful for charging batteries available in the late 19th century after Guelcher's works were tested and approved in Germany in 1891-1893 (Koller 15b: 560-561).

Besides Seebeck's metallic plates, spheres, prisms, cylinders and other geometrical shapes of thermopiles, Koller also discussed Julius Conrad von Yelin's (1771-1826 Edinburg) observation of prism at Munich academy, printed as (Koller 15b: 562; Yelin, *Neue electro-magnetische Versuche: die magneto-motorische Wirkung der flüssigen Säuren, Basen und Salze mittelst einfacher metallischer Leiter: und eine einfache Ladungssäule mit trennbaren einpoligten Elementen* (*New electro-magnetic experiments: the magneto-motoric effect of liquid acids, bases and salts by means of simple metallic conductors: and a simple charged pile with separable single-pole elements*, published in Munich by Ignatz Joseph Lentner on 7 March 1823). The Habsburgians were quick enough at least to endorse if not to invent the novelties, as Koller read about the electrical currents in cold metals described by Nobili in 2nd or 4th volume of Baumgartner's *Zeitschrift*. By examining volcans by Ampère's ideas for his notes in *Ann.Phys.* in 1826, Nobili declared the volcanos as biggest problem of physics and of geomagnetism of those days, when Kreil and his friend Koller also embraced the research of the new

Gaussian-Weberian geomagnetism as their main fields of research (Koller 15b: 563-564).

As the last third of his preparations for lectures about electricity, Koller described his "C) Electrochemical appearances", also by contact-electricity of chemical compound reactions of metals modified by the electric currents (Koller 15b: 566, 608 as a clear copy). H. Davy proposed to the admiralty of his Royal Navy the protection of stern (back of a ship) by the metals at his publication for RS in London. By Davy's advice, the parts of boats should be pasted with copper against oxidation of iron-plates in water where the copper surfaces provide the protection in seawater also by their alloys with zinc (Koller 15b: 65, 461, 463, 568-570, 610; Baumgartner 1826 # 340 page 464 2nd part). Dumas in Paris got the similar results for his boats on the river Siena: coal-plate was also useful instead of iron or copper, as was the porcelain. Jean Baptiste André Dumas (1800-1884) and Davy therefore advanced the old research of Gabriel Gruber who used to deal with impregnation of wooden boats earlier in Ljubljana. Koller certainly praised Davy's electrochemical theory. Koller preferred German or French translation of Davy's Bakerian Lecture (Koller 15b: 571; Davy, On the Relations of Electrical and Chemical Changes, *Phil.Trans.* 1826). Koller quoted from the French translation (Davy, *Annales de chimie et de physique* volume 33 in 1826). Therefore, Koller scarcely used English publications as he mastered French, German, Latin and Italian languages better than English language, and English language journals were not that frequent in Habsburgian monarchy in Koller's times. Koller did not use many English language originals because they were very scarce in Habsburgian libraries or because of Koller's linguistic inabilities, at least before his Londoner travels in autumn 1838. Later in Stefan-Boltzmann's era English language readings of sciences was more common in Habsburgian circles.

In previous lectures-paragraphs, Koller discussed magnetic influence on electrical states of metals and other bodies, including the involved chemistry. Then Koller embraced Yelin's comment on changing of magnetic meridian and magnetic poles (Koller 15b: 572). Avogadro, Oersted and Becquerel discussed (Peltier's) electricity caused by chemical influences. Jean Charles Athanase Peltier (1785-1845 Paris) was originally a watch-dealer. He reversed Seebeck's effect in 1834, introduced the concept of electrical induction in 1840 and researched Koller's favorite atmospheric electricity. Koller might have met him in 1838 although Koller did not mention that event in his travelogue. Peltier's reversion of Seebeck's effect resembled three years earlier Faraday's experiments of 29 August 1831 as a reversion of Oersted's discovery of electromagnetism noticed on 21 April 1820.

As a strict Francophile, Koller noted how the Parisian music teacher Nicholas Gautherot's (1753-1803 Paris) used platinum for "secondary" current flow with electrophoresis occurring in a drop of water (Koller 15b: 576). The effect was used for later lead-acid storage battery in 1801.

In his clear copy, Koller or his assistant repeated their illustrated Schweigger's experiment in similar drawing which was not quite identical to Koller's first picture (Koller 15b: 523, 525, 586, 588). Koller's or his assistant's clear copy avoided lengthy calculations and even omitted some supposedly less important illustrations (Koller 15b: 526, 533/588, 532, 537/591, 597/541-542, 553/601). In his clear copy, Koller or his assistant also provided some comment as annotations not included in Koller's older variant, for example about a native of Reggio Emilia L. Nobili's metal slices used for the electrical current (Koller 15b: 607). Koller had two of his paragraphs excluded from his clear copy (Koller 15b: 566/608). Almost no illustration was included in Koller's clear copy of his part C), maybe made by one of Koller's assistants (Koller 15b: 616).

Summary: Thermo-magnetism and thermoelectricity as clear copy of Koller's previous notes

Koller and Baumgartner discussed their version of thermo-magnetism by Faraday's experiments with polar wire in mercury to highlight the interactions between electricity and magnetism, as both are agents of nature (Koller 15b: 647; Baumgartner 1826 # 392 page 496). In their research of multiplicators, Haüy and Brewster found that in crystal the force is changing by temperature according to Seebeck's data of 1821, as well as following Oersted and Fourier's experiments (Koller 15b: 651; Baumgartner 1826 # 395 page 497). Seebeck's experiments were also the result of an induction of electrical current in his circuits. Oersted reacted to Seebeck's discovery with his own paper On Some New Experiments in Thermoelectricity Performed by Baron Jean-Baptiste-Joseph Fourier and Oersted (*Annales de chimie*, 22 (1823), 375-389). Oersted reported to academy the results of their ten experiments. Oersted and Fourier heated several connected metals while Seebeck has proved that an electrical current may be produced in a circuit formed of solid conductors only by disturbing the equilibrium of the temperature. That enabled a new class of electric circuits called thermoelectric circuits. Seebeck distinguished them from galvanic circuits, which from now on may conveniently be called hydroelectric by Seebeck's opinion. That Seebeck's term did not catch. On the other hand, very interesting Fourier's collaboration with Oersted puts Fourier in the frame of Naturphilosophie of Seebeck, Oersted, or Goethe, which additionally explains Fourier's Egyptian Napoleonic opposition to Berthollet-Laplace's society of Arcueil, although Berthollet was also Fourier's fellow member of Napoleon's Egyptian expedition. The silent quarrel between Fourier's fans and Laplacians was based on their deeply different somewhat hidden philosophical views and not so much on their published mathematical physics. That antagonism resembled Bernoulli-Leibniz's opposition to Newtonians: Leibniz just knew all about then widely known hidden alchemy behind the Newtonian formulas, which are hardly observable by the modern reader. In any case, Koller was able to praise equally the French and the British researchers (Koller 15b: 662-665), including de la Rive's hypothesis, Faraday's experiments with torsion-twisting, Faraday's mentor Davy, as well as Ampère's force of polarity distinguished by Koller's acclamation **Vive Ampère!** as the sign of Koller's Francophonic youthful devotion (Koller 15b: 665).

Heat

Koller constructed his lectures about heat by following Baumgartner's 2nd edition of 1826, while Koller still kept his quotations of Baumgartner's 1st edition of 1824 in brackets (Koller 15b: 273). Koller's friend Baumgartner published his textbooks in 1824 and 1826 with his introduction signed in January 1826, while his 3rd edition of 1829 was not considered yet by Koller's course about heat, which means that Koller wrote that part of preparation for his lectures in late 1826 when he still had in mind his friend Baumgartner's 1st edition of 1824. Koller commented on specific heats and aggregation states with Dulong-Petit's capacity and results valuable for all gases, in opposition to Delaroche and Bérard's conclusion provided in 1824 (Koller 15b: 273, 276; Baumgartner, 1826 p. 387). The relevant data were published by the Doctor of medicine of Saint Mary Islington, William Tutin Haycraft († 1856). Haycraft had his paper printed by *Transactions of the Royal Society of Edinburgh.*

Another relevant research was Avogadro's in 1825, but Koller did not mention those among his two methods for the determination of specific heats. Koller noted Prevost's heat rays researched in 1809 with ideas of reflection of heat as resembling the relations between light and heat (Koller 15b: 277-278; Baumgartner 1826 388-389; Baumgartner 1826 400-401, 405-406). Koller discussed Lavoisier-Laplace's table of calorimetric measurements (Koller 15b: 283-286; Baumgartner 1824 2: 151-160) as Koller was commenting just those Baumgartner's data which contained some mathematical formulas that Koller loved to develop! (Baumgartner 1824 part 2nd paragraphs).

Koller added note on calorimeter of Rumford (Koller 15b: 286; Baumgartner 1824 2: 161). Koller discussed Black's methods like Baumgartner did (Koller 15b: 291-293; Baumgartner 1826, 385-386), except that Koller at his 2nd case at the end noted Adair Crawford's (1748-1795) method of cooling, while Baumgartner preferred to discuss Delaroche and Bérard's results which Baumgartner probably preferred. Koller even noted the revolutionary physician Jean-Paul Marat (1743-1793) and Auguste-Arthur de la Rive (Koller 15b: 293-295). As well, Koller loved Dulong-Petit's connection between the atomic gravity of simple bodies and their heat capacity. Koller also described the drips of fluids under heavy pressure of 36-40 atmospheres and connected their behavior with astronomical cosmic air in space under pressure and its latent heat. Koller discussed the data of Franklin, then of the naturized Viennese Jan Ingenhousz, and Desperetz's work of 1825, as Koller borrowed from Baumgartner's textbook (Baumgartner 1826, 392-393; Koller 15b: 297-299). Koller included Lambert's research, Biot's different kinds of metals, de la Rive, the transport of heat as researched by Fourier and the data of Genevan botanist Augustin Pyramus de Candolle (Alphonse Pyrame, 1778-1841) or his son (* 1803) baptized with the same name. Candolle blackened his equipment to determine the conduction of heat. Koller noted Gay-Lussac's *Leçons de Physique* published illegally by Grosselin in Paris in 1828 while two years earlier Baumgartner mentioned just Rumford, Leslie, and Howard's differential thermometer (Koller 15b: 302; Baumgartner 1826: 394-395). **Therefore, Koller wrote those notes about heat in 1828!**

Description with an engraving of a Differential Thermometer by Dr. W. Howard appeared in *Quarterly journal of science, literature…* in 1820. It resembled Leslie's features. According to Koller the refraction of heat rays was first noted by Du Fay, then by the other Frenchmen Delaroche and Bérard, while Baumgartner in 1826 noted just Herschel and Scheele (Koller 15b: 303). Koller discussed Dulong and Petit's data which later became Stefan's favourite topics (Koller 15b: 308-311). Koller included La Roche (Delaroche), Dalton, and Dulong-Petit's merits described by César-Mansuète Despretz. As the mathematician, Koller loved the ideas of the geometric progression representing the speed of cooling, the cooling in free air released through the pipe of balloon, and the cooling in vacuum noted as probably Koller's own text which was not included in Baumgartner's textbook of 1826 (Koller 15b: 313-320). In 1852 it was reinvented as Joule-Thomson effect. Koller also explained the cooling by loosing of heat by radiation in emptied balloon-vessel related to later Stefan's and Dewar's approaches. Koller discussed Dulong-Petit's experiment by Newtonian approach, and praised Dulong-Petit's measurements of the speed of cooling in vacuum (Koller 15b: 318). Koller again postulated that the cooling happens in geometrical progression, while the distance between the observed objects increases in arithmetic progression; therefore, the exponential power law was involved. Koller noted in his comment column the cooling of air and gases, as the cooling possibilities of gases are different compared to the atmospheric air. He noticed the cooling in hydrogen as quicker and better than in CO_2 or other gases, because it depends on the smaller or greater mobility of the particles of a gas, therefore on the molecules in modern notation (Pouillet, *Élémens de physique expérimentale et de météorologie*. Paris 1828, volume 1, pt. 2; Koller 15b: 320, 325). Based on the results of Dulong-Petit, the cooling also happens because of the radiation which was soon endorsed and updated by J. Stefan.

Koller and Baumgartner were excited about Davy's safety lamp as described in Gilbert's *Annalen* in 1817 (Koller 15b: 321; Baumgartner 1826, page 407 # 245). It was useful in flammable atmospheres after it was invented by Sir Humphry Davy in 1815. Like Stefan later in the era when Galileo's duration of time was suddenly much easier measurable as very important experimental fact, Koller tried to determine the speed of cooling by changing the pressure in gases. He changed their cooling capacity by modifying their partial pressures. Probably there Koller used a concept borrowed from Dalton although Koller did not mention Dalton's name (Koller 15b: 322-324). Dulong-Petit also noted how the changing of surface influenced the speed of cooling as its mathematical function. Different surfaces' areas mean different heat capacity for metals like gold, silver, or copper. The idea was endorsed by the most of physicists including Dulong and Petit as their law brought a lot of clearness to the mechanism of heat transfers.

Optics of heated lamps

Koller was enthusiastic about the Brewster's monochromatic lamp described in 1828 in the Parisian *Annales de chimie et de physique* and in *Ann.Phys.* in 1829 (Koller 15b: 325-326). Koller noted that the lamp gives homogenous white light with traces of green and red in form of rings in the water and in the prism, as stated by Brewster. Next British step was an invention of yellow light of platinum put in a candle. The designer was William Henry Fox Talbot (1800-1877) who used to be an English pioneer of photography and of the spectral analysis of stars. In 1826, John Herschel introduced Talbot to David Brewster as Brewster and Talbot's research on light frequently overlapped. Brewster began publishing Talbot's scientific articles in his journal and both forged an unusually close lifelong friendship. Koller met Brewster in 1838 but did not mentioned Talbot in his travelogue.

Koller noted spreading of heat in heated air (Koller 15b: 327; as extension of Baumgartner 1826, *# 225* page 394). It was researched in Vienna in 1823, but only Baumgartner mentioned the chemist Paul Traugott Meißner (Meissner, * 1778 Medias (Medias⬜h) by Sibu in Transylvania; † 1864 Vienna-Hernals) who taught in Viennese Polytechnic from 1815, from 1842 as director of the department of general chemistry. Meißner's hot-air central heating system became famous, but as a free-thinker he was opposed by Prechtl and Liebig, and evidently also by Koller while Baumgartner eventually did not go that far. Koller did not mention Meißner even if Koller wrote about the heated air and doubled insulation, probably because Meißner's free-thinking could not be Koller's favourite.

Koller noted colour of the flame of substances of boron acid, alcohol, and burning air in Brewster's inflamed air at monochromatic lamp (Koller 15b: 329-330; Baumgartner 1826 # 246 p. 408). The flames of different bodies of Talbot and Henry Home Blackadder were described in Baumgartner's *Zeitschrift* in its volume 1. Their conclusions resembled Faraday's later experiments with coloured flames, presented in *Chemical History of a Candle* as a series of Faraday's six lectures about the chemistry and physics of flames at the Royal Institution in 1848, two decades after Koller's notes. Koller described the source of light visible from the distance of 126 miles developed by the professor in Prague Adolf Martin Pleischl (1787 in Horní Planá (Hossenreith), Bohemia-1867 Dorf an der Enns). In 1815 the chemist Pleischl obtained his medical doctorate from the University of Prague, where he then served as a professor of general and pharmaceutical chemistry in 1821-1838 (Koller 15b: 332-333). Later he lectured at the Viennese university where he taught Johann August Natterer (* 1821) who successfully continued Pleischl's attempts to liquefy the CO_2. Pleischl also analysed mineral waters like Koller did in Kremsmünster. In 1780 Aimé Argand (1850-1803 Geneve) invented a lamp with output 6 to 10 candelas, which was brighter than earlier lamps. It enabled completer combustion of the candle wick and oil than in other lamps. It also required much less frequent trimming of the wick as the top of the first flame heated another. Certainly, Prechtl's gaslighting described in his Viennese monograph in 1817 had not disfavoured candles of his times, while Edison's bulb was just J. Stefan's future for Koller before March Revolution.

Heat & Chemistry of optical lamps

Koller relied on then currently valid theory of the compounds of two or more bodies, but still had to note phlogiston which Lavoisier's fan the Viennese baron Jacquin already buried and replaced by caloric during Koller's youth; a generation later, Stefan was free to ignore them both, the obsolete phlogiston and the caloric (Koller 15b: 335-337; Baumgartner, 408, # 447). In his 1830s description of combustions, Koller did not mention atoms by name as he only noted Lavoisier's caloric, while mentioning of Lavoisier's ideas by itself distinguished Koller from four decades younger Stefan for whom Lavoisier was just a history. For Koller in 1830s, all decaying-burning bodies have in their insides some substances, therefore still the caloric, while the phlogiston proved to be unusable explanation after the beginning of burning developed and the phenomenon of fumes predominated. So, any initial pleasure by the theory of phlogiston does not explain why it was wasted during burnings. It was replaced by the oxygen by Lavoisier's fans after Lavoisier proved the suspicious cases where phlogiston should have a negative weight. The concept of burnt up oxygen is wiser, as it explains what is lost. Koller's initial dependence on this view was caused by the simple influence of multitudes of Lavoisier's explanations of nature and it won the case. According to Lavoisier, all bodies regarding to their losses of weight during burning were divided into two classes. Lavoisier named his new stuff caloric. Berzelius and Davy researched those burnings where the mass greatly enlarges, or the weight is lost from body's volume while the behavior of air might be in principle explained by its electricity. So, Koller greatly extended Baumgartner's discussion into the historic Lavoisier's attacks against phlogiston which were soon dogged by Davy and Berzelius. By his own evidence, Berzelius revised and generalized the acid/base chemistry chiefly promoted by Lavoisier, while Davy proved the existences of acids without the oxygen, like HCl. Koller afforded that historical discussion as a prelude into his repeated description of Davy's safety lamp called Miners' friend which operated without any open flame, based on the reflection of rays of heat (Koller 15b: 339; Baumgartner 1826 page 406 # 243).

Koller's Atmosphere

After his clear copy Koller again resumed prevailing style of Koller's own big letters to discuss one of his favourite topics, the height of atmosphere, just like his fellow Benedictine Robida did later (Koller, 15b: 584-616). Koller used the known formula for the pressure of air and temperature. In Koller's formula for height

$h = \alpha D / d$

the force is determined by the speed c of sound. Koller also provided his marginal column note to underline that the tension is determined by the temperature (Koller 15b: 619; Baumgartner 1824 # 278 old edition; Baumgartner 1826 2nd edition page 635 # 231 in a short paragraph, with no Baumgartner's edition providing the paragraph number quoted by Koller).

Koller might not have been aware of Lambert's optical research which was not widely read yet at those times because of Lambert's Latin language publication, but Koller knew how Lambert determined the heat of atmosphere while the temperature at the surface of earth proportionally diminished by the height, also according to the timing of simultaneous temperatures in different places (Koller 15b: 621-622). Koller supported logarithmic equation which Schmidt designed by Humboldt's observations (Koller 15b: 624). Schmidt explained it by Lambert's hypothesis, somehow resembling Lambert's law where the loss of intensity of light propagated in a medium is directly proportional to intensity and the distance. Much later in 1852, August Beer discovered another attenuation relation.

Koller certainly provided a derivation of good old Laplace's formula for the height-measurements to find the pressure of air at the known height (Koller 15b: 627, 630). That enabled Koller to use the changing temperature in the air by Laplace's theory for determination of height of atmosphere as 18336 meters (Koller 15b: 631). That was a good approximate result as today's Earth's atmosphere is about 480 kilometres thick, but most of it is within 16 km above the surface while the air pressure decreases with altitude. Koller did not directly use Laplace's escape velocity which is a clue to determine how high and far could Earth keep its molecules of air from escaping into cosmos.

After his calculations of the height of atmosphere, Koller discussed the atmospheric electricity with historical intermezzo about electrical effects in air researched by Cavallo, Humboldt and Pouillet (Koller 15b: 635; Baumgartner 1826 page 676 # 296). For Koller, the lightning is an electrical spark bringing thunder and colour, which goes through good conductors, oxides the metals, and kills animals (Koller 15b: 637, 671; Baumgartner 1826 # 296 # 299 pages 676-677). In Koller's and Baumgartner's times, Franklin's lightning rod was still interesting scientific topics with the relevant opinions of celebrities like Charles and the French lightning rods described in *Ann. Phys.* in 1824. The especially important pamphlet was the work of the son of influential deist Hermann Samuel Reimarus (1694 Hamburg-1768 Hamburg), Johann Albert Heinrich Reimarus (1729 Hamburg-1814 Rantzau, Holstein). He published his *Regulations for the erection of lightning rods on all buildings designed after reliable experience (Vorschriften zur Anlegung einer Bliz-Ableitung an allerley Gebäuden nach zuverlässigen Erfarungen entworfen)* in Hamburg in 1778 (Koller 15b: 639; Baumgartner 1826 # 301 page 678).

Koller praised Lichtenberg, Gay-Lussac, Biot and especially Volta's theory of hail involving the sunlight, temperature and electricity, which was criticized by the abbot Angelo Bellani (1776 Monza-1852 Milano). Bellani opposed Alessandro Volta's theory of the formation of hail. In his Defense of the supposed letter to Volta in Milano in 1823, Bellani argued that the heat of the sunrays increases the temperature of the total mass of the clouds, so that the outer surface evaporates more quickly, but does not cause the initial freezing of the cloud. That was contrary to Volta's opinion. To prove his thesis, Bellani had frozen his two thermometers, one exposed to the sunlight and one in the shade, highlighting how the second froze faster. Therefore, hailstorms occur more frequently at (afternoon and approaching) night when the air is cooler. In 1826 Bellani also claimed the uselessness of the anti-hail tool called paragrandini. He criticized the theses of Paolo Beltrami of 1825. Together with Bellani other famous contemporary scientists supported the uselessness of prevention systems against lightning and hail, among which were Gaetano Melandri Contessi, Giuseppe Demongeri, Alexandre Lapostolle, Le Normand, Giovanni Majocchi, Pietro Molossi, Francesco Orioli, Charles Richardot, Antonio Scaramelli, Charles Tholard, and probably even Baumgartner and Koller as Koller also cited Bellani's letter addressed to Volta against Volta's theory in 1806. The meteorologist Koller eventually at least considered oscillations of Bellani's clouds forcing the downfalls of solid bodies from time to time. The idea was highlighted together with Arago's work in *Annuarie des Bureau de Longitudes* in 1826 and in Baumgartner's *Zeitschrift* (Baumgartner's *Zeitschrift* volume 4/3: 324-333f; Koller 15b: 643-644; Baumgartner 1826 # 304 pages 679-681; Volta, *About the Hail (Sopra la Grandine)*, Bologna 1816; Bellani, *Il propagatore dei paragrandini* 1826; Bellani (letter addressed to Volta) *Nuove sperienze ed osservazioni fisico-chimiche: istituite cogli elettro-motori*, Milano, July 1806).

Koller finished his lecturing notes by explaining the electrical phenomena in air, claiming that up to now we discussed the artificial electricity – now we discuss the natural atmospheric electricity, thunderstorms, discharges in metallic wires and sparks (Koller 15b: 669; Baumgartner 1826 page 675 # 295). Koller again provided some early history by describing the pointed lightning rods of physicists in Europe. Franklin's fan Thomas-François Dalibard (1709-1778) performed his experiments in Merly-la-Ville in 1752, which was probably not important for Baumgartner's textbook anymore. To scare his students from any unauthorized experiments, Koller described the tragedy of G.W. Richman (1711–1753) in Petersburg on 6 August 1753 while his companion engraver to the Royal Academy at Petersburg named Sokolow survived. Koller did not mention their collaborator Lomonosov who soon became even more famous after the successful Russian propaganda (Koller 15b: 670-671).

Summary of Koller's lecturing

Koller's notes of his preparations for lecturing about nonmechanical phenomena were not properly arranged by the branches of physics and chemistry, unlike his notes about the mechanics. He occasionally included into them even some acoustics, then again electrodynamics (electromagnetism) on 15b: 151f then heat on 15b: 269f and again magnetism and electromagnetism on 15b: 387f. That was certainly not the order governing Koller's lectures, but more represents his notebooks' preparations for lectures put stochastically in the same notebook! Koller's 15a manuscript preparations for courses of mechanics and his notes from Stefan's lectures were put in order, but not Koller's manuscript 15b.

Like his friend Baumgartner and the other Habsburgian members of Koller's Benedictines' order including Stefan's teacher Karl Robida (1857-1862) and Karl Puschl, Koller probably initially advocated William Robert Grove's wave theory of heat highlighted at Grove's lecture of 1843 and did not yet endorse proper kinetic theory before the Spring of Nations. Grove's wave theory of heat analogous to optics was certainly still somewhat inside then modern mainstream theory of heat termed as "thermodynamics" for the first time at Baumgartner's extract *Anfangsgründe der Naturlehre* in 1837 (Brush 1976, pp. 322, 332; Baumgartner 1837 page 132 # 226; Kangro 1976, p. 229). Baumgartner coined that new name by analogy with electrodynamic, aerodynamic, geodynamic, dynamic, and hydrodynamic as another aspect of Law of motion of heat. In 1850s, at the gymnasiums in Ljubljana, Novo Mesto and J. Stefan's Klagenfurt, they used that Baumgartner's abridged textbook. Robida's obsolete wave theory of heat might be the main reason why Stefan so soon abandoned the ideas of his class teacher Robida. Stefan soon preferred the modernized Ettingshausen's kinetic theory of atoms as Ettingshausen became Stefan's teacher at university. Later favorite but today obsolete mechanical theory of heat (*mechanischen Wärmetheorie*) was also advocated by Baumgartner, certainly afterwards. A quarter of century after his promotion of "thermodynamics" he described new mechanical theory of heat in his paper *Die mechanische Theorie der Wärme: Vortrag gehalten in der feierlichen Sitzung der Kaiserlichen Akademie der Wissenschaften* on 30 May 1864. For his lectures, Koller alternatively used first five editions of Baumgartner's textbooks published in 1824-1836. In his *Naturlehre* of 1836, Baumgartner used the same subtitle Law of motion of heat (Gesetz der Bewegung der Wärme) like in 1837, but not a term Thermodynamik followed in parenthesis of his abridged earlier edition (Baumgartner 1836, page 438 # 134), although the text was quite similar at the beginning, explicitly based on analogy with optics and stating that the heat is omnipresent like gravitation. Nomen est Omen. In fact, Baumgartner based his ideas on the data of Leslie and Macedonio Melloni. In 1837, Baumgartner also promoted the Law of equilibrium of heat (Baumgertner 1837 pages 138-148 # 236-#247; Baumgertner 1836 page 451 (# 199).

Koller also quoted the textbook of the Piarist Remigius Döttler (1741-1812) which Koller used as a student (Koller, 15a: 212, 220, 224-225, 233). At Viennese university, Döttler already abandoned or at least neglected Bošković's theory of forces represented by the famous Bošković's curve even if Koller still quoted Bošković's Viennese-Carniolan fan Jurij Vega and Baumgartner cited Bošković as one of his main references in 1826 (Koller, 15a: 243, 534, 542).

Koller left for us thirteen hundred pages of his preparations for lecturing of physics-chemistry-meteorology. No astronomy was included as the professor of astronomy in Kremsmünster was the other monk and not Koller who headed the observatory anyway. Many of those Koller's notes were repetitions as clear copies, some of them probably written by his assistant. Most of citations were vague, but some chapters of electromagnetism provided footnotes clearly designed for a publication of his eventual textbook (Koller 15b: 475 on Magneto-electric effects), which never succeeded, probably because Koller soon lacked sufficient time as a head of observatory and later as even more busy state counsellor. His Habsburgian friends Baumgartner (1824) and Baumgartner's brother-on-law Ettingshausen as Baumgartner's buddy at 7th edition of textbook in 1842 already prepared their own textbooks of physics-chemistry, while Joseph Johann von Littrow did the same for popular and academic audiences of astronomers in 1824 and 1832-1834. Koller's travel companion Kunzek published optical textbook in 1836 and Gymnasium-Real school textbook in 1864 while many others followed. The translation of French Pouillet's meteorology was hard to surpass and Koller's friend Karl Kreil published Italian and German instruction for magnetic observations in Habsburgian monarchy in 1856, which was reprinted in 1858. After his father Nicolaus' success, Joseph Franz Jacquin (1766 Schemnitz (Banská Štiavnica); † 1839) authored chemical textbook in Vienna in 1836. So, there was a sharp competition preventing more Habsburgian textbooks, differently from the Theresian Jesuitic era half of a century earlier when the (Hungarian) university professors were obliged to author their textbooks yearly.

Koller certainly relied on first Baumgartner's editions published in 1824, 1826, 1829 and 1832. The first part of Koller's preparation (15a) on 694 pages focused on chemistry (15a: pages 2-175), mechanic (pages 176-570, 667-668, 679-694) and acoustic (pages 15a: 571-666, 671-675, 15b: 123 on pendulum clocks, 15b:

147-150 elasticity as part of acoustic or theory of heat). No equivalent independency was given to the course of hydrodynamic or hydrostatics (Koller 15a: 257-270). Koller's 174 pages of chemistry, 418 pages of mechanics, and 101 pages of acoustic meant that mechanics was far the most important of all branches of his physics. Koller's first lecturing part (15a) was clearly divided by topics except of last twenty-eight pages. He relied on then modern Dalton's atomism including Dalton's law of partial pressures of individual gases. Koller included Dalton's tables of gas' temperatures and pressures with four kinds of imponderables: electricity-magnetism and light-warmth stuff (Koller 15a: 3, 97, 382-402, 453-454). As advancement of Lavoisier's ideas but not yet reaching Mendeleev, Koller offered his students fifty-eight kinds of ponderable stuff, eleven kinds of gases not yet brought into fluid state as non-metallic stuff including oxygen, hydrogen, iodine. Only a decade after Koller's death, Louis Paul Cailletet in France and Raoul Pictet in Switzerland succeeded in producing the first droplets of liquid air in 1877. Koller also listed forty-one kinds of metals, beginning with the lighter aluminium and ending with a heavier mercury (Koller 15a: 19). Koller even favoured a kind of early kinetical theory including the collisions between the smallest particles of bodies of similar kind (Koller 14b: 125-126). They make the chemical affinity in electrical current as they produce chemical moment which is enabled in body by the collisions of smallest particles as the function of their mass and their overall distribution. That was pretty much the kinetic atomism, two decades later professed by Clausius, Maxwell and Stefan as Koller's diminished cohesion of the smallest particles liquified bodies.

Like Jeremias Benjamin Richter (1762 Jelenia Góra in Polish Silesia–1807), Koller was fascinated by Lavoisier's promotion of the role of mathematics in chemistry and by the atomism. In his course, Koller included Berthollet-Proust's controversy (Koller 15a: 42), dimension of atoms (Koller 15a: 46) and Berzelius' water still denoted as HO (Koller 15a: 62). After 175 pages of chemical introductions, Koller switched to physics-mechanics by his analytical determining of the location of a point (Koller 15a: 176). Koller's statics was followed by his dynamics of gases, liquids, or solids (Koller 15a: 495). Next in row was acoustics (Koller 15a: 571) with Tartini's combinations tone included (Koller 15a: 647). Koller finished his first volume of lectures (15a) by almost thirty (in fact twenty-eight) pages of mixture of dynamic (15a: 667-668, 679-684) and acoustic (Koller 15a: 671-675), filled by additional occasional notes which followed his temporal lecturing interests.

Koller's optics was somewhat scattered through his manuscript (15b: 82-110, 15b: 143-144 as inserted notes on diffraction of light, 15b: 179-272). Koller's beginning of optics was later repeated (Koller 15b: 223) with his inserted notes on speed of light, followed by discussions about the goniometer and other optical instruments, magnetism, and electricity (electrodynamics at Koller 15b: 4-80, 115-142; chemical affinity with electrolysis like an intrusion at pp. 125-130, 151-178, 387f; heat at 15b: 269-386).

Koller's second volume (15b) of preparations for lectures in concluding 8th grade of Lyceum Kremsmünster's philosophical studies in 1826-1839 consisted of 672 pages. His lyceum of those times separated Gymnasium from the university as the school lower than university proper, but the lyceum courses of physics, chemistry and biology were the higher available in then Habsburgian monarchy except for some lectures at the faculties of the faculties of medicine. But the lyceums transformed into concluding classes of Gymnasiums after the Spring of nation of 1848. Therefore, Koller's lectures in 1830s were on slightly lower lever than J. Stefan's university courses three decades later in 1860s, after Stefan absolved his lower-level chemistry-physics in the Benedictine Karl Robida's Klagenfurt's class in early 1850s.

Koller's science of heat was not considered as equally important title like the others, even if Baumgartner in 1826 in his 2nd edition discussed heat at independent paragraph of his 2nd part after optics and before electricity, while magnetism was Baumgartner's last 4th paragraph. For both Koller and Baumgartner, 1st part was mechanics which is mostly still today in elementary courses of physics.

Baumgartner's 3rd part was divided into 1) Physical Astronomy – 2) physical Geography – 3) Physical meteorology. Koller apparently did not discuss those except some meteorological heights of atmosphere at the end of his 15b preparations for lectures. Koller's 672 pages of 15b preparations consisted of four A3 leaves (Halbbogen) of small format, and additional fifteen A3 leaves (Halbbogen) of great format were noted by his later Kremsmünster editor who probably examined just few paragraphs as he (wrongly) termed all volume 15b of Koller's lectures as discussions of Oersted and Faraday's discoveries about the electrical current. In fact, the manuscript contains today many more pages, which may suggest that the additional ones were inserted later. That title of Koller's manuscript is misleading as the mathematician Koller's principal hero was the mathematician Ampère, praised by Koller's acclamation Vive Ampère!

(Koller, 15b: 665, 667). Besides those Frenchman, which Koller learned to respect during his teenaged studies in Napoleonic Ljubljana, Koller probably noted so many now almost forgotten (north) Italians because as a Benedictine he had many connections there, especially in then Habsburgian Padua where Koller was a member of a learned society as an honorary member of the "Academia di scienze, lettere ed arti di Padova". Koller was equally an honorary member of the "Ateneo di scienze, lettere ed arti di Bergamo" at then Lombardian Habsburgian city of Bergamo 200 km west of Padua.

Koller divided his electrodynamics at his several cleared copies into three parts: A) Electromagnetism proper or Electromagnetic phenomena (Koller 15b: 4f, 508f, 589f (with some discussions on Koller's favourite geomagnetism of stable earth's hemispheres including Koller's own experiences with measurements of geomagnetism at Koller 15b: 43-47)); B) thermoelectricity as Thermoelectric appearances (Koller 15b: 53f, 553f, 601f); and C) Electrochemical appearances (Koller 15b: 63f, 115f as electrolysis, 566f, 608f). Koller's first variant of his course was interrupted by his chapters about optical double refraction (Koller 15b: 82f) and by the polarizations focused on Fresnel and Brewster's optics.

SECOND PART:

Koller's greatest discovery was Stefan

It is no wonder that Stefan's talent exited Koller. Talent, however, is not enough, it is at most a prerequisite for success. Therefore, Koller acted as Stefan's chief mentor in 1850s and 1860s.

Stefan's Courses at Viennese high school

On Koller's recommendation Stefan got unpaid job at Viennese university. To enable Stefan's survival in expensive metropolis which Koller himself was denied in his early days, Koller had to arrange Stefan's paid teachings of mathematics and physics including his administrative duties as deputy schoolmaster at Viennese real school.

As early as on 17-18 May 1860, Koller attended Stefan's lectures at the Viennese Real School at Bauernmarkte, founded in 1586, which Koller promoted from a lower to a higher secondary school by decrees of the Ministry of Education and Worship on April 2, 1858 no. 1192 and February 6, 1859 no. 1422. That private Real School employed the best young lecturers. Stefan received his first paid position there in 1857/58 as a physics lecturer four hours a week in the 4th grade, and in 1858/59 as a fifth-grade class teacher, while at the same time teaching mathematical physics at the university three hours a week and hydromechanics one hour a week. From 1857 to 1860 Stefan resided at Alservorstadt no. 288 north of his high school in the center of the town, where Slovenes, Croats and Slovaks have lived for a century and a half. Certainly, it was not really a prestigious apartment.

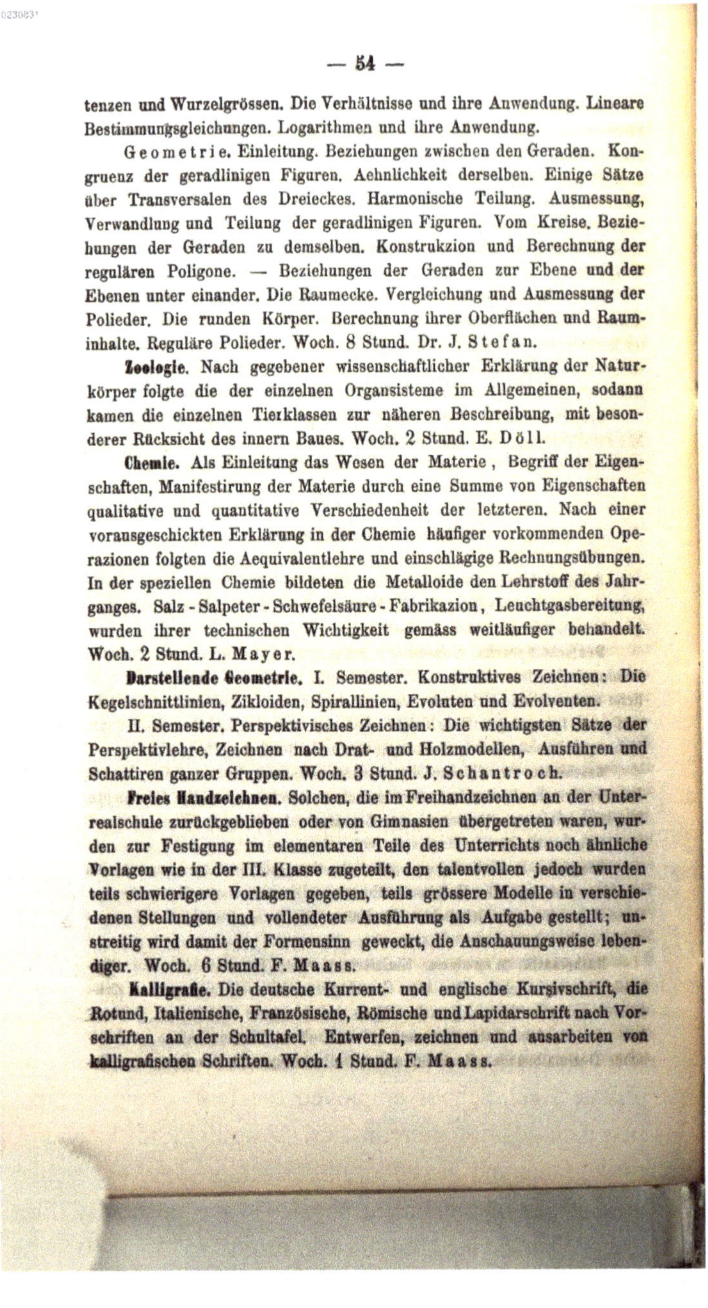

tenzen und Wurzelgrössen. Die Verhältnisse und ihre Anwendung. Lineare Bestimmungsgleichungen. Logarithmen und ihre Anwendung.

Geometrie. Einleitung. Beziehungen zwischen den Geraden. Kongruenz der geradlinigen Figuren. Aehnlichkeit derselben. Einige Sätze über Transversalen des Dreieckes. Harmonische Teilung. Ausmessung, Verwandlung und Teilung der geradlinigen Figuren. Vom Kreise. Beziehungen der Geraden zu demselben. Konstrukzion und Berechnung der regulären Poligone. — Beziehungen der Geraden zur Ebene und der Ebenen unter einander. Die Raumecke. Vergleichung und Ausmessung der Polieder. Die runden Körper. Berechnung ihrer Oberflächen und Rauminhalte. Reguläre Polieder. Woch. 8 Stund. Dr. J. Stefan.

Zoologie. Nach gegebener wissenschaftlicher Erklärung der Naturkörper folgte die der einzelnen Organsisteme im Allgemeinen, sodann kamen die einzelnen Tierklassen zur näheren Beschreibung, mit besonderer Rücksicht des innern Baues. Woch. 2 Stund. E. Döll.

Chemie. Als Einleitung das Wesen der Materie, Begriff der Eigenschaften, Manifestirung der Materie durch eine Summe von Eigenschaften qualitative und quantitative Verschiedenheit der letzteren. Nach einer vorausgeschickten Erklärung in der Chemie häufiger vorkommenden Operazionen folgten die Aequivalentlehre und einschlägige Rechnungsübungen. In der speziellen Chemie bildeten die Metalloide den Lehrstoff des Jahrganges. Salz - Salpeter - Schwefelsäure - Fabrikazion, Leuchtgasbereitung, wurden ihrer technischen Wichtigkeit gemäss weitläufiger behandelt. Woch. 2 Stund. L. Mayer.

Darstellende Geometrie. I. Semester. Konstruktives Zeichnen: Die Kegelschnittlinien, Zikloiden, Spirallinien, Evoluten und Evolventen.

II. Semester. Perspektivisches Zeichnen: Die wichtigsten Sätze der Perspektivlehre, Zeichnen nach Drat- und Holzmodellen, Ausführen und Schattiren ganzer Gruppen. Woch. 3 Stund. J. Schantroch.

Freies Handzeichnen. Solchen, die im Freihandzeichnen an der Unterrealschule zurückgeblieben oder von Gimnasien übergetreten waren, wurden zur Festigung im elementaren Teile des Unterrichts noch ähnliche Vorlagen wie in der III. Klasse zugeteilt, den talentvollen jedoch wurden teils schwierigere Vorlagen gegeben, teils grössere Modelle in verschiedenen Stellungen und vollendeter Ausführung als Aufgabe gestellt; unstreitig wird damit der Formensinn geweckt, die Anschauungsweise lebendiger. Woch. 6 Stund. F. Maass.

Kalligrafie. Die deutsche Kurrent- und englische Kursivschrift, die Rotund, Italienische, Französische, Römische und Lapidarschrift nach Vorschriften an der Schultafel. Entwerfen, zeichnen und ausarbeiten von kalligrafischen Schriften. Woch. 1 Stund. F. Maass.

Figure 6: Stefan's mathematical lectures at the secondary school - geometry curriculum in 1861/62 (*Jahres-Bericht der Öffentlichen Ober[1]Realschule (Bauernmarkt(e) Nr. 11) zu Wien*. Wien: Anton Schweiger, 1861/62, pp. 53-54, here page 54.

The library of Stefan's high school was subscribed to Schlömilch's journal of physics-mathematics, where Stefan published nearly a dozen papers on integrals, hydrodynamics, kinetic theory of gases, acoustics, and optics (*Schlömilchs Zeitschrift für Mathematik und Physik* 7 (1862): 356-359-11 (1866); Jahresberichte Bauernmarkte 1859, 39, 53-54, 62; Jahresberichte Bauernmarkte 1860, 101-103, 107).

On September 30, 1858, the school councilor and inspector of secondary schools, former Liechtenstein's private teacher Moriz Alois Becker (Moritz, * 1812 Staré Město (Altstadt) in Moravia; † 1887 Lienz) organized a special meeting of Stefan's colleagues secondary school teachers, for which the principal thanked him heartily. The principal also praised Koller, the auxiliary Viennese bishop, canon-cantor of St. Stephen church Franz Ksaver Zenner (* 1794 Vienna; † 19 October 1861 Vienna) and Commissioner for Viennese Secondary Schools Joseph Scheiner (Josef, * 1798 Česká Lípa; † 1867 Vienna). Scheiner was a Roman Catholic priest, theologian, and university professor, from 1855 a canon in the Cathedral of St. Stephen in Vienna, the Archbishop's Consistory Councilor, Commissioner for Catholic Educational Institutions, Viennese Gymnasiums and Secondary Schools. Together with Koller, he was a consultant to the Ministry of Education (Jahresberichte Bauernmarkte 1860, 1859, 66, 67, 70, 72).

Becker then organized a couple of meetings at Stefan's high school. On May 18, 1860, he and Koller inspected the high school and attended eventually even Stefan's lectures, for which the principal thanked them both again. Becker made rapid progress: immediately after Stefan's departure from high school, he became a private tutor to the unfortunate heir to the throne Rudolf and his sister Gisele in 1864, and in 1867 he was knighted as a topographer and correspondent for the Viennese daily *Die Presse*.

Stefan's physics courses at Viennese High School

In 1858/59 Stefan began his real school lectures on mechanics in the 5th grade with thermometers and finished them with the vacuum pumps. His final examination was extended to an hour and a half from 10:30 a.m. on July 29, 1859 on Friday. Stefan examined twelve of his altogether fourteen students of 5th class aged fifteen to eighteen years. Among them, in addition to the Catholic Germans, there was one Protestant, two Jewish candidates, two Slavs, and three Hungarians. Vienna of those days was certainly a real conglomerate of languages and religions.

The final sixth grade was introduced in the following year, when Stefan lectured similarly in the fifth grade with somewhat less emphasis on thermometers. In the final 6th grade, Stefan began four hours of weekly lessons with the waves on water. In his subsequent course of acoustics, he described resonance and the ear, and in optics he explained the eye, optical devices, interference, deflection, and polarization. He then tackled magnetism, electric force, electrolysis, polarization, Ohm's law, the connection (Wechselbeziehung) between electricity and magnetism, electrodynamics, the physiological effects of electricity, thermo-electricity as thermal effects, electricity in air, as well as the use (Anwendung) of electricity and magnetism. Next, he focused on thermal expansion, thermometry, calorimetry, mechanical theory of heat, phase transitions, evaporating water, heat sources, connections between thermal phenomena and chemical composition of bodies, heat transfer, radiation, and heat distribution along the Earth. He did not highlight the third possibility of heat transfer by convection in a separate group of lectures, but he certainly left no doubt in the kinetic atomistic mechanical theory of heat, which he otherwise somewhat agnostically denied, at least according to Koller's notes two years later.

In the concluding lectures of his high school course Stefan determined the points on Earth and in the sky, rotation of the axis (Axendrehung) of the Earth, measurements of time, daylength, seasons, and classifications of the celestial bodies. The semester ended on February 17 and after a week of vacation they continued with classes in the second semester on February 23, therefore with a one-month delay compared to today's winter holidays and much earlier than at the then universities, where they finished the first semester just before the Easter holidays.

Stefan's mathematical courses at Viennese High School

Between 1859/60 and 1862/63, Stefan also taught mathematics in the 4th grade at the Viennese High School for eight hours a week as a full professor of 6th rang, which was higher than most of his associates and also entailed higher incomes: therefore, the advancing Stefan was able to move into a more respectable apartment. He lectured a total of 16 hours a week, in addition to his university courses. In math class, he started with algebra and finished with geometry. He first introduced algebraic symbols and notations, and then narrated their history. He discussed basic operations with general numbers, the number of greatest common divisors and least common multiples (Gemeinschaftliche grossen Mass und kleinsten Vielfache), divisibility of numbers (Teilbarkeit), numeral systems, decimal numbers, periodic decimal fractions (Dezimalbrüche) with their calculations, powers and roots, probability calculus (Verhältnisse), linear equations and finally the logarithms. In geometry, he first considered lines (Geraden), then the construction of rectangular shapes, the Triangle Proportionality Theorem (laws of transversals of triangles), harmonic division, measurements, rectangular shapes, a circle, the construction of rectangular shapes, the intersection in space (Raumecke), calculating the circumference and volume of convex bodies, and finally the regular polyhedral.

The first semester of this course did not end until February 26, and the second semester began on March 6, 1860, while Moriz A. Becker re-inspected the school. Stefan donated Franc Ksaver Moth's algebra textbook to his school library, which he used while still studying (I. Šubic, 1902, 78; Moth, 1827; Moth, 1852; *IV Jahres-Bericht Bauernmarkte*, 1862, p. 60). Apparently, Stefan no longer needed that textbook, as he ended his career as a mathematics lecturer forever in 1863. That principal's report was followed by the principal's thanks to Koller, Becker, economic councilor Paul Schmidt, and J. Scheiner as a Commissioner for Viennese Secondary Schools.

Meanwhile, Stefan became Skřivan's deputy director (Stellvertreter) at Real School (Wisniak, 2006, 188), an assistant professor of mathematical physics at the university, and a corresponding member of the academy. However, as his income rocketed, he moved to a better apartment at Laimgrube no. 16, while the principal-director Skřivan resided nearby at Laimgrube no. 214 as a professor of mathematics in 5th and 6th grade (Jahres-Bericht der Öffentlichen Ober = Realschule am Bauernmarkt Wien 1862, 53-54, 56-58, 60-61).

Koller arranged the commendation of the director Gustav Skřivan (Skrivan, 1831 Krucemburk (Kreuzberg in the Czech Republic-1866 Prague). Skřivan published upon Stefan's final departure for the university: Full Regular senior lecturer dr. J. Stefan, corresponding member of k.k. Academy, was appointed with the highest determination a full professor of mathematics and physics at the University of Vienna. So, at the end of the 1st semester he left our Real School educational institution. His excellent teaching performance, successful work in the field of science, extremely loving behavior towards students, strength of character and true collegiality have largely earned him recognition and respect of all (Der Wirkliche Oberlehrer Dr. J. Stefan, Korrespondierte Mitgliede der kk Akademie, wurde mit Allerhöchste Entschliessung zum ordentlicher professor der Mathematik und Physik an der Wiener-Universität ernannt, und trat in Folge dessen an Schluss dem 1. Semester aus der Verbände der Lehrkörper dessen Lehranstalt. Seine ausgezeichnete Wirksamkeit als Lehrer, seine erfolgreiche Tätigkeit auf dem Felde der Wissenshaft, sein äusserst liebevolles Benehmen gegenüber der Schülern, seine Charakterfestigkeit (Karakterfestigkeit) und wahre Kollegialität, haben ihm die Anerkennung und Hochachtung aller im vollsten Masse erworben).

The 1st semester ended with Stefan's farewell on February 11, 1863, and the 2nd semester began without Stefan at the high school on February 19, 1863.

Stefan and Skřivan had spent nearly the same term at their Viennese Real School. On April 2, 1858, a member of the Engineering Society, Geographical and Secondary School Association the new principal Skřivan replaced the owner of the school of botany-physicist Karl Schelivsky (* 1814 Kyjov (Gaya) in southern Moravia), who has since run only the neighboring main school. As a student, Skřivan took part in the Prague demonstrations at the barricades during the Spring of Nations in 1848; together with Stefan he left the Viennese High School and became the first professor of mathematics at the Prague Polytechnic in 1863. At his Viennese Real school, he was replaced by the new principal Georg Kahrer, who specifically dealt with Ferdinand Redtenbacher's equations in 6th grade course of mechanical engineering, although Stefan no longer took Koller's friend Redtenbacher's Dynamides seriously (*Jahres-Bericht Bauernmarkte,* 1859: pp. 39, 53-54, 62; 1860: pp. 101-103, 107).

So, it is no wonder that Koller's eye flickered at Stefan's talent. Talent is not enough. It is at most a prerequisite for serious success: that is why the established professor Koller went straight to the student benches of the beginner Stefan to enable Stefan's full professorship and membership in the academy. Bullseye!

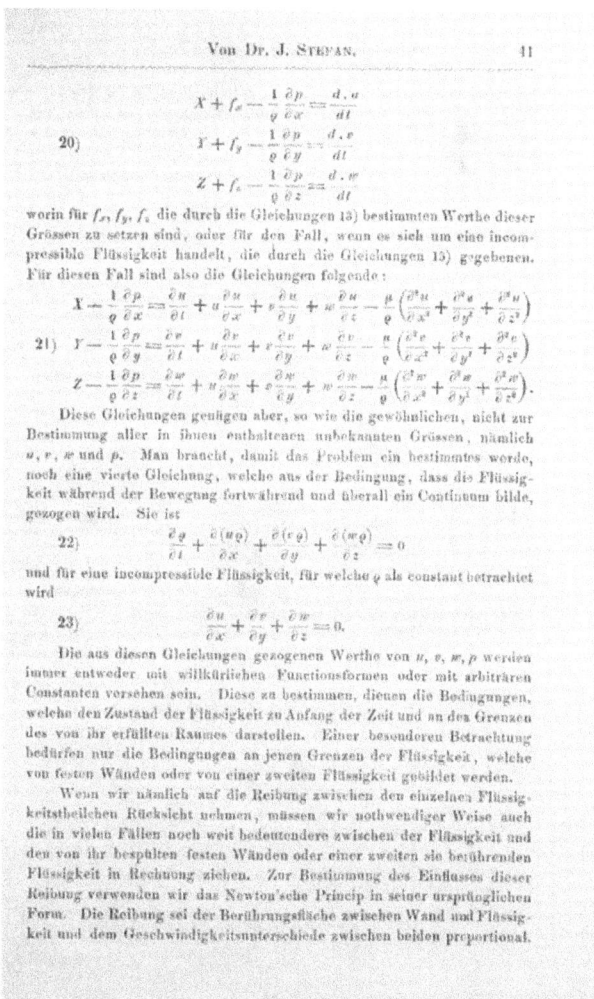

Von Dr. J. Stefan. 41

$$20) \quad \begin{aligned} X + f_x - \frac{1}{\varrho}\frac{\partial p}{\partial x} &= \frac{d \cdot u}{dt} \\ Y + f_y - \frac{1}{\varrho}\frac{\partial p}{\partial y} &= \frac{d \cdot v}{dt} \\ Z + f_z - \frac{1}{\varrho}\frac{\partial p}{\partial z} &= \frac{d \cdot w}{dt} \end{aligned}$$

worin für f_x, f_y, f_z die durch die Gleichungen 13) bestimmten Werthe dieser Grössen zu setzen sind, oder für den Fall, wenn es sich um eine incompressible Flüssigkeit handelt, die durch die Gleichungen 15) gegebenen. Für diesen Fall sind also die Gleichungen folgende:

$$21) \quad \begin{aligned} X - \frac{1}{\varrho}\frac{\partial p}{\partial x} &= \frac{\partial u}{\partial t} + u\frac{\partial u}{\partial x} + v\frac{\partial u}{\partial y} + w\frac{\partial u}{\partial z} - \frac{\mu}{\varrho}\left(\frac{\partial^2 u}{\partial x^2} + \frac{\partial^2 u}{\partial y^2} + \frac{\partial^2 u}{\partial z^2}\right) \\ Y - \frac{1}{\varrho}\frac{\partial p}{\partial y} &= \frac{\partial v}{\partial t} + u\frac{\partial v}{\partial x} + v\frac{\partial v}{\partial y} + w\frac{\partial v}{\partial z} - \frac{\mu}{\varrho}\left(\frac{\partial^2 v}{\partial x^2} + \frac{\partial^2 v}{\partial y^2} + \frac{\partial^2 v}{\partial z^2}\right) \\ Z - \frac{1}{\varrho}\frac{\partial p}{\partial z} &= \frac{\partial w}{\partial t} + u\frac{\partial w}{\partial x} + v\frac{\partial w}{\partial y} + w\frac{\partial w}{\partial z} - \frac{\mu}{\varrho}\left(\frac{\partial^2 w}{\partial x^2} + \frac{\partial^2 w}{\partial y^2} + \frac{\partial^2 w}{\partial z^2}\right). \end{aligned}$$

Diese Gleichungen genügen aber, so wie die gewöhnlichen, nicht zur Bestimmung aller in ihnen enthaltenen unbekannten Grössen, nämlich u, v, w und p. Man braucht, damit das Problem ein bestimmtes werde, noch eine vierte Gleichung, welche aus der Bedingung, dass die Flüssigkeit während der Bewegung fortwährend und überall ein Continuum bilde, gezogen wird. Sie ist

$$22) \quad \frac{\partial \varrho}{\partial t} + \frac{\partial (u\varrho)}{\partial x} + \frac{\partial (v\varrho)}{\partial y} + \frac{\partial (w\varrho)}{\partial z} = 0$$

und für eine incompressible Flüssigkeit, für welche ϱ als constant betrachtet wird

$$23) \quad \frac{\partial u}{\partial x} + \frac{\partial v}{\partial y} + \frac{\partial w}{\partial z} = 0.$$

Die aus diesen Gleichungen gezogenen Werthe von u, v, w, p werden immer entweder mit willkürlichen Functionsformen oder mit arbiträren Constanten versehen sein. Diese zu bestimmen, dienen die Bedingungen, welche den Zustand der Flüssigkeit zu Anfang der Zeit und an den Grenzen des von ihr erfüllten Raumes darstellen. Einer besonderen Betrachtung bedürfen nur die Bedingungen an jenen Grenzen der Flüssigkeit, welche von festen Wänden oder von einer zweiten Flüssigkeit gebildet werden.

Wenn wir nämlich auf die Reibung zwischen den einzelnen Flüssigkeitstheilchen Rücksicht nehmen, müssen wir nothwendiger Weise auch die in vielen Fällen noch weit bedeutendere zwischen der Flüssigkeit und den von ihr bespülten festen Wänden oder einer zweiten sie berührenden Flüssigkeit in Rechnung ziehen. Zur Bestimmung des Einflusses dieser Reibung verwenden wir das Newton'sche Princip in seiner ursprünglichen Form. Die Reibung sei der Berührungsfläche zwischen Wand und Flüssigkeit und dem Geschwindigkeitsunterschiede zwischen beiden proportional.

Figure 7: The page of Stefan's paper about integrals in the yearly of his Real School (Stefan 1862, 41)

155

's ancestors.

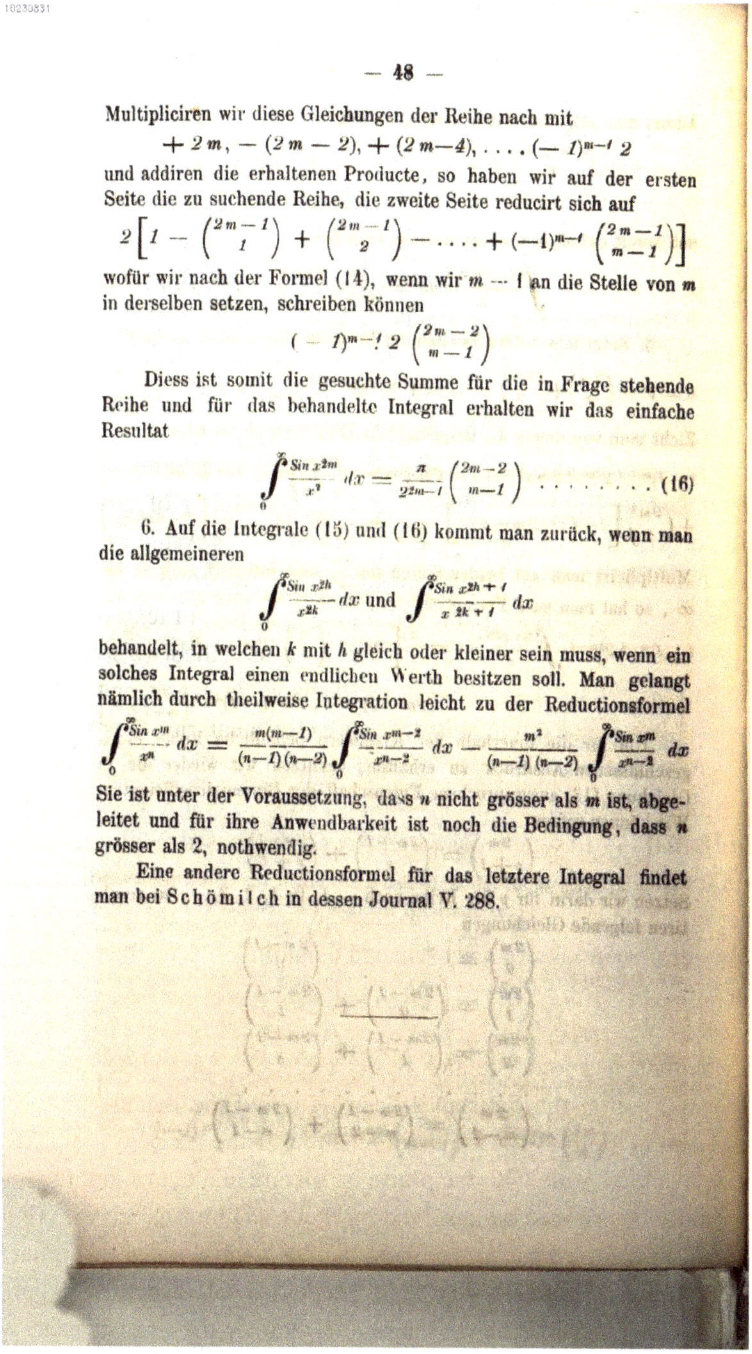

Multipliciren wir diese Gleichungen der Reihe nach mit

$$+ 2\,m, - (2\,m - 2), + (2\,m - 4), \ldots (-1)^{m-1}\,2$$

und addiren die erhaltenen Producte, so haben wir auf der ersten Seite die zu suchende Reihe, die zweite Seite reducirt sich auf

$$2\left[1 - \binom{2\,m-1}{1} + \binom{2\,m-1}{2} - \ldots + (-1)^{m-1}\binom{2\,m-1}{m-1}\right]$$

wofür wir nach der Formel (14), wenn wir $m - 1$ an die Stelle von m in derselben setzen, schreiben können

$$(-1)^{m-1}\,2\,\binom{2\,m-2}{m-1}$$

Diess ist somit die gesuchte Summe für die in Frage stehende Reihe und für das behandelte Integral erhalten wir das einfache Resultat

$$\int_0^\infty \frac{\sin x^{2m}}{x^2}\,dx = \frac{\pi}{2^{2m-1}}\binom{2\,m-2}{m-1} \quad \ldots \ldots \ldots (16)$$

6. Auf die Integrale (15) und (16) kommt man zurück, wenn man die allgemeineren

$$\int_0^\infty \frac{\sin x^{2h}}{x^{2k}}\,dx \quad \text{und} \quad \int_0^\infty \frac{\sin x^{2h+1}}{x^{2k+1}}\,dx$$

behandelt, in welchen k mit h gleich oder kleiner sein muss, wenn ein solches Integral einen endlichen Werth besitzen soll. Man gelangt nämlich durch theilweise Integration leicht zu der Reductionsformel

$$\int_0^\infty \frac{\sin x^m}{x^n}\,dx = \frac{m(m-1)}{(n-1)\,(n-2)}\int_0^\infty \frac{\sin x^{m-2}}{x^{n-2}}\,dx - \frac{m^2}{(n-1)\,(n-2)}\int_0^\infty \frac{\sin x^m}{x^{n-2}}\,dx$$

Sie ist unter der Voraussetzung, dass n nicht grösser als m ist, abgeleitet und für ihre Anwendbarkeit ist noch die Bedingung, dass n grösser als 2, nothwendig.

Eine andere Reductionsformel für das letztere Integral findet man bei Schömilch in dessen Journal V. 288.

Figure 9: Note about Koller's friend Schlömilch which concluded Stefan's paper about integrals in the yearly of his Real School (Stefan 1862, 48)

As soon as he became a full university professor, Stefan cashed in on his experience as a high school lecturer to become a critic and reviewer in a Habsburg grammar school newspaper. He first assessed the textbook of the then director of Aachen School of Commerce, a later professor of the Aachen Polyte hni , Stefan's peer Adolf Wüllner (1835 Düsseldorf; † 1908 Aa hen). Despite the title, Wüllner's textbook was based on Gilles Celestine Jamin's (* 1818; † 1886) work only in the introdu tion and in the first hapter on the balance and motion of liquids and solids. This was followed by Wüllner's explanation of atomism, which Stefan particularly approved of. In determining the average density of the Earth, he cited both H. Cavendish's experiments and the measurements of Koller's friend Airy (Stefan 1864, 161). In dealing with capillarity, Wüllner studied the experiments of Koller's friends Savart and Magnus. In the third chapter Wüllner dealt with the vacuum pump, Regnault's experiments, and Bunsen's investigation of absorption of gases resembling Stefan's doctoral dissertation; in the chapter on waves, Wüllner considered the experiments of Koller's acquaintance Charles Wheatstone (Stefan 1864, 162) and Helmholtz's acoustic experiments. In his next part of textbook focused on optics, Wüllner considered Cauchy's wave dispersion theory (Stefan 1864, 163) against partial optics in predicting the decisive Foucault and Fraunhofer's experiments. Wüllner examined Brewster's experiments and Gustav Kirchhoff's law of equivalence of absorption and radiation of light. He also discussed Fresnel's mirrors, which were an essential part of Stefan's lectures at the time (Stefan 1864, 164). In conclusion, Stefan not only praised Wüllner, but even the heir and son-in-law of his Leipzig publisher, the Freemason Benedictus Gotthelf Teubner (1784 and Grosskrausnik in Luckau in Lower Lusatia-1856 Leipzig), who also printed books of Clebsch and Schlömilch's mathematical journal (Stefan 1864, 165-166).

Wüllner remained among Stefan's basic references for another decade and a half, especially after the Parisian academy offered the Bordin prize (prix *de* Charles-Laurent Bordin) for the determination of the temperature of the Sun. The prize went to Violle with the praise of Vicaire and Creva, but the question was officially determined as unsolved by the academicians anyway (Rossetti, 1879, 325). In the third part of his most famous treatise of 1879, Stefan began with a report on Wüllner's textbook adaptation of Tyndall's experiments consistent with then new Stefan's law; that half of the printed page was an essential part of Stefan's contribution. However, Stefan did not mention either Wüllner or Tyndall in the earlier manuscript summary of his work. Even in his published article, their result seemed less important than the measurements of

John William Draper and John Ericsson, with which Stefan continued the third chapter of his contribution as indicated by Violle: "Stefan can no longer find certain benchmarks to compare his T^4 formula to experience. The interesting, but not very precise, research of Draper and Ericsson provides only indications which Stefan considers to be in accordance with the law of the fourth power. I must add that my personal observations do not confirm Stefan's view".[10] The nationalistic Violle defended his old French compatriots Dulong and Petit as well as de la Provostaye and Desains against new Ericsson's (1876) and Stefan's (1879) critiques. Lehnebach 410, César-Mansuète Despretz 418-419, Pouillet 420, 426, Wüllner, Tyndall, Draper (1847) 422, Ericsson's *Contributions to the Centennial Exhibition*, New York: The Nation Press September 1876 II, Radiation at different temperatures 17-51, against Dulong-Petit on pages 48-51 here Stefan cited p. 49 (Stefan 1879 422-423), Charles Soret 427, Stefan's fellow Viennese student in 1854-1857 promoted to Regnault's Parisian collaborator in 1864 and professorships of physics at Liceo di Santa Caterina in Venetia and University of Padua Francesco Rossetti (Rosetti, 1833 Trento-20 April 1885 Padua) Sur la température du Soleil, recherches expérimentales, *Annales de Chimie et de physique,* June 1879 17: 177-228, Translated by F. Engels' acquaintance the crippled social reformer John I. Watts (Ph.D. of Giessen in 1844, 1818 Coventry–1887 Manchester) as a secretary to the Owens College in Manchester in LIII. Experimental Researches on the temperature of the sun *Phil.Mag.* October 1879 8/49: 324-332, 8/51 (December 1879) 438-449. The original Rossetti's Italian language article was published a year earlier as Sulla temperatura del sole in *Nuovo Cimento (giornale di fisica)* Serie 3, tom 3 (1878), pp. 238-256 and in a book form as *Indagini sperimentali sulla termperatura del sole : memoria,* From *Reale Accademia dei Lincei,* Anno CCLXXV (1877-78) Serie 3a.--*Memorie della Classe die scienze fisiche, matematiche e naturali,* Volume IIa.--Seduta del 6 gennaio 1878, Roma: Coi Tipi del Salviucci, 1878 in 34 pages (Stefan 1879 427-428).[11] Rossetti also used Wüllner's textbook which was evidently very popular in Viennese institute. It is very probable that Stefan found out about Wüllner's publication of Tyndall's results only after he had already written his manuscript draft, but before the final publication. Therefore, Stefan added Wüllner's data to his work, but more as a

---

[10] Violle, 1881, 318.

[11] Rossetti, Sulla Temperatura delle Fiamme, *Atti del R. Istituto Veneto,* vol. iii. 1877; Rossetti, *Di alcuni recenti progressi nelle scienze fisiche e in particolare di alcune indagini intorno alla temperatura del sole : orazione inaugurale dei corsi accademici dell'anno 1877-78 letta nell'aula magna dell'Università di Padova il 19 novembre 1877,* Padova : G.B. Randi, 1877.

minor piece of evidence, as it was obvious that Tyndall did not measure very accurately. In later development, it was Tyndall's contribution that most supported Stefan's ideas; Tyndall was certainly happy to see his contributions praised there, even though everyone knew they were not accurate.

In 1865, Stefan evaluated Schellbach's Berliner textbook of elliptic integrals and (Riemann-Carl Ludwig Siegel (1896 Berlin-1981 Göttingen)) theta functions in the journal of Austrian grammar schools (Stefan 1865, 16: 747-748). Karl Heinri h S hellba h (* 1805 Eisleben; † 1892 Berlin) studied mathematics and electricity in Halle with Johann Salom Christoph Schweigger (1779 Erlangen-1857 Halle) after Schweigger assembled his famous sensitive galvanometer in 1820. Between 1855 and 1889, Schellbach presented the difficult art of teaching to young mathematicians in his Berliner Mathematical-Pedagogical Seminar. Schellbach's seminar exercises were attended by Clebsch, Carl Neumann, Koenigsberger, Georg Cantor, and many others. Schellbach's lectures appeared in a book on the doctrine of elliptical integrals and theta functions with an emphasis on the practical side, its applications in various tasks in mechanics, astronomy, and physics. In the first part of his book, Schellbach gave the theory of elliptical integrals, and in the second part he listed the possibilities of their use that interested Stefan the most. Of course, Stefan recommended the work of master Schellbach to all lovers of mathematics. In the same magazine, Stefan sharply criticized the textbook of Albert Ferdinand Trappe, the professor and vice-rector of the Zwinger High School in Wrocław. Trappe retired after 38 years of teaching mathematics and physics at the Zwinger First Class High School on Easter 1876.[12] Stefan was critical of Trappe's mathematically overly complicated explanation, which was supposed to go beyond the students 'options and put professors in a time crunch due to the overabundance of the material taught. Even more sharply, Stefan criticized the loose definitions of a momentum as a force, the superfluous enumeration of machines including their manufacturers (as a kind of advertising), the incomplete explanation of Kepler's laws and the mechanical work with kinetic energy. Thus, Stefan was not only critical of the textbooks of his Slovenian compatriot Simon Šubic, but he was also otherwise strict. In the same volume of the journal of the Habsburg Gymnasiums (Zeitschrift für die österreichischen Gymnasien, 16: 865-866), the University of Graz mathematics professor,

---

[12] Stefan J. (1865). Book review: Trappe, Albert Ferdinand. 1865. Die Physik für den Schulunterricht bearbeitet von Albert Ferdinand Trappe, 3rd edition on 296 pages. Breslau (Wrocław): Ferdinand Hirt. *Zeitschrift für die österreichischen Gymnasien*. 16: 670-673.

Johann Fris hauf (* 1837; † 1924), rather sharply assessed the dis ussion of the Czech professor of the Ljubljana grammar school Nejedli,[13] saying that he did not sufficiently follow the Harriot-Descartes' rule of signs from Descartes' La Géométrie printed in 1673 and the work of the English founder of algebra Thomas Harriot (* 1560 Oxford; † 1621 London) printed posthumously in 1631 as *Artis Analyticae Praxis*. The fan of Slovenes and Alpinism Frischauf was Ettingshausen's Viennese student soon after Stefan, and he also privately taught Boltzmann's half-Slovenian bride Jetti. In the reports of the Ljubljana grammar s hool, Josip Nejedli (Johann Josef, * 1821 Prague; † 1919 Ljubljana) published numerous discussions on algebraic analysis. In 1863 he dealt with Euler's procedure for solving indeterminate first-order equations. Močnik's preliminary discussion of Cauchy's methods was concluded in 1865 with Nejedli's discussion of Budan's and Horner's algorithm for solving numerical equations of higher species (Frischauf 1865, 865-866). The English mathematician William George Horner (1786 Bristol-1837 Bath) is better known today than the French amateur mathematician-physician Ferdinand François Désiré Budan de Boislaurent (1761 Limonade, Haiti-1840 Paris), who developed his method in 1803 with a final publication in 1807.

In the same volume of that Grammar School journal, Stefan praised the work of the then lecturer in mathematics at the University of Halle, Carl Gottfried Neumann (Karl, 1832 Kaliningrad (Königsberg)-1925 Leipzig), the son of famous mineralogist Franz Ernst Neumann (* 1798; † 1895). Carl Neumann was educated as a student of the Schellbach's Berliner Seminar; he became one of the initiators of the theory of integral equations. Three years after Stefan's review, in 1868, together with Alfred Clebsch, Neumann co-founded the journal *Mathematical Annals (Mathematische Annalen)* in 1868, which is still published today as one of the twenty most important mathematical magazines in the world. Carl Neumann drew on Wilhelm Weber's ideas about the importance of the velocity and acceleration of electric particles, which together with the distance between them define force: of course, as was the custom at the time, he thought of Ampère's magnetic molecules and particles of the luminiferous. Eventually, Maxwell's approach based on Carl Neumann's mathematics supplanted Weber's approach also with Stefan's help; Stefan tried to merge Weber's and Faraday's data in his course noted by Koller, and Stefan

---

[13] Frischauf J. (1865). Recension: K. Nejedli, Elementäre Ableitung der Budar-Horner'schen Auflösungsmethode Höherer Zahlengleichungen (22 pages), *Zeitschrift für die österreichischen Gymnasien.* 16: 865-866.

gave his best pupil Boltzmann an English dictionary so that he could read Maxwell's works. Of course, Stefan knew well the authority he was dealing with, so he warmly recommended the work of Carl Neumann to friends of the mathematical sciences (Stefan 1865, 16: 305-306).

Further in the same volume of that journal, Stefan's promotion to the post of full Viennese academician was reported (*Zeitschrift für die österreichischen Gymnasien*, 1865, 16: 472). That success and Koller's death probably ended Stefan collaborations in the Viennese pedagogical journals.

Koller Promoted Stefan at Viennese University and Academy

Stefan's collaborations in Viennese pedagogical journals promoted by Koller's ministry further advanced Koller's support of Stefan. Already famous professor Koller went into the students' benches of the beginner Stefan. Koller listened carefully and recorded at least six semesters of his lectures, the last semester together with Ludwig Boltzmann whose own study materials were never examined. Koller wrote his impressions to show his colleagues at the ministry that they had to employ Stefan as a full professor and academy member. What did those records look like?

dignes d'en faire usage dans le Catalogue de la Société Astronomique, je vous prie de me notifier aussi dorénavant les travaux que vous souhaiterez exécutés par les astronomes, et je ne manquerai pas de faire mon possible pour seconder vos vœux. . . . .

M. KÖLLER,

Kremsmünster, en Autriche, 1838, *Decemb.* 17.   *Astronome.*

A.S.C. No.	Star.	Right Ascension, 1838.	No. of Obs.	Declination, 1838.	No. of Obs.
2	11 Cassiopeiæ β ..	0 0 34·503	21	+58 15 22·47	7
41	15 Cassiopeiæ κ ..	23 50·636	18	+62 2 15·06	9
55	Ceti	29 0·950	6	....	..
91	Cephei.......	0 47 49·186	6	+85 23 0·33	6
164	Ceti	1 22 3·237	7	−26 27 28·15	7
194	53 Ceti χ	1 41 37·786	4	−11 29 28·15	4
268	Ceti	2 27 12·303	4	+ 6 6 33·63	2
271	Arietis præced..	27 37·987	2	+23 56 18·76	2
	sequens.	27 41·002	1	+23 56 15·23	2
279	Ceti	31 43·893	6	+ 5 24 41·43	1
283	34 Arietis μ	33 14·656	6	+19 19 1·06	5
340	Persei	2 57 24·752	5	+48 59 17·07	5
356	14 Eridani	3 8 44·867	5	− 9 45 30·26	5
389	20 Eridani	28 54·670	5	−18 0 27·41	5
399	41 Persei ν	34 12·655	6	+42 3 36·28	6
424	28 Eridani	3 40 41·683	6	−24 22 47·89	6
482	41 Eridani	4 11 45·937	4	−34 11 49·62	4
503	71 Tauri	17 7·382	4	....	..
506	43 Eridani	17 57·345	2	−34 23 44·83	4
515	80 Tauri	4 20 54·701	4	+15 16 39·16	3
616	5 Leporis μ......	5 5 39·379	3	....	..
635	22 Orionis ε	13 29·569	3	− 0 32 52·65	1
641	Eridani	15 7·414	1	....	..
672	Columbæ ν	25 27·814	2	−35 35 30·26	2
685	40 Orionis φ²	28 0·405	1	+ 9 11 46·47	1
693	49 Orionis d......	5 31 2·773	2	− 7 18 31·90	3

Figure 10: Koller's publication in London, page 374 with signature.

Stefan as University Teacher

How did Stefan lecture? In our days, there are not yet online videos of Stefan's lectures, but we could animate them without difficulty. We have detailed available Koller's synopses of Stefan's courses that can be combined with Stefan's photos, familiar descriptions of his appearance and the way he moved with one shoulder somewhat sagging because of his juvenile carrying of his father's heavy baker's bags filled with flour.

Koller initially visited Stefan's courses at a high school in Vienna as the inspector. Next, Koller did the same in Stefan's university classrooms outside of Koller's official duties. At the Viennese university, Koller first attended the privatdocent Stefan's course entitled: Theory of solid-state bodies (Theorie der Elasticität fester Körper). Stefan taught that course in the winter semester 1861/62 and Koller noted those lectures on thirty-six folios (= 144 pages) where the additional last thirty-six pages were in fact devoted to the course of acoustic in the next summer semester of 1861/62. All those early Koller's notes of Stefan's courses were different from the later notes as there Koller did not reserve his left half of the pages for eventual appendixes and illustrations. In that winter and summer semester 1861/62, Koller illustrated his notes directly between the text which means that he made almost no additional notes. The idea of blank left columns occurred to him only later while he attended additional Stefan's courses.

Stefan read his first lecture on 22 October 1861 within his part no # 1. On Koller's second page Stefan defined the stresses (tensions) of particles of matter without explicitly referring to atoms or molecules.

Stefan read his second lecture on 29 October 1861 within his part no # 5 with underlined subtitle Progressive movement accompanied by the illustration of a parallelogram of forces (Koller-Manuskripte, 18: page 7).

In his part no # 6 Stefan described any pair of forces illustrated by the tetrahedrons (Koller-Manuskripte, 18: page 9). Next, Stefan switched to the problem of conditions for rotation in space defined by three coordinate axes and discussed the center of gravity (Koller-Manuskripte, 18: page 10).

Stefan read his third lecture on 5 November 1861 with a part no # 7 entitled Relations between stresses for different planes in one and the same point. Koller illustrated them again by the stress tetrahedrons (Koller-Manuskripte, 18: page 13). In 1822, Cauchy presented the idea of traction vector that contains both the normal and tangential components of the internal surface forces per unit area and used his tetrahedron to prove the existence of stress tensor as the main part of the modern foundation of continuum mechanics. Koller and Stefan did not use Cauchy's name in that context.

Next, Stefan read his part no # 8 (Koller-Manuskripte, 18: page 18). Stefan read his fourth lecture on 12 November 1861 with a part no # 9 about mechanical stresses (tensions) while he constantly used the term normal stress (Koller-Manuskripte, 18: page 21). A normal stress occurs when a member is loaded by an axial force. For any prismatic section, the value of the normal force is the force divided by the cross-sectional area as a normal stress occurs when a member is placed in tension or compression.

Next, Stefan read his part no # 10. He also dealt with stress ellipsoid alias elasticity ellipsoid and radius vector of the stress. Soon after Cauchy's idea of tetrahedron published in 1822, Lamé with the help of his companion Clapeyron in Petersburg described his stress ellipsoid as an alternative to the later Christian Otto Mohr's (1835 Wesselburen in Holstein-1918) circle for the graphical representation of the stress state at a point, as any surface of Lamé's ellipsoid represents the locus of the endpoints of all stress vectors acting on all planes passing through a given point in the continuum body (Lamé, Mémoire sur l'équilibre intérieur des corps solides homogènes (reviewed by Poinsot and Navier in 1828) *Extrait des Mémoires des savans étrangers*, Tome 4, 1833; Koller-Manuskripte, 18: pp. 23-24).

Stefan read his part no # 11 to answer a question: which are the main stresses inside a body (Koller-Manuskripte, 18: page 24). Stefan read his fifth lecture on 19 November 1861 with a part no # 12 (Koller-Manuskripte, 18: page 28).

Stefan read his part no # 13 to discuss again his equation no. 12 (Koller-Manuskripte, 18: page 30). Stefan read his part no # 14 by solving cubic equation (of 3rd order) no. 14 (Koller-Manuskripte, 18: page 32). Stefan concluded the discussion with two illustrations showing sinus functions (Koller-Manuskripte, 18: page 36).

Stefan read his sixth lecture on 26 November 1861 with a part no # 15. He discussed the shifting of the points with the illustration of sinusoids (Koller-Manuskripte, 18: pp. 36, 38).

Stefan read his part no # 16 to discuss the pair of differentials of his equation no 18 (Koller-Manuskripte, 18: page 39).

Stefan read his part no # 17. His treated surface might be elliptic or hyperbolic as an extension surface (area). He also used the orthogonal system (Koller-Manuskripte, 18: pp. 44-45, 49).

Stefan read his seventh lecture on 3 December 1861 with a part no # 18, using the illustration of an orthogonal parallelopiped (Koller-Manuskripte, 18: pages 51, 53).
Stefan read his part no # 19 and afterwards his part no # 20 with the Extension proportional to the pressure (distancing). He obtained thirty-nine coefficients in so called homogenous body (Koller-Manuskripte, 18: pages 55-56, 59). Even under the assumption of linear elastic behavior of the material, the relation between the stress and strain tensors is generally expressed by a fourth-order stiffness tensor with twenty-one independent coefficients.

Stefan read his eighth lecture on 10 December 1861 with a part no # 21. Stefan read his part no # 22 about stresses-pressures for any arbitrary chosen system of coordinates, also for the tangential stresses (Koller-Manuskripte, 18: pages 59, 61, 62).

Stefan read his nineth lecture on 17 December 1861 by a part no # 23 (Koller-Manuskripte, 18: page 65) with notes about the axes of prism in direction of its three axes to determine the stress along the principal axis:

$$P = X + Y + Z = \lambda \cdot (a + b + c) + 2 \cdot \mu \cdot a = ((3 \cdot \lambda + 2 \cdot \mu)/3) \cdot (a + b + c) = ((3 \cdot \lambda + 2 \cdot \mu)/3) \cdot \theta$$

$$\theta = P. (3/(3 \cdot \lambda + 2 \cdot \mu))$$

where

$$\alpha = 3/(3 \cdot \lambda + 2 \cdot \mu)$$

is a coefficient of elasticity on the surface (Koller-Manuskripte, 18: page 66).

Stefan (or maybe just Koller) added special annotation about the measurements of μ and λ of the researcher of critical points and acoustic of siren, the baron Charles Cagniard de la Tour (Latour, 1777 Paris-1859 Paris). From his equation for the prism, Cagniard de la Tour got the relation $\lambda = \mu$. Guillaume Wertheim (1815-1861) offered the relation $\lambda = 2 \cdot \mu$ for three different punches to find the relations for all kinds of bodies. With the sharpened pencil while using the same ink, Koller added the additional remarks about Clapeyron, who found for iron or ice: $\lambda = \mu$. His method deserved several valuations. With these uncertain results he was only left with the assumption that λ and μ would have different values in different bodies, which will be the same in some bodies (Koller-Manuskripte, 18: page 68; Clapeyron, Sur le travail des forces élastiques dans un corps solide déformé par l'action de forces extérieures, *C. R. Acad. Sc.*, t. 46, 1858, p. 208; Lamé & Clapeyron, Sur l'équilibre intérieur des corps solides homogènes, *Académie des Sciences*, 1833; Guillaume Wertheim, Recherches sur l'élasticité: 2nd De l'élasticité et de la cohésion des alliages, *Annales de chimie et de physique*, 3e série, t. 12. - Présenté à l'Académie des sciences le 8 mai 1843). That early Stefan's use of Clapeyron and his friend Lamé's data might have been symptomatic for later Stefan's extensive uses of their treatments of movable borders in phase transitions.

Stefan read his part no # 24 by the underlined title: Transformation of the equations of equilibrium (Koller-Manuskripte, 18: page 69).

Stefan read his part no # 25 to conclude with a remark about Clausius ground-breaking new book *Potentialfunction und das Potential*, published in 1859 (Koller-Manuskripte, 18: pages 72, 78). Certainly, Stefan was already a fan of kinetic atomism which Clausius already described a couple of years earlier.

Stefan read his tenth lecture on 7 January 1862 with a part no # 25 filled by notes about the work of the force denoted by the letter P (Sic!, it should be #26, Koller-Manuskripte, 18: page 74). Increase of the force enables the motion of a body (Koller-Manuskripte, 18: page 77).

Stefan read his part no # 26 by the underlined title: Work of the force defined by stresses (tensions, pressure, Koller-Manuskripte, 18: page 78).

Stefan read his part no # 27 about the elasticity force applied to the surface, explained by Clapeyron's equilibrium of forces (Koller-Manuskripte, 18: page 80). The applied force P acts on the surface of the wire by the force p (that is, as considered in wire). It acts as elasticity in the interior of the body. Clapeyron has put forward a theory about the equality of these forces, which Stefan now wanted to present somewhat shortened. We have found the following expressions for equilibrium equations of elasticity of a body noted in Stefan's formula No. 4. Let us multiply these equations' relationship with coefficients (Koller-Manuskripte, 18: pages 80-81).

Stefan continued: We will discuss work of the stress (tensions forces) on the surface of the body. The normal stress is constant, so we get Clapeyron's theorem. Then Stefan or Koller added one of his frequent annotations finished by Lamé's equations for stress along the principal axis A, B, C (Koller-Manuskripte, 18: pages 84-86).
Stefan read his eleventh lecture on 14 January 1862 with a part no # 28 to check Clapeyron's theory of Stefan's equation no 27. From the equation no. 27 we get by the development of Clapeyron's theory the integral equation no. 41. We only want to make use of the expression already found and seek the relationship between the rotational speed around the axis of a rod attached to one end and the rotating force acting on the rod (Koller-Manuskripte, 18: pages 86-88).

Stefan read his part no # 29 about his equation no 41 (Koller-Manuskripte, 18: page 91). Stefan read his part no # 30 focused on the work function at the areas pressed by that force (Koller-Manuskripte, 18: page 92).

Stefan read his twelfth lecture on 21 January 1862 with a part no # 31 by notes about the equilibrium of elastic bodies based on his equation no. 4 (Koller-Manuskripte, 18: page 95).

Stefan read his part no # 32 about the equation for μ, involving the components of the stress containing 21 constants while consulting his previous lecture at part no. 20 (Koller-Manuskripte, 18: pages 97-98). He used his own system of major shifts-stresses.

Stefan read his 13th lecture on 29 January 1862 with a part no. # 32 by notes about the elasticity of suspended burdened prism which Koller explained by his illustration (sic!, in fact no. # 33, Koller-Manuskripte, 18: page 99).

Stefan read his part no # 33 to solve his previously discussed problem on another way, while seeking additional answers (Koller-Manuskripte, 18: page 101).

Stefan read his part no # 34 to solve the equation for equilibrium on surface of elasticity (Koller-Manuskripte, 18: page 104).

Stefan read his 14th lecture on 4 February 1862 with a part no # 35 dealing with his previous equation no 55=11 for ξ to achieve the given shape of a function f (Koller-Manuskripte, 18: pages 109-114).

Stefan read his 15th lecture on 11 February 1862 with a part no # 36. He discussed the algebraic function f and the equation no 60 (Koller-Manuskripte, 18: page 114).

Stefan read his part no # 37 by notion of the previous solving of his equation no 64 with its constant a. The discussion ended with Koller's personal appendix-annotation, stating: In my opinion the constant K in this equation is = 0, because for x = a, y = 0 and z = 0 one takes over the point of the axis (Koller-Manuskripte, 18: page 118-119).

The older theory, which takes the number of shifts by its panoramic view, gives a value for ξ which is therefore by a degree too small (Koller-Manuskripte, 18: page 120).

Stefan read his 16th lecture on 18 February 1862 with a part no # 38 (Koller-Manuskripte, 18: page 120). He focussed on the right-angled prism of forces in the case of equilibrium. Next, Stefan read his part no # 39 (Koller-Manuskripte, 18: page 124).

Stefan read his part no # 40 to determine the constant C_m (Koller-Manuskripte, 18: page 127). Then Koller added his own explanatory note as annotation to the 1st line of part 4 of his lecture: This force, with which a material point is kept in its state of rest (equilibrium), decreases with the mutual approaching of their material particles closer to each other, therefore with their diminished distance. It must be regarded as handy turning of these particles, depending on their distance and on their mass (Koller-Manuskripte, 18: pages 127, 135).

Next followed a note, probably about the lecture no 4 (or maybe Stefan's part #4, or equation no 4). Lamé used in his theory a tension dydz noted with a letter X (Koller-Manuskripte, 18: page 135). Lamé denoted by f_x, f_y, f_z the components of external force which Stefan noted as X_o, Y_o, Z_o, at least by Koller's narration. In fact, in his 4th part Stefan used notation X_x, X_y, X_z and continued with his similar symbols in continuation of his lecture no 4 (Koller-Manuskripte, 18: page 4).

In that Koller's explanatory annotation, Koller certainly had in mind Lamé's edition of *Theory of elasticity* (Lamé, *Leçons sur la théorie mathématique de l'élasticité des corps solides, par M.G. Lamé* Paris: Bachelier, 1852). Stefan and Koller obliviously read Lamé's work in its French original just like the newly printed Lamé's *Leçons sur la Théorie de la Chaleur* published in Paris by the same (Mallet-)Bachelier few months before those Stefan's lectures in 1861. Lamé's lectures about heat, acoustic, and optics as a second part of his physics course at the Parisian Polytechnic were translated by the translator of Arago, Poisson and Cauchy's works, Christian Heinrich Schnuse (1800-1878). The professor of mathematics in Carolinum in Braunschweig in 1835/36, Schnuse, translated Lamé's work into German language in 1838 and soon added the translation of Lamé's course on electricity, magnetism, and electrodynamics, published in Darmstadt by Leske in 1841. Eventually, nobody translated two important later Lamé's books, *Leçons sur la Théorie de la Chaleur* or *Leçons sur la théorie mathématique de l'élasticité,* where the eventual German title might have been *Theorie der Elastizität.*

Koller's note about the data no. 5 (probably Stefan's lecture no 5) was also focused on Lamé's works according to Lamé's formula needed for the equilibrium (balance). The balance was designed as parallelepiped of forces. For further annotations and comments about Stefan's data no=5, Koller advised his readers-students to see the end of the following page which was eventually crossed out. Koller's comments about # 15 focused the displacement and cosine functions (Koller-Manuskripte, 18: pages 136-139).

No matrices or tensors were used by Stefan and Koller, except at Koller-Manuskripte, 18: page 140 and on the very end of Koller's manuscript no 18, written on the separate leaf. The term matrix was introduced by the English Jewish mathematician-lawyer James Sylvester (1814 London-1897 London) in 1850, but only his friend similar mathematician-lawyer Arthur Cayley (1821-1895 Cambridge) developed the algebraic aspect of matrices in two papers in 1858. Therefore, Stefan and Koller probably knew about that matrix novelty in early 1860s. The term tensor (and scalar) was already used by the Irishman William Rowan Hamilton (1805 Dublin-1865) during Irish first year of mouldy (blight) potato famine leading to starvation in 1846, as an extension of his quaternions later promoted and recommended by Thomas Andrews (1813 Belfast-1885 Belfast) in Belfast and by Kelvin's Scottish collaborator Peter Guthrie Tait. Among most physicists, the tensors became widely known only after Einstein's use of them for his general theory of relativity in Prague and

Zürich just before the First World War, so that they remained foreign even to Stefan's best student Boltzmann with his bulky publications which frustrated Maxwell. Only several years after Stefan's death, in March 1900, Gregorio Ricci Curbastro (1853 Lugo di Romagna-1925 Bologna) and his former student Tullio Levi-Civita (1873 Padua-1941 Rome) first published about their tensors.

Also, probably as a part of (Koller's) appendix notes there was the underlined title with no new part stated. It was entitled: Relations of pressures on the same point involving different surfaces. The text was written by the other hand which was not Koller's but eventually of one of his assistants sufficiently trained in physics, or even Stefan himself might be there as the handwriting clearly resembles his earlier Slovenian sonnets and later manuscript papers used for his plead for habilitation at Viennese university on 3 September 1858. That now unknown person alias Stefan wrote much smaller letters slightly inclined forwards on the top and used different ink. There was no text crossed out as it was a clear copy on one folio (Koller-Manuskripte, 18: pages 147-148).

Also, probably as a part of (Koller's) appendix notes there was the underlined title with no new part # entitled: Geometric relations of pressures of the same point of different surfaces. The text was again written by the other hand which was not Koller's. The now unknown person again wrote much smaller letters and used different ink. That handwriting was followed by another underlined title, third in the row: Geometric Relations of normal pressures of the same point (Koller-Manuskripte, 18: pages 149-150). The writer quickly finished that text after half of a page.

Koller or the later editor of his manuscripts added inserted unrelated leaf which noted the Prime-Convict in Fiume (Rijeka) and in Castel Nuovo del Cattaro (Herceg Novi at the entrance of Kotor Bay, Koller-Manuskripte, 18: page 151). The first Benedictine nunnery in Rijeka was built in the 17th century in the Old Town next to the still existing church of St. Rochus that entered its composition. In fact, Sancti Petri was the name of the site where the Benedictine monastery and church were located, and today that site is called Rake in Herceg Novi. As a Benedictine, Koller certainly had relations with that monastery which became the part of Habsburgian monarchy after the Napoleonic downfall, and Koller might even paid them a visit. There is nothing about Stefan's lectures at that peace of paper as it was indeed inserted there by the error.

Therefore, Stefan's 16th lecture on 18 February 1862 contained parts no # 38-40. It was followed by three blank pages, by new appendixes, and by two additional unnumbered parts which we may consider as numbered 41 and 42. No further parts were numerated after the number 40 which was probably the sign that the following text was not proof-edited like the other Koller's notes were. No notes were dated on 25 February 1862 or on 4 March 1862.

Next note dated on 18 March 1862 was not written with ink, but with the graphite pencil. There, Stefan's discussed the equilibrium in his prism of forces. It might be considered as his 18th lecture about mechanics. The manuscript again assumed the usual Koller's handwriting. The text was not underlined while it began as: Equilibrium on triangular (three-corner) prism. At first, we got ... stresses (tensions) at points. On next page there was an underlined subtitle: We research our prism of forces in the elliptic coordinate while Stefan was computing a differential equation for the coefficient ξ (Koller-Manuskripte, 18: pages 153-154). It resembled a pressure prism as a way of visually describing the variation of hydrostatic pressure within a volume of fluid.

Next note was dated on 11 March 1862, a week before the previous one! It might be considered as 17th lecture. It was again not written by ink, but by the graphite pencil. The problem focused was again the same equilibrium on prism. At the top of page 164 Koller noted a kind of reference to a hardly readable name (Koller-Manuskripte, 18: pages 161-164). Koller took no notes on 25 March 1862 as no preserved note was dated then. Next, he added additional text with no separate date provided.

In following, Koller added his dated note on 1 April 1861, again about triangular prism with a separate pagination of folios reading as no 1, with just one folio included, comprising of four pages, still focused on equations involving ξ (Koller-Manuskripte, 18: pages page 173). It might be considered as Stefan's 19th lecture. Clearly, the text was still about the elasticity and not yet about the acoustics. Koller probably planned to make its clean copy later, but he never did. Those 16 or even 19 dated lectures in 40 or even 45 parts of winter semester were too numerous as the usual number of winter semester weekly lectures was only 14. On the other hand, the lectures in course about acoustic in the following summer semester of 1861/62 were far less numerous as only seven of them with dated entrances were noted instead of usual twelve, while even those were not enumerated as separate lectures or divided into parts. If we sum together the lectures read in winter and summer semester, we get something like the expected 26 lectures for both, but less than third of them were read during the summer semester course of acoustics. In 1862, the Easter Good Friday was on 18 April, followed by the Easter Saturday, Sunday, and Monday on 21 April 1862. Therefore, the holidays between semesters were certainly at about the same space of time as the dates in Koller's manuscript.

Motions of Waves as Acoustics (and Optics)

Next, Koller dated his note about a lectured course on 6 May, probably in 1862 (Koller-Manuskripte, 18: page 177). It was entitled: Static and Dynamic of Elasticity. Koller probably connected it with acoustic as a wave theory including diffraction of waves and wave surface. Koller wrote his separate pagination of folio no. 1 with just part of folio included, comprising of three pages. The notes might be from Stefan's course, somebody else's course, or they included Koller's own ideas. Koller's travel companion Kunzek taught the course entitled "the static of fluids" in summer semester 1864 with Boltzmann among his students. Again, Koller used only graphite pencil and his handwriting shows kind of a hurry involved. Those notes were probably connected with Stefan's course of mathematically treated acoustic, which Boltzmann attended two hours per week two years and a half later in winter semester 1864/65 (Höflechner, 1994, 15).

Next Koller wrote a dated note on 13 May 1862 again in three pages full of computations but without any underlined title (Koller-Manuskripte, 18: page 180).

Koller dated his next entry on 20 May 1862 entitled "About the laws of sound in particular cases (instances)". There he dealt with his formulas noted by squares (Koller-Manuskripte, 18: pages 184, 187).

Koller therefore attended at least two Stefan's mathematical courses about mechanics in two consecutive semesters of 1861/62. The first one was devoted to Clapeyron-Lamé's theory of elasticity (Koller-Manuskripte, 18: pages 1-177). It contained forty notes and several additional parts, all of them written between 22 October 1861 and 1 April 1861. The other course comprised less carefully edited seven dated entries where the lectures were not enumerated or divided in parts-paragraphs. Those data were focused on static and acoustics as they were put down after Easter holidays between 6 May 1862 and 8 July 1862 (Koller-Manuskripte, 18: pages 177-189, with some wave optics up to page 213/215). In late April there might have been some introductory lecture on acoustics which Koller did not note or preserve. The elasticity was still the main topics in that new course of static and acoustics, but it was not clear copied inside Koller's notes. Stefan noted the measurements of λ of the researcher of critical points and acoustic of siren the baron Charles Cagniard de la Tour (1777 Paris-1859 Paris). Stefan also liked Guillaume Wertheim's (1815-1861) Recherches sur l'élasticité: 2nd (Part) De l'élasticité et de la cohésion des alliages, *Annales de chimie et de physique*, 3e série, volume 12, Presented to Académie des sciences on 8 May 1843. And, certainly, Stefan praised Clapeyron in Stefan's lectures about elasticity focused on work of force connected to the law on conservation of energy (Koller-Manuskripte, 18: page 68). On 7 January 1862 Stefan or maybe just Koller cited a reference in parenthesis on the end of his part 26 with the data about Rudolf Clausius's *Potentialfunction und das Potential: ein Beitrag zur mathematischen Physik*. Leipzig: Barth, 1859 (Koller-Manuskripte, 18: page 78 in concluding comment to part # 25). Clausius' book had its introduction dated in Zürich on 23 January 1859, which put it among the extremely new sources in Stefan's course. Clausius generalized the law of conservation of energy into the concept of potential as the sum of vis viva and ergal in his criticism of Helmholtz's approach (Clausius, 1859, 9; Elizabeth Garber, *The Language of Physics: The Calculus and the Development of Theoretical Physics in Europe, 1750–1914*, Birkhäuser, 1999, 301-302). As early protegee of the physician Ludwig raised in German areas, Stefan at least initially preferred the physician turned physicist Helmholtz, while Koller was closer to the great mathematical physics of Clausius based on Hamilton, Dirichlet, Gauss, and the principle of least action. Clausius also loved Lagrange–d'Alembert principle as the dynamic analogue to the principle of virtual work for applied forces in a static system, while Clausius generalized Hamilton's principle (Clausius, 1859, 82). For Clausius (and Coulomb), there was no difference between gravitational, electrical, or magnetic forces' inverse square law of distance, except for the involved constants. That early reading of Clausius' work certainly contributed to Stefan's kinetic theory and probably Koller also accepted it soon. Stefan clearly relied on book of Lamé published in 1852 (Koller-Manuskripte, 18: pages 86, 135-136) and on Lamé's friend Clapeyron's theorem (Koller-Manuskripte, 18: pages 68, 80f, 85-86). In Stefan's course about acoustic noted by Koller no references were added, but the main source was probably still Lamé. Lamé spent eleven years in Russia and returned to Paris together with Clapeyron in 1831, immediately after July revolution, which forced Cauchy to go into his voluntary exile. Lamé believed in connections between branches of physics which he also highlighted in 1860/61 as he tried to join the theory of elasticity and theory of translation of heat after 1833. That Lamé's effort was probably as much fundamental as Maxwell's joining of all physics except the gravity, finalized in his *Treatise* in 1873 (Bachelard, 1973, 112). The ideal unity of all branches of physics might still be a fair Lamé's goal before the speedy diversifications among the physicists after the Spring of Nations, while among the mathematicians the unity of their sciences was still promoted by the influential universal mind of David Hilbert (1862-1943) even much later.

Lamé believed that Descartes' coordinates, his own Lamé's curvilinear coordinates related to the ellipsoid of elasticity, and others' changing of the coordinate system always brings new knowledge. In was like observing the same problem from the different angle, from another perspective. Lamé also believed in (Fresnel's) aether, as did Koller and Stefan (Voronina, 1987, 172).

Cauchy and Poisson discussed the generalized case where the elasticity changed with a direction around the given point, which involved the treatment of angle besides the dealing with the distance; but they were unable to propose their mathematical-physical theory as a pure set of equitation without the involved hypotheses, without any proposed law (Lamé, 1861, X, XI), resembling later Kelvin's problems with mechanical modeling of Maxwellian theories. It was the hardly reachable Newtonian ideal: "Hypotheses non fingo (I feign no hypotheses)". In means that everything nonmechanical is hypothetical, which was a possible stance in Newton's era, but not anymore later. Cauchy and Poisson ware personally great mutual antagonists dealing with the same problems as antagonists. Clausius and Poisson's elasticity changed by the angle and by the distance, while their time was almost not involved as far, until Stefan introduced his movable barriers. The analytic theory of Cauchy-Poisson-Fourier, as advanced by Lamé, was a mathematical theory of fields while the mechanical theory of heat was an experimental physics' theory without clear parameters, which enabled them to coexist in the same branch of physics without apparent clashes. They nearly ignored each other tragicomically because they were researched by two different types of scientists even if both groups relied on atoms (Truesdell, 1980, 143 footnote). Only statistical mechanic finally joined them for their mutual theory of translation of heat and electricity. Lamé had a great idea about the distances between the ponderable particles comparable to the wavelength of light although none of them was experimentally measurable during next decades (Lamé, 1861, V). The scientific work of those days has often limited itself to the proper use of Fourier's series under given boundary conditions, regardless of the physical differences between the problems considered. Truesdell sarcastically called it Duhamel's "linearization before thinking" (Truesdell, 1980, 146). Their research of the thermal properties of gases also included pneumatic chemistry. Lamé used Fourier's theory needed for Lamé's own theory of elasticity in 1860/61, while he was admiring the crystallography of Fresnel's fan Henri Hureau de Senarmont (Sénarmont, 1808 Broué-1862 Paris) who coedited the completion of Fresnel's works (Voronova, 1987, 173; Lamé, 1861, XIX).

Duhamel forever connected the theory of elasticity and theory of the conduction, but he was no match for thermodynamics which needed a hundred and thirty-six years from 1837 to 1963 to include the thermo-elasticity (Truesdell 1980). As early as October 31, 1857, Stefan read a treatise on the motion of heat in moving fluids authored by the heir to Fourier-Poisson's theories, Jean Marie Constant Duhamel (* 1797 Saint-Mallo; † 1872). It was published in *Comptes rendus de l'Académie des sciences*, and Stefan used the original French printing as Stefan already had a good command of the French language, just like his mentor Koller. Duhamel was best known for his research of the conduction of heat in crystals and he finally replaced the deceased Siméon-Denis Poisson at the French Academy. In those days, Stefan also read Poisson's Sur le mouvement des ondes from *Mémoires de l'Académie des sciences* (in fact, Poisson, Mémoire sur la théorie des ondes, *Mémoires de l'Académie des Sciences*, 1 (1816), 1818, pp. 71-186). Stefan even studied Weber's *Wellenlehre*. He certainly used the famous Weber's book *Wellenlehre, auf Experimente gegründet, oder über die Wellen tropfbarer Flüssigkeiten, mit*

Anwendung auf die Schall- und Lichtwelen. Von den Brüdern Ernst Heinrich Weber, und Wilhelm Weber. The physician Ludwig's advice led Stefan to the work of more medically oriented among both brothers, Ernst Heinrich Weber. Ernst Heinrich Weber's book published in Leipzig in 1850 was needed for Weber's application to the teaching profession; he dealt with the circulations of blood (*Kreislauf des Blutes*). Stefan also used Poisson's Mémoire sur l'équilibre et le mouvement des corps élastiques, read in front of Parisian academy on 14 April 1828. A reprint followed in 1842. Or Stefan might have in mind Poisson's speech on 12 October 1829 at the Académie des sciences in Paris, later published in the *Journal de l'École polytechnique* 20th cahier, Tome 13 under the title: Memoir sur les equations générales de l'équilibre et du mouvement des corps solides elastiques et des fluids. Stefan perhaps used its German translation *Abhandlung über die allgemeinen Gleichungen für das Gleichlwicht der elastischen Körper als auch der Fluida; übersetzt durch F. Ott*, Neisse: JA Müller, 1847. Stefan also studied Navier's Mémoire sur les lois de mouvement des fluids, read before the Parisian Academy on 18 March 1822, and Gabriel Lamé's *Traité de Physik*. Stefan probably meant Lamé's *Cours de physique de l'École polytechnique*, Paris, Bachelier, 1836-1837, reprinted in 1840. Or perhaps, Stefan in his diary mentioned German translation of Lamé's work, published in Darmstadt by Carl Wilhelm Leske's (* 1784 Leipzig; † 1837 Darmstadt) son Carl Friedri h Julius Leske (* 1821 Darmstadt; † 1886 Darmstadt) in 1841. Karl Marx tried in vain to publish his Parisian *Kritik der Politik und Nationalokonomie* by Leske in Darmstadt in February 1845. Lamé's book contained the first reports on the solidification of the Earth with regards to its latent heat. In 1831, Lamé and Clapeyron published about the solidification of the Earth under the influence of latent heat after they read their paper in front of Parisian Academy on 10 May 1830 (I. Šubic, 1902, 79; Šarler, 2011, 21; Truesdell, 1980, 143-147; Lamé and Clapeyron, 1831, 250-256).

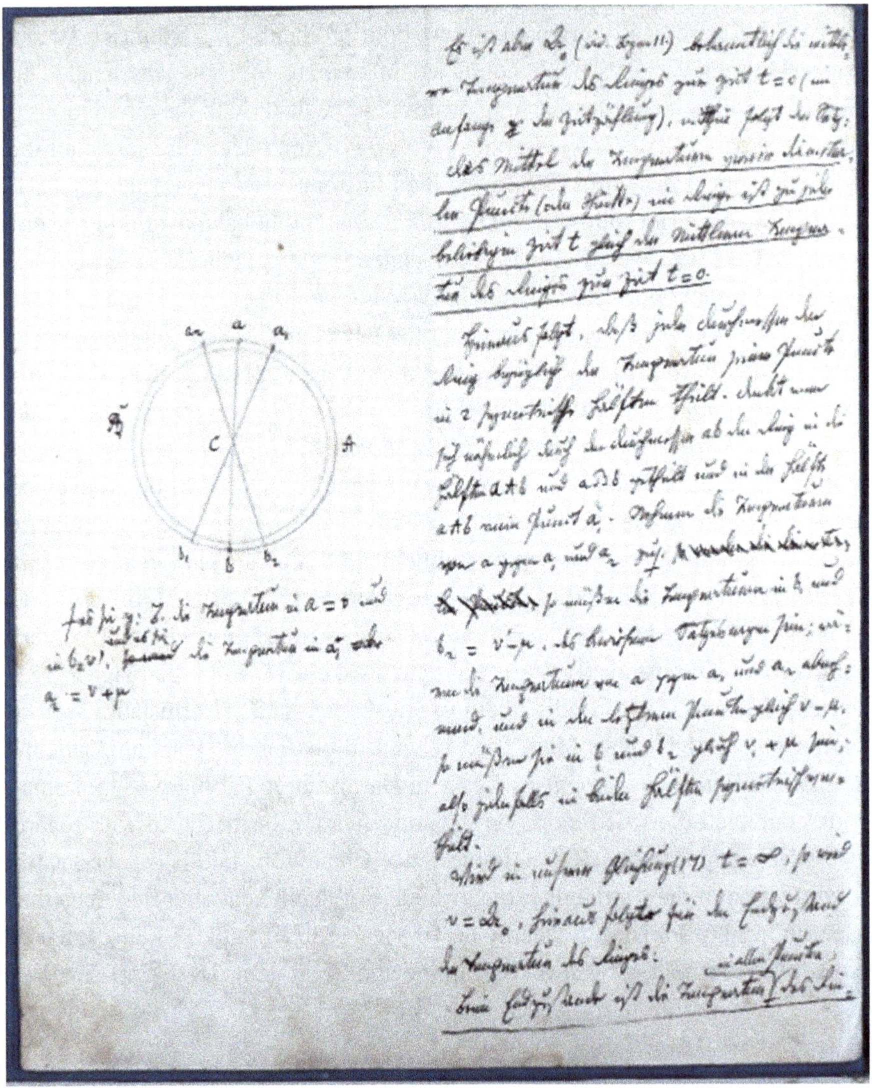

Figure 11: Stefan's Theorem: Average temperature of two points at any moment t equals the average temperature of all their area at the initial time t = 0. Koller illustrated it by a sketch of two concentric circles with points at the outer circle and the centre marked with letters. Koller draw his picture in the left column reserved for notes, corrections, comments, and illustrations.[14]

---

[14] Marian Koller Stefan's lecture *Über die Theorie der Wärme*, Koller-Manuskripte, 31, page 52, photographed by p. Amand Kraml.

Stefan's most notable successes interfered with the movable border between different states of aggregation, theory of the heat of the Sun, and the radiation of the black body. Koller wrote down a dozen of Stefan's altogether fourteen lectures in the winter semester of 1862/63, with the semester slightly shifted from the current ones in USA, as it was not ending in January but rather just before Easter.[15]

In February 1863, Koller described the last 14th Stefan's lecture on the theory of thermal conduction. Unfortunately, the last lectures did not contain any conclusions similar to the initial first lecture full of kinetic atomism of the modified Fourier's approach, at least according to Koller's notes. Those Koller's records perhaps predominantly illustrated Koller's somewhat outdated beliefs basically different from Stefan's modernized views, which also included critique of infinite velocity of propagation of Fourier's disturbances of temperature as radiated heat travels with the velocity of light, while the conducted heat has the velocity of sound (Truesdell, 1980, 73; Stefan, Über die Fortpflanzung der Wärme 4 April 1863, 331, 344). In any case, the winter semester was then completed for the sake of the Easter holidays, even if it seems that the missing lecturer's conclusion with a summary of the achievements discussed during his course of lectures would somehow complete Stefan's introduction of October 1862. After forty days of fasting, the Easter Sunday was on 5th April 1863, which means that at the end of March 1863 there were certainly no lectures.

Stefan's lecturing style: Thermodynamics

Koller began his writing of Stefan's lectures by comments on temperature and measurements of body temperature. On 21 October 1862, Stefan began his first lecture about heat with a supposedly agnostic statement that contradicted his and Boltzmann's beliefs.

These introductions were followed by many Stefan's equations, which presented his pedagogical efforts mainly as exercises in applied mathematics,

---

[15] Marian Koller Lecture notes Dr. Josef Stefan's lectures, 1862-1864.

the topics which Koller loved the most. Certainly, Koller also preferred to take mathematical notes. Stefan's lectures on heat theory were not much different from what first-year physics student hear today as a freshman, for example in Ljubljana. Of course, today, the professor does not initially raise any doubts about the true nature of heat, after the victory of Stefan's kinetic atomism. Stefan's integrations and differential equations might be mathematically slightly above the level of modern students, which may have been the fruit of the exaggeration of the recorder Koller, who wanted to impress his colleagues at the court with Stefan's abilities.

It is not an issue that Koller wanted to learn much of that Stefan's basic-elementary calorimetry for his own education, since he already learned enough earlier. Koller mostly endorsed all Fourier's theories as a student at Linz lyceum and than lectured on his own advanced colorimetry as the professor of physics in Kremsmünster for a decade and a half. As a professor of physics, he taught basic and advanced calorimetry.

It is much more likely that Koller wrote his booklets filled with Stefan's lectures to show them off and persuade his colleagues at the Court Ministry to assign Stefan to the most important chair at the University of Vienna, even if Koller's colleagues and collaborators at the administration did not understand many of these long mathematical formulas, or precisely because of it. Anton Peterlin used the same approach a century later to persuade Boris Kidrič and Kraigher to give more money to his projects in Ljubljana. Anton Peterlin tried to lure the rulers to approve more money for Peterlin's research. Oppenheimer probably used the same approach to humiliate and simultaneously attract the general Leslie Groves (1896 Albany, New York–1970 Washington DC) in early 1940s.

Stefan's mathematical equations illustrated his modernized atomistic Fourier's theoretical ideas. Instead of Fourier's series, he preferred to use seemingly more handy differential calculus to define the coefficient of heat transfer (thermal conduction). Because he did not use a tensor calculus, Stefan's derivations were extremely long.

Broader view of Koller's notes on Stefan's lectures: Koller on Stefan's Thermodynamics

Koller took notes of private-docent Stefan's lectures "Über die Theorie der Wärme (On theory of Heat)" during winter semester 1862/63 on 35 leaves which formed 140 pages. L. Boltzmann attended those lectures a year and a half later in summer semester 1863/64 (Höflechner, 1994, 15). That course was related to Stefan's recent publications: Notices on absorption of gases (Bemerkungen über die Absorption der Gase), *Sitzungsberichte der Kaiserlichen Akademie der Wissenschaften in Wien (Wien.Ber.)* 27: 375–430 (1858); On Dulong-Petit's Law (Über das Dulong-Petit'sches Gesetz), *Wien.Ber.* 36: 85–118 (1858); On phenomena of absorption in gases (Über der Erscheinungen der Gasabsorption), *Programm der Ober-Realschule, Vienna* (1859); On specific heat of water (Über die specifische Wärme des Wasserdampfes), *Ann. Phys.* 110: 593 (1860); Notices about the theory of gases (Bemerkungen zur Theorie der Gase, 3rd session on 22 January 1863), *Wien.Ber.* 47, II: 81–97; On the speed of sound in gases (Über die Fortpflanzungsgeschwindigkeit des Schales in gasförmigen Körpern, 14th April 1863), *Ann. Phys.* 118: 494-496 (1863) which was probably a part of Stefan's two days later publication: Über die Fortpflanzung der Wärme, 10th session on 16th April 1863, *Wien.Ber.* 47, II: 326–345 & *Ann.Phys.* 125 (volume 201 by overall counting (der ganzen Folge)): 257-265 (1865). Stefan discussed Fourier's, Clausius', Newton's, and Laplace's theory of the speed of spread of heat in bodies focused on the Newtonian velocity of sound on pp. 344-345. It was published just before Marian Koller's Bericht über ein nahezu vollendetes handschriftliches Werk der Verstorbenes Akademikers Dr. K. Kreil (died 21 December 1862) über die Klimatologie in Böhmen, 13th session on 15th May 1863, *Wien.Ber.* 47, II: 427–428. Koller's reminiscences on Kreil were reprinted in *Die feierliche Sitzung der kaiserlichen Akademie der Wissenschaften*, in Vienna on 30 May 1863 pp. 145-147. Koller's writing was a part of General-Secretary of Academy after 1851 Anton Schrötter von Kristelli's obituary of Kreil on pp. 118-152. There, Koller especially highlighted Kreil's research of the history of Bohemian meteorology focused on the Jesuits Joseph Stepling (1716-1778) and Antonin Strandt (Strnad, 1746-1799), the Premonstratensian Aloys Martin David (1757 Dřevohryzech u Teplé-1836 Teplé) and the Piarist Cassiano Hallashka (Hallaschka, Franz Ignaz Cassian, František Ignác Kassián Halaška, 1780 Budišov nad Budišovko

(Bautsch)-1847 Prague). Certainly, Koller and his young protégé Stefan read each others works even if their main research fields were not the same.

Koller began his notes of Stefan's lectures about the theory of heat with the idea of temperature of a body, including its measurement. Stefan's 1st lecture was on 21 October 1862, 11th lecture on 20 January 1863 (Koller-Manuskripte, 27: page 27a) and the last noted by Koller was 14th lecture which he wrote down on 10 February 1863.

From October to December 1862, Koller attended nine lectures on theory of heat, while he missed 8th and possibly also 10th lectures. He then attended four more lectures from January to February 1863, until the semester ended just before Easter.
At those times, the Habsburg and German semesters were different than in the United States or Slovenia, where the winter semester ends soon after Christmas. By modern timetable, Koller would attend even a part of summer lectures after Christmas and missed about six lectures of final part of summer semester.

But the modern German and the Habsburgian *Wintersemester* (WiSe), during which most students start university, often goes from 1 October until 31 March, with lectures starting around 15 October and lasting fourteen weeks, which was exactly the number of Stefan's fourteen weekly lectures which Koller attended while noting twelve among them and providing the blank pages for additional 10th lecture with no notes survived or available data presented. Koller probably missed at least Stefan's lecture in mid-December for the sake of other Koller's commitments.

There was usually a two-week break around Christmas and New Year which is not considered a p part of those fourteen weeks of a semester. The Habsburgian and modern German *Sommersemester* (SoSe, summer semesters) consequently usually went from 1 April until 30 September with lectures starting after Easter and lasting twelve weeks which made summer semester lectures for two weeks shorter than the winter semester. The Habsburgian and modern German summer semesters usually took place from Easter until 30th June. Their lectures began few days after Easter. There were the holydays enjoyed between semesters. Two lecture-free periods between the twelve to fourteen weeks long semesters were needed to take exams, practice doing internships, laboratory courses, and the like.

By Koller's narration, Stefan began his 1st lecture on 21 October 1862. Stefan started with the agnostic statement which seemingly contradicted his and Boltzmann's later convictions even if Stefan in following lectures countered that supposed agnosticism by using the term molecules in kinetic sense all the time. Initially, Koller narrated: "Theory does not make any conclusions about the nature of heat. The theory was set and completed by Mr. Fourier." "According to Joseph Fourier (1768 Auxerre-1830 Paris), theory makes no conclusions about the nature of the heat". That might even be Koller's own opinion not endorsed by Stefan, as Stefan continued to refer to molecules by modern kinetic ideas and spoke about them, also to his high-school students.

Joseph Fourier (1768 Auxerre in Burgundy 160 km southeast of Paris-1830 Paris), of course, really avoided much of the conclusions about the nature of heat in his approach. In their final form, he published his novelties in 1822 as the *Théorie analytique de la chaleur* during his placement as permanent secretary of the Paris Academy in that same year 1822; a position that Stefan soon took over at his own Viennese Academy.

In fact, Fourier was the only authority which Stefan quoted in his lectures about the theory of heat, at least according to Koller's notes.

The ideas of Stefan's older contemporaries William Thomson Kelvin (1824-1907), James Clerk Maxwell (1831-1879) and maybe even some novelties of Rudolf Clausius (1822-1888) may have seemed too new for undergraduate students' benches, which was always a most common approach of teachers of all eras.

Few decades later Ernest Rutherford also somewhat masked his ideas about the warmth of Sun because Koller's acquaintance, the almighty Kelvin, was (napping) in the front row of his audience.

Koller did not publicly discus his own otherwise firm beliefs in atoms. Koller never published his own preferences about kinetic atomism as he almost exclusively issued experimental researches, but his fellow Viennese scientist certainly knew and were aware about his atomistic opinions expressed and preserved in his Kremsmünster manuscripts: in 1825-1830 Lectures on Natural history read at philosophical studies in Kremsmünster, and most of all in 1826-1839 Lectures about physics read at philosophical studies in Kremsmünster with some kinetical atomism involved. There was also Koller's unpublished

Treatise "Determination of the path of a homogeneous light beam through several different surfaces causing refractions".

The last undated manuscript paralleled and complemented Stefan's lectures about optics, which Koller attended in summer semester 1863/64. That Koller manuscript was probably straightforwardly related and connected to Stefan's data and lectures, as the optical part of physics was most closely related to Koller's own professional research of astronomical observations.

Fourier's antagonist and opponent, Laplace, used to mention both caloric and kinetic atomistic variants of then popular ideas. But Laplace certainly preferred the theory of the caloric of his deceased boss Lavoisier.

The kinetic atomism was a favorite approach of Stefan and Stefan's teacher Andreas von Ettingshausen, although there were also some hostile renegades who categorically refused atoms among the students of Ettingshausen, including Ernst Mach.

Koller attended and noted Stefan's lectures about heat little more than five years after Clausius's fundamental treatise on the motion (of molecules) called heat, written in Zurich on 5 January 1857 and printed in Leipzig in the same year.

Few weeks later appeared influential system of Dynamides of Koller's juvenile friend and student, Ferdinand Redtenbacher, signed at Karlsruhe on 5 July 1857 and printed in Mannheim in the same year.

Maybe Stefan still preferred to present his kinetic atomistic beliefs somewhat carefully, but only for a few more years as he had already introduced them in his High School lectures. At least Koller pretended that some agnosticism was still involved.

More likely, that agnosticism might be just Koller's own old-fashioned interpretation of Stefan's data. Of course, Stefan knew well that Koller would be attending his lectures, so he might have diplomatically and politely preferred to avoid criticism of Koller's most influential fundamental core supporters, the Upper Austrian family of Redtenbacher.

Stefan continued, at least by Koller's notes: The temperature is determined by the measurements of stretching of a body. The bodies are brought into contact with one another for temperature comparisons, based on experience. The heat ousted from the body is measured by comparing of the temperatures.

To quantify temperature, we use a body whose volume can be easily measured, and these devices are called thermometers. They enable the measurements of temperature.

One relies on the fact, that both bodies gradually achieve the same temperature and then we no longer meet hotter outbreaks. In a state of stability, both bodies balance their temperatures, so that we no longer encounter any hot eruptions."

Then Stefan offered a subtitle of his first lecture: "Exchange of the heat. Fine structure of the bodies and testing of their ability to conduct heat measures their thermal conduction".

Regarding to the exchange of warmth, it has been found through derivation from experimental that the warmth given out is proportional to the temperature of the body - this is also used with the analogy in consideration (related to the hydrodynamics). In respect of heat exchange, the experiments proved that the output of heat is proportional to the body temperature - this also follows from the analogue of hydrodynamics equations, which was Stefan favorite approach in the following decades.

Then Stefan went even further, saying that this law of exchange also partly follows from the composition and constitution of the bodies, therefore from the relationship between molecules.

Stefan therefore described the emission output of heat of the bodies, without explicitly mentioning or declaring its three mechanisms, probably because Stefan's model Fourier was mainly interested in the conduction of heat in his days.

A decade later in 1872, as the first in Europe, Stefan measured the coefficients of heat transfer (thermal conduction) of gases with his newly invented diathermometer.

One of the most prominent proponents of Bošković's physics and the atheism of David Hume was John Leslie (1766 Largo in Fife in Scotland-1832

Edinburgh). Already in 1819 as the newly appointed Edinburgh professor of natural philosophy (physics), Leslie separated all those three ways of heat transfer: radiation, convection (stirring, mixing), and the thermal conduction. Unfortunately, he theorized without distinguishing between specific heat at constant volume and the specific heat at constant pressure.

With Leslie's cube and his differential thermometer as predecessor of Stefan's diathermometer, Leslie established basic laws of heat radiation, while reviving the debate about the physical composition of heat. He announced his novelties by his tractate *An Experimental Inquiry into the nature, and propagation of heat*, published in Edinburgh in 1804. There, he revitalized the research of the substantial nature of heat, which Fourier tried to ignore in attempt to be as exact as possible by Newtonian "Hypotheses non fingo (I feign no hypotheses)".

In his introduction to his textbook, Koller's Ljubljana professor and the cofounder of Viennese Polytechnic Johann Philipp Neumann (1774-1849) recommended the thermoscopic Leslie's cube as a great new success of the professor Leslie.

Stefan continued by telling Koller and other attenders on 21 October 1862: Each body has the property to take up space or give it up (by absorption capacity or by the emitting capacity). He claimed that the body could take up a larger space by expansion or give it up by contraction. This is achieved by absorbing or by transmitting heat according to the capacity of a body.

There the emitting (radiation) capacity is determined through the amount of heat, which is given off by the curse of time because of the temperature difference, while the temperature is measured in degrees Celsius. The determination of the temperature is performed by means of the air thermometer.

He noted the ability of body to radiate (or receive) the heat while emitting flow = 1 at the time = 1 and at temperature difference i = 1 measured in degrees Celsius. The temperature is measured by the air thermometer: certainly, more theoretically than in practice.

By means of hydrodynamic analogies Stefan therefore approached his main goal of his next decades: he was trying and attempting to determine the coefficients of heat transfer (thermal conduction) of gases. Soon, he had on his disposition his newly invented diathermometer.

Place a solid body of imaginary completely homogeneous substances between two unlimited levels of the surfaces marked as A and B. Lower plane A will stabilize by its own causes only at a new even temperature (due to the same or different reason) and therefore a > b applies between the temperatures those planes. It is clear by itself that all points of this arrangement have the same temperature in equilibrium. The temperature of the lower plane A over time became the same as that of the upper surface B when they are both in balance.

Koller noted Stefan's second lecture which Stefan read a week after his first lecture on 28 October 1862 under the title: Temperature (i.e. heat condition, warmth) of a body located between two unlimited flat surfaces when both those planes have inconsistent temperatures (Koller-Manuskripte, 27: folio 1b (= page 2)).

Let us put a solid body of imagined totally homogeneous substance between two unlimited flat surfaces. Designed as A and B, the lower plane A will stabilize itself due to its own causes only at a new enough uniform temperature v (due to the same or the different cause) and therefore (the temperatures of planes A and B follows the relation) a > b.

We now ask ourselves what temperature will be found in the body for whatever things are in planes A and B? It is by itself clear, that all points of this arrangement would have the same temperature (in equilibrium). If the distance between planes A and B is AB = e, we take Aα = a as the temperature of the lower, and Aβ = b as the temperature of the upper surface as the same.

Koller continued with a kind of comments about Stefan's second lecture entitled "Fourier's theorem" (Koller-Manuskripte, 27: folio 1e (= page 5)). Koller there filled four additional unpaged leaves before his folio no. 2 which meant that first folio had 8 pages instead of four, unlike all the others! Those additional somewhat inserted four pages dealt with Fourier's theorem.

By Koller's narration of Stefan's course, Fourier based his mathematical theory of heat on the following theorems:

1. Among two molecules that are extremely close to one another, in the unit of time the warmer molecule only sends quantity of heat q to the colder molecule. That transferred amount of warmth gradually decreases with the diminishing

distance t between these particles and becomes equal to 0 if this distance t between those two particles is reduced to an extreme smallness.

For this matter, a person needs to regard the amount of heat as a function, f (t), of the density of substance made of these molecules.

2. This amount of heat q is also proportional to the temperature difference v_I - $v_{II\ between}$ these molecules.

After that, Stefan expressed his (and modified modernized atomistic Fourier's) ideas in mathematical form of equations. He adapted Fourier's novelties to his own atomistic kinetic theory. Instead of Fourier series, Stefan preferred to use a seemingly more convenient differential calculus to define the heat transfer coefficient K' (Koller-Manuskripte, 27: folio 1h = page 9).

Stefan in Koller's narrations continued his notes about the coefficient of heat transfer, focussed on thermal conduction (Koller-Manuskripte, 27: 2a = page 9). Koller presented the temperature (i.e. conditions of warmth) in a very long and thin rod (Koller-Manuskripte, 27: 2b = page 10).

By Koller's narrations Stefan provided third lecture of his course on 3 November 1862 under the title: Temperature (i.e. Heat condition) of a thin rod of known length (Koller-Manuskripte, 27: 3a = page 13). He discussed the determination of the ratio of the heat transfer coefficient of different substances by means of such rods. His aim was a calculation of the ratio of coefficients of thermal conductivity of different substance inside the rod.

As noted by Koller, Stefan's fourth lecture on 11 November 1862 was entitled: Heat condition of a thin one-dimensional rod at a given time (Koller-Manuskripte, 27: Folio 5d = page 24).

Stefan's next subtitle was a Heat condition (state of warmth) of a in-dimensional body at a given time (not having extension in space, Koller-Manuskripte, 27: Koller 6c = page 27). Stefan's conditional equation was only true for some determined values of the variables used for the temperature v (Koller-Manuskripte, 27: 7b = page 30). Koller added his own annotated remarks about those equations (Koller-Manuskripte, 27: folio 7d = page 32).

Stefan's next fifth lecture on 18 November 1862 had the title: Integration of the differential equation of the temperature v for a very thin rod (Koller-Manuskripte, 27: 9a =37), neglecting the heat flowing away from the substance (medium) while providing application to an example of a circle (Koller-Manuskripte, 27: 10a = 41).

Stefan's next sixth lecture on 25 November 1862 was the Conclusion (Koller-Manuskripte, 27: 12b = 50). Stefan declared his theorem: The mean temperature of two points (or shells) in a range at any given time t equals the mean temperature at that range at time t = 0. Koller illustrated it by a sketch of two concentric circles with the annotated points of the outer circle and the center indicated by the letters, all of those in Koller's left column reserved for comments, reparations, titles and illustrations (Koller-Manuskripte, 27: Folio 12d = page 52). Then followed the integration of the equation and notes on heat conduction between two separated masses (Koller-Manuskripte, 27: 13b = 53, 13a = 54). Koller noted the equations found for the true nature of $d\alpha$ and $d\beta$ used for heat conduction between several (n) separate masses (Koller-Manuskripte, 27: 14c = 59, 14d = 60).

Stefan explained the conduction of heat between equal masses separated from each other by equal mutual distances on the periphery of a body. Some kinds of boundary conditions were therefore taken into consideration, first presented by the self-taught miller George Green's electric potential at *An Essay on the Application of Mathematical Analysis to the Theories of Electricity and Magnetism* in Nottingham in 1828, and more rigorously derived by Peter Gustav Lejeune Dirichlet, at least for the ball, in Dirichlet's 1850 paper submitted to the Prussian academy (Koller-Manuskripte, 27: 16a = 65).

Stefan's next seventh lecture on 2 December 1862 exceptionally had no underlined title (Koller-Manuskripte, 27: 16c = 67). Stefan explained the Integration of equations (Koller-Manuskripte, 27: 16d = 68), and discussed the problem of a Transition to a generalized approach of an infinitive (∞) large number of material points (Koller-Manuskripte, 27: 21a = 85).

Koller missed Stefan's 8th lecture on 9 December 1862. Stefan's next ninth lecture on 16 December 1862 again had no underlined title (Koller-Manuskripte, 27: 22c = 91). Koller's folios 106-108 were left blank, probably for Stefan's 10th lecture which was never included. Koller probably attended Stefan's 10th lecture (on 23 December 1862 or more probably on 13 January

1863) and took notes but never transcribed them into his ordinary now preserved notebook. Between the Christmas on 25 December and Holly Three Kings' days (Epiphany) on 6 January there were no lectures because of official Church holidays.

Stefan's next eleventh lecture on 20 January 1863 again had no underlined title (Koller-Manuskripte, 27: 27a = 109). Stefan began with a list of symbols including the temperature coefficient denoted as K', and its specific heat marked as C'.

Stefan's next twelfth lecture on 27 January 1863 again had no underlined title (Koller-Manuskripte, 27: 29a = 117). After short introductions of just few words, Stefan began with definite integrations and limits. No experiments were noted by Koller as probably Stefan and his assistants presented them at their special exercises during the same week or in special experimental lab courses for higher grades which Koller did not attend as he and his colleagues at ministry were more interested in theoretical questions and in Stefan's lecturing abilities. Apparently, Stefan did not provide experimental demonstration lectures in Faraday's style, at least there are no traces about those in Koller's notebook. Stefan's lecture room used to be crowded while some students also stayed just near the lecturer, with Koller as a special guest certainly sitting in the first row.

Stefan's next 13th lecture on 3 February 1863 again had no underlined title (Koller-Manuskripte, 27: 31c = 127). In Koller's notes many equations were crossed out and repaired. Koller ceased to note the titles of Stefan's lectures after the first ones which were probably written with better care. If Koller presented his booklets of notes to his ministerial colleagues, he certainly expected that they will just examine the bulk of it and few of its first pages, because the higher mathematics used would prevent their further reading as they were mostly not trained mathematicians.

Stefan's next 14th lecture on 10 February 1863 again had no underlined title (Koller-Manuskripte, 27: 34d = 140). Koller ceased to attend Stefan's late February lectures or maybe noted them elsewhere, or, most probably, 14th lecture might end the course even if it provided no conclusions comparable to the kinetic atomistic variations of Fourier in Stefan's introductory 1st lecture, at least in Koller's narration. In any case, the winter semester was over then in mid-February or a little later because of the Easter Holidays. Compared to the

October 1862 introduction, there was no equivalent conclusion of Stefan's course as a summary of achievements of his lectures about the theory of heat. After the forty days of Lent fasting and Holly week the Easter Sunday (Resurrection Sunday) was on 5 April 1863, which means that there certainly were no lectures in late March 1863.

Table of contents of Koller's notes on fourteen Stefan's lectures about heat, which were not divided into special paragraphs as were Stefan's lectures on optics. Index of fourteen lectures on heat, unlike the subsequent Stefan's optics:

First Lecture, October 21, 1862, Temperature of Body and its Measurements, page 1.
Second Lecture, October 28, 1862, Thermal state (i.e. Temperature) of the body placed between two unlimited flat surface levels, when both planes initially had inconsistent temperatures (Koller-Manuskripte, 27: Folio 1b = page 2). Supplement to Lecture II (Koller-Manuskripte, 27: 1e = 5): Fourier's theorem, containing four inserted pages not considered in the counting, as an additional unnumbered folio explaining Fourier's theorem, pages 2-12.

3rd lecture, November 4, 1862 (Folio 3a = page 13): Thermal state of a thin rod of known length. On the same page Koller added a subtitle: Determination and calculation of the ratio of the conductivity coefficients of different bodies in the middle of the rod, as the heat transfer coefficient of different substances, pages 13-24.

Fourth lecture on 11 November 1862 (Folio 5d = page 24): Temperature describing the thermal state of thin one-dimensional rod at a given time, pages 24-37.

5th lecture, November 18, 1862 (Folio 9a = page 37): Integration of the differential equation in terms of temperature v for a very thin rod, ignoring the heat flowing from the medium, therefore neglecting the heat which is running out, pages 37-50.

Sixth lecture on 25 November 1862 (Folio 12b = page 50): Conclusions. On page 52 the specific theorem: The average (mean) temperature of two points in the range is equal to the average temperature of their area, illustrated by a sketch of two concentric circles with points of the outer circle and its centre, marked with the letters. Koller put it in the left column reserved for purposed notes, corrections, comments and illustrations, pages 50-67.

Seventh lecture on 2 December 1862 (Folio 16c = page 67), without underlined title; on page 85 a subtitle of that lecture: Generalization of the approach for an infinitely large number of material points, pages 67-91.
Notes of the eighth lecture are missing.

Ninth lecture on 16 December 1862 (less likely January 16, 1863), no underlined title at folio 22c = page 91, integrations needed to determine the temperature (heat conditions) of a particular point X according to conductivity λ, pages 91-105.
Koller's folios 26b-26d (pages 106-108) were left blank and empty, probably reserved for Stefan's 10th lecture, which was never included. Anyway, Koller is believed to have attended that tenth lecture on December 23, 1862, or more likely on January 13, 1863, pages 106-108.

Eleventh Lecture on 20 January 1863, without underlined title (Folio 27a = page 109). Stefan first noted a list of symbols which he used and then continued to integrate the cosine function of heat transfer (thermal conduction) according to the proposals of Oscar Xaver Schlömilch's *Compendium der höheren Analysis*, pages   109-117.

Twelfth lecture on 27 January 1863, without underlined title (Folio 29a = page 117), resolving the definite integrals and limits of Fourier series of sinuses determining the heat transfer (thermal conduction), pages 117-127.

Thirteenth lecture on 3 February 1863, without underlined title (Folio 31c = page 127): Definite integrals from the previous lecture as odd trigonometric-exponential functions of the heat transfer (thermal conduction) dealing with the volumes and surfaces of the body, pages 127-140.

Fourteenth final lecture on 10 February 1863, without an underlined title (Folio 34d = page 140), Integration of Laplace's exponential equation for heat transfer (thermal conduction), in the right-hand column marginal notes containing the later crossed-out generalization of D. Poisson equations on potential theory as a partial differential equations of 2nd order, pages 140-142.

Further Generalizations of Broader Themes of Stefan's theory of heat (thermal conduction):

I Lecture, October 21, 1862, Temperature of Body and its Measurements, page 1.

II Lecture, October 28, 1862, Temperature of a body placed between two unbounded flat surfaces when surfaces have inconsistent temperatures, by Fourier's theorem modified for Stefan's (hidden) molecular kinetic, pages 2-12.

III-IV lectures, November 4, 1862-November 11, 1862: Thermal state of a thin one-dimensional rod of known length with the conductivity coefficients in given time, pages 13-37.

V-VI Lectures, November 18, 1862-November 25, 1862: Integration of the differential equation of temperature for a very thin rod, mean temperature of two points in the range, pages 37-67.

VII-X lectures, December 2, 1862-January 13, 1863, Generalization of the approach for an infinitely large number of material points: the thermal state of a thin one-dimensional rod of infinite length with the conductivity coefficients in given time & Integrations determining heat at point X with respect to conductivity λ, pages 67-108.

XI-XIV Lectures, January 20, 1863-February 10, 1863, list of symbols, integrating the cosine function according to Oscar Xaver Schlömilch & Definite Integrals and Limits & Definite integrals from the previous lecture as odd trigonometric-exponential functions of heat transfer from body volume and surface & Integration of Laplace's exponential equation for heat transfer, in the right-hand column of the strikethrough generalization in D. Poisson's equation of potential theory as a partial differential equation of the 2nd order, pages 109-142.

Stefan lectured on Interference, Diffraction, and Polarization with Koller in his audience. Koller noted the first five of supposedly twelve Stefan's lectures taught at that summer semester course in 1863/64 with L. Boltzmann also among Stefan's students after his matriculation in October 1863 (Höflechner, 1994, 15; Koller's notes on 10 (15) leaves forming 40 (60) pages in Koller-Manuscript no. 27).

First page of Stefan's optical lectures was Koller's title page noted by the modern handwriting on the bottom as: "15 quarto pages by the hand other than Koller's enclosed (von fremden Hand 15 Quarterseiten beiliegend)", while in fact it encloses just a little more than 10 folios and not 15 of them (Koller-Manuscript no. 27: 1a=page 1).

On 12 April 1864 Stefan gave his 1st lecture with no underlined title. It began by descriptive title put in Koller's left column, just like Koller's notes on Stefan's course of theory of heat which was left without any modernized title page (Koller 27: Page 2). Stefan-Koller's 1st paragraph point stated: "According to the undulation theory, a luminous point turns into an oscillating movement. This propagates itself in all directions. The wave surface, apart from the constant movement, is based on the path through which the movement is seen during an oscillation time (T). We can measure the wavelength (λ), and the speed of this propagation (v). The wave surface has a different shape depending on the speed of the light in different directions. If this speed is the same in all directions, then it is a ball, etc., if the distance between the shining points (of lighting) is very great."

As in the case of Fourier's theory of heat, Stefan (and Koller with him) certainly accepted Augustin-Jean Fresnel's (1788-1827) undulatory theory of light (and heat) instead of William Robert Grove's and Stefan's class teacher Karl Robida's (1857) intermediate vibrational hypotheses of electricity criticized by Stefan's colleague Ettingshausen's son-in-law Wilhelm Joseph Grailich. Fresnel's 137-page essay *De la Lumière* (*On Light*), was finished in June 1821 and published by February 1822, almost simultaneously with Fourier's final theory of heat entitled *Théorie analytique de la chaleur*. Both fundamental books mostly originated in Paris where their authors certainly endorsed many resemblances of then French Bourbon restoration's ideas, following the humiliating Napoleonic downfall. Like today, French nation was deeply divided in those times two centuries agi with Stendhal praising Napoleon as at consular post in Trieste in 1830 which Metternich refused. The nobility coursed Napoleon and most of Robespierre's fans were hiding.

In his 2nd paragraph-point, Stefan stated: Each of the oscillating movement propagates in the medium, while a point moves further by the time of oscillation (period) designed as T when illuminating points provide their transfer at a certain time t. The motion becomes a function of T and t, i.e., its ability for transfer:

$$\xi = f\,(T, t)$$

Koller added his own remark as a formal annotation (27: folio 1c = page 3). He wrote that the form of the function's limit is determined in the following way: If one thinks in the frame of aether-particles, his projection of the equations is only focused on the distance x which namely assume that the (force) opposes acceleration. The force leads a particle back to its old position, as it strengthens. Therefore, it is proportional to the distance x, hence equal to Ax. In the limit of great values, we then have the formula for the second derivation:

$$d^2x\,/\,dt^2 = Ax\;\ldots$$

The equation (I) shows that v = o for x = a, which means that a is the largest distance between our aetheric particles. The same equation also gives the formula for dt (Koller 27: folio 1d = page 4). You could add the time measured when the aether-particle is in the equilibrium position, so that you have the series resembling Fourier's series (Koller 27: folio 1e = page 5). If the distance of a particle from the origin of its movement is equal to Δ, we find further that the oscillation movement makes the path λ during the time T. Then we define the time that the media needs to come to a few nearby points after $\Delta = T$. Δ / λ at time t will be determined for the aether-particles by the smaller distance Δ between the same positions. Both of aether-particles are at the origin of the movement at time $t = T$. Δ / λ wannish by their motion. Hence the position of the particle at the distance Δ at time t is namely $X = a\,\sin\,(2\,\pi\,(\ldots$ where we have the t defined by the last equation, in addition to the light which has its own oscillation. If v is the period (oscillation-time, freqiancy) of propagation of light in our medium, we have $\lambda = v\,T\;\ldots$

There Δ is the distance of the aether-particle, whose relocation X at time t is propelled by the basic movement.

Koller next noted Stefan's ideas about the interference in his 3rd paragraph: If point at time t follows the prescription 1) $\xi = a \cdot \sin\ldots$, a second main point has therefore the prescription 2) $\xi = a \cdot \sin\ldots$ (Koller 27: folio 2a = page 6).

Koller continued: ...It is well known that the time-related quantities are measured by the square of their amplitude; so, this was widely approved, and such assumed measure is also the most natural (Koller 27: folio 2c = Page 8). Therefore, the professor Stephan (sic!, Koller's misspelled Stefan's name which was not rare in his times) just gave proofs following the data from theory of George Biddell Airy (1801 Alnwick in Northumberland-1892) of Greenwich. Koller did not state which work of his later befriended Airy he had in mind, but most probably it was Airy's newest English Edition or his German translations, mostly provided by Karl Ludwig Littrow (George Biddell Airy, *Mathematical tracts on the lunar and planetary theories, the figure of the earth, precession and nutation, the calculus of variations, and the undulatory theory of optics: designed for the use of students in the university*, Cambridge: Macmillan and Co., 1858. Maybe also German translations of Karl Ludwig Littrow: George Biddell Airy, Ueber die Diffraction eines Objectivs mit kreisrunder Apertur, *Annalen der Physik und Chemie*, 121/9 (1838): 86-95; George Biddell Airy, *Abriss einer Geschichte der Astronomie im Aufange des neunzehnten Jahrhunderts (1800-1832). Aus dem Englischen übersetzt von Karl Ludwig* Littrow. Wien: Gerold, 1835; G.B. Airy, *G.B. Airy's directors der königlichen sternwarte zu Greenwich, Populäre physiche astronomie übersetzt von Karl Ludwig Littrow*, Stuttgart: Hoffmann, 1839).

On 19 April 1864 Stefan gave his 2nd lecture on Fresnel's mirrors. Fresnel bi-mirror (double mirror) consist of two plane mirrors forming a dihedral angle with a few angular minutes less than 180° used for the observation of interference of coherent beam of light (Koller 27: folio 2c = Page 8). Fresnel proposed it in 1816. Light (preferably monochromatic) is reflected by two mirrors with the extremely small but adjustable angle between perpendiculars to those mirrors' planes. The superposition of the light waves produces an interference, which is observed with a magnifying glass.

In his 4th paragraph Stefan discussed Fresnel mirror experiment: It is well known that Fresnel rays, which were reflected by two mirrors placed against each other under extremely small angles, caused bending and the observed interference. As experience proves, the rays must come from the same source of light to produce the interference (knitting), of which more later (Koller 27: folio 2d = page 9).

Next followed mathematical derivations. The light starkens itself in those points M defined as follows: $\xi = \ldots$ (Koller 27: folio 3c = pages 10, 12). In white light you see the white middle strip, then the second one black strip, then the colored strips (Koller 27: pages 10, 12 (folio 3c), 14 (folio 4a). The intensity is very much lower that it can be, because of the original impacts of the vibrations of the ray, or rather its polarization.

In his next 5th paragraph Stefan discussed the interference prism as related to Newton Rings, in which two glass surfaces, one flat and one convex, are pressed into close contact, forming a thin film with circular symmetry between glass and air. Stefan discussed the central points of circular waves (Koller 27: pages 14 (folio 4a), 15 (folio 4b)).

On 26 April 1864 Stefan delivered his 3rd lecture (Koller 27: 4d = Page 17). In his 6th paragraph with no formal title noted, he discussed the rays in prism as spheres where the velocity of light in air was noted by v, and in glass by v' (Koller 27: bottom of page 17). The waves of coloured (light) did not fell of Fresnel mirrors (Koller 27: Folio 5b = Page 19 bottom). The wavelength was not changed by Fresnel mirror. Stefan also used the middle of second mirror which formed a magnifying glass (Koller 27: 5c = Page 20).

Stefan's 7th paragraph was about the Colors of Thin films (Koller 27: folio 5d = Page 21). When white light, which consists of a range of different wavelengths, falls on the film, certain wavelengths (colors) are intensified while others are attenuated. The interference on thin-film explains the multiple colors seen in light reflected from soap bubbles and the oil films on water.

On 3 May 1864 Stefan presented his 4th lecture (Koller 27: 7b = Page 27). There, his 8th paragraph had no special title noted. Koller mostly wrote down the computations with sinuses (Koller 27: 8a = Page 30). The division of the wave provides the relaxation on the border between air and glass (Koller 27: 8b = Page 31). After crossing through several thin films, the wave is shifted by the half of its original wavelength (In the destructive interference two waves with same amplitude are shifted by exactly half of their wavelength when they merge together, then the crest of one wave will match up perfectly with the trough of the other wave, and they will cancel each other out). That discovery is analogous to the rule which governs two elastic spheres of different masses, when the smaller mass is approaching and contacting the larger mass in the state of motion. That was certainly an analogy between the optics of light and mechanics, but it did not yet involve any hydrodynamics, which soon became Stefan's favourite analogy.

Koller denoted his and Stefan's angle of incidence by α and discussed the intensity of the reflected and refracted light rays. That intensity designs the whole surplus of that division through reflection and refraction. One can say: the phenomena of the reflected light and of the transmitted (let through) light are complementary. The wavelength or an effective area of the length of waves is only determined as the same distance, which is also true for the color (Koller 27: folio 9b = Page 35 top). The reflected light might disappear if it has a small density, so that color can be lost in the process. If extremely clear prism is imposed, then the stroking is observed by pressure switched on the mica, where the light falls and the reflected light decomposes through the prism (Koller 27: folio 9a = Page 34 on the top).

In his 10th paragraph, Stefan discussed Newton's colour rings as a series of concentric light- and dark-coloured bands observed between two pieces of glass when one convex is put by its convex side on another piece having a flat surface. Thus, a layer of air exists between them. Stefan also described the lens with a spherical curvature (Koller 27: Folio 9c = Page 36 almost on top).

On 10 May 1864, Stefan held his 5th lecture, beginning by 11th paragraph about Appearances in diffuse scattered light as the scattered images (Koller 27: 10b = page 39). The dispersed appearances in diffuse light center with the reflection like the light reflected from a surface. It was already observed by Newton. He took a concave mirror and sprinkled it with dust. With new mirrors made exclusively from glass, the light emits colored rings. If the image and the opposite strand coincide (which, as is well known, does not affect the object in the middle of a mirror), then colors remain the same. In the case of a convex or covered mirror, this phenomenon does not occur. To explain these phenomena, consider two dust particles at incoming rays, as the surface of dust particles is reflecting. But it had to fall from dust particles, because light was diffused and refracted and gets into the eye, which on the real surface naturally occurs (Koller 27: 10c = page 40).

The last written page (Koller 27: 10c = page 40) contains just few words on the top, while the next (last) page was left blank. Therefore, Koller's introductory title-page remark about 15 folios included is false as we have just 10 or maybe 11 folios preserved. At his 39 pages and one additional line, Koller recorded five lectures divided in 11 paragraphs:

1st lecture 12 April 1864 (0.5 page, top of the folio 1b=Page 2), 1st paragraph: No underlined title, it begins: According to the undulation theory, a luminous point turns into an oscillating movement. This propagates itself in all directions. Followed by discussion at one untitled paragraph at 3.5 pages, total 4 pages as Introduction to aether theory of undulation optics.

2nd Paragraph, 1st lecture (3.5 pages, middle of folio 1b=Page 2), no underlined title, it begins by: Each of the oscillating movement propagates in the medium, then becomes a point inside the further oscillation.

3rd Paragraph, 1st lecture (2.5 pages, folio 2a = page 6): Interference.

19 April 1864 2nd lecture (5.5 pages, middle of folio 2c = Page 8), 4th paragraph: Fresnel mirror experiment: determined by well known Fresnel rays and by the theory of Airy.

5th paragraph 2nd lecture (3 pages, folio 4a = Page 14): Interference Prism, followed by calculations at one untitled paragraph consisting of 4 pages, total 7 pages.

6th paragraph 3rd lecture 26 April 1864 (4 pages, folio 4d = Page 17), no formal title noted, begins with: Like previous things, by letting beam into the prism.

7th paragraph 3rd lecture (6 pages, folio 5d = Page 21): Colors of Thin films, followed by calculations at two untitled paragraphs at 8.5 pages, total 14.5 pages.

8th paragraph, 3 May 1864 4th lecture (4 pages, folio 7b = Page 27), no special title noted, begins with: About adding up the sums.

4th lecture, 9th paragraph (4.5 pages, the middle of folio 8b = Page 31): No separate title, begins with: Determining the intensity of motion.

4th lecture, 10th paragraph (3 pages, Folio 9c = Page 36 almost on top): Newton's colour rings.

5th lecture, 11th paragraph (One page, folio 10b = page 39, up to additional single line at next page, 10 May 1864): Scattered images in diffuse light.

Koller in fact attended just first five Stefan's lectures focused on interference, and not the following probably seven lectures focused on diffraction and polarization of light even if Stefan mentioned some polarization and used diffraction inside his teachings about the interference. Index with untitled paragraphs full of aether theory or calculations belonging to the titled paragraphs may be divided into seven topics:
(Title Page)
Introduction with aether, pages 2-6,
Interference 6-8,
Fresnel's mirrors by Airy 8-14,
Interference Prism (related to Newton's Rings) 14-21,
Colors of Thin Films 21-36,
Newton's Color Rings 36-39,
Scattered images in diffuse light 39-40.

That division in seven parts is the following:

1) 1st -2nd lecture 12 April 1864 (4 pages 2-6), Introduction to aether theory of undulation Optics,

2) 3rd Paragraph, 1st lecture (2.5 pages, pages 6-8): Interference,

3) 19 April 1864 2nd lecture (5.5 pages, Pages 8-14), 4th paragraph: Fresnel mirror experiment explained by theory of Airy,

4) 5th -6th paragraphs 2nd lecture, followed by calculations of 6th paragraph 3rd lecture 26 April 1864 (7 pages, 14-21): Interference Prism,

5) 7th - 9th paragraphs, 3rd lecture & 3 May 1864 4th lecture with calculations at two untitled paragraphs (Pages 21-36): Colors of Thin Films,

6) 4th lecture, 10th paragraph (3 pages 36-39): Newton's color rings,

7) 5th lecture, 11th paragraph (One page, 39-40, 10 May 1864): Scattered images in diffuse light.

Soon after Easter Sunday, which was celebrated on 27 March 1864, Koller attended just five Stefan's lectures about optics from 12 April 1864 to 10 May 1864. He visited much less than fourteen lectures of Stefan's course about heat which he heard a year and half earlier from 21 October 1862 until 10 February 1863, just before Stefan became the youngest professor of physics-mathematics at Viennese university and the co-director of the Physics Institute on 9 March 1863. In that time, Koller published his paper Über das Passage-Instrument in *Yearly of Natural History Association in Brünn (Brno)* in 1861 as Koller was an honorary member of that society together with Encke, Ludwig Redtenbacher, Kreil, and Rokitansky, while the former Rijeka professor Franjo Mathon and Ferdinand Lippich were ordinary members (Verhandlungen des naturforschenden Vereines in Brünn, Brno 1861 (issued in 1862) 1: XIV). Koller began to attend Stefan's optical lectures in Stefan's classroom with Boltzmann among the students two weeks after Easter Sunday of 1864 but ceased to appear anymore just before the middle of altogether probably twelve Stefan's summer semester lectures. He might have had myriads of reasons for his abrupt finishing of attendance of Stefan's course, including Koller's own knowledge of practical astronomical optics far surpassing Stefan's trainings and maybe even Petzval's knowledges. Stefan might have begun his lectures about optics in winter semester 1863/64 with theoretical part of optics which Koller and Boltzmann did not attend. Kunzek taught the science of light and heat in winter semester 1863/64 with Boltzmann but not Koller in his class. Stefan continued Kunzek's or maybe even his own general course in summer semester of 1863/64 while lecturing about special cases of optics with Boltzmann among his students, and during almost whole first half of course also with Koller in the audience. Therefore, Stefan's lectures about the theory of heat in winter semester 1862/63 were also probably continued with some special cases including entropy, but Koller did not preserve any manuscript of those even if he attended them, which is doubtful. The second Stefan's course of theory of heat might have been more sophisticated with some entropy involved.

Stefan probably lectured for two hours in a row as he taught two weekly hours about all his topics except his course of practical experimenting. He usually taught two different topics per semester, mostly about heat and optics, as his courses delivered in the same semester. In summer semester 1863/64 and in the winter semester 1865/66, Boltzmann attended Stefan's lectures about the theory of heat, and also Kunzek's lectures on more experimental calorimetry in summer semester 1863/64 with Koller not in the audience anymore.

Petzval also taught Science of Light and Heat in winter semester 1863/64 with Boltzmann but not Koller in his class, while Loschmidt taught Molecular Physics in winter semester 1866/67 with Boltzmann but not Koller in his class. Koller attended five Petzval's lecture just in 1850/51-1855/56 to promote Petzval's talent just like Koller highlighted Stefan's abilities a dozen years later. Petzval's optical manuscripts were bobbed in 1859 and he gradually ceased teaching that topics. Like Stefan later, in 1869 at the age of 62 Petzval married his housekeeper, but she died four years later.

Stefan lectured on Introduction to physical experiments ten hours weekly in summer semester 1866. He probably delivered that course in his labs where he recently became a boss with Boltzmann but not Koller attending as it was a kind of introduction to Boltzmann's final exam. A year earlier Boltzmann attended ten hours per week of the similar Ettingshausen's course entitled "Introduction to physical experiments" in summer semester 1865 (Höflechner, 1994, 14-15). Certainly, the introductory courses about heat and optics were full of Stefan's mathematics, but Boltzmann attended Loschmidt's course about the molecules only in Boltzmann's last 4th class in winter semester 1866/67. In 1837 at Prague's Charles University, Loschmidt became a personal reader of the philosophy professor Franz Serafin Exner. In his 1861 booklet, *Chemische Studien ("chemical studies")*, Loschmidt proposed two-dimensional representations for over 300 different molecules in a style resembling modern chemists. Among these were aromatic molecules such as benzene (C_6H_6), and the related triazines. Loschmidt symbolized the benzene molecule by a large circle suggesting its cyclical structure, four years before the similar Kekulé's French and German papers were published in 1865 and 1866. Kekulé never admitted any borrowing from Loschmidt as he preferred his story about the dreaming of a snake catching its own tail. In 1865, Loschmidt was the first to estimate the size of the molecules that make up the air: his result was only twice the true today accepted size, just before Boltzmann attended his classes to become his friend. Certainly, the molecules and atoms were far smaller than the ancient Greeks ever imagined! Loschmidt became professor of physical chemistry at the University of Vienna in 1868 as a leader of second physical institute where he shared private rooms with Stefan; they also shared their love to their naughty female housekeeping cleaners as each of them soon married one of them, certainly not the same one.

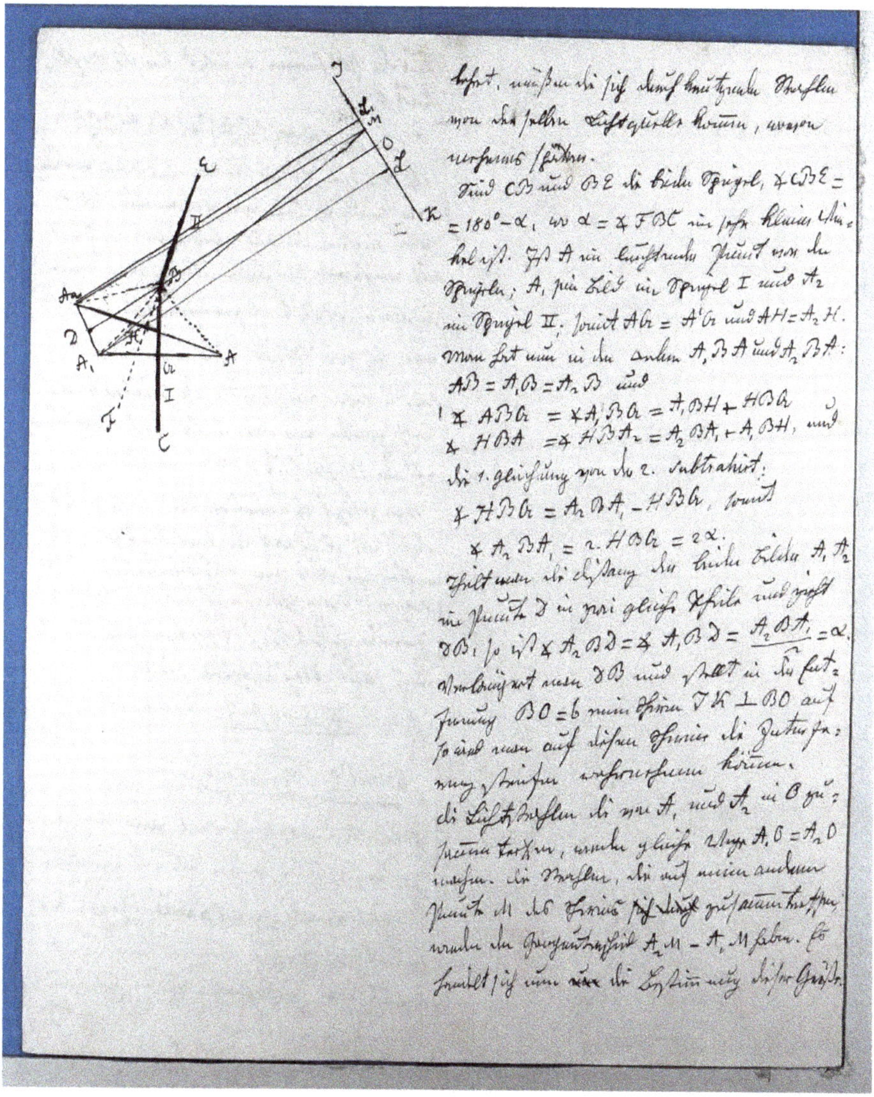

Figure 12: Koller's notes on Stefan's lectures on the optics of refracted light (Koller-Manuskripte: Stefan, Über Interferenz, Beugung, Polarisation, summer semester 1863/64, folio 2d, Directions-Archiv der Sternwarte Kremsmünster, Photo: Amand Kraml).

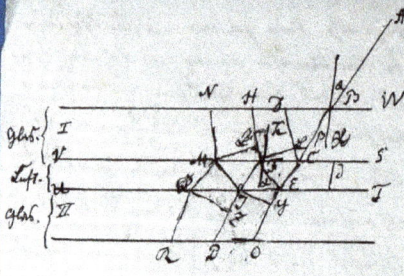

Figure 13: Koller's notes on Stefan's lectures on the optics of colored thin film (Koller-Manuskripte: Stefan, Über Interferenz, Beugung, Polarisation, summer semester 1863/64, folio 5d Directions-Archiv der Sternwarte Kremsmünster, Photo: Amand Kraml).

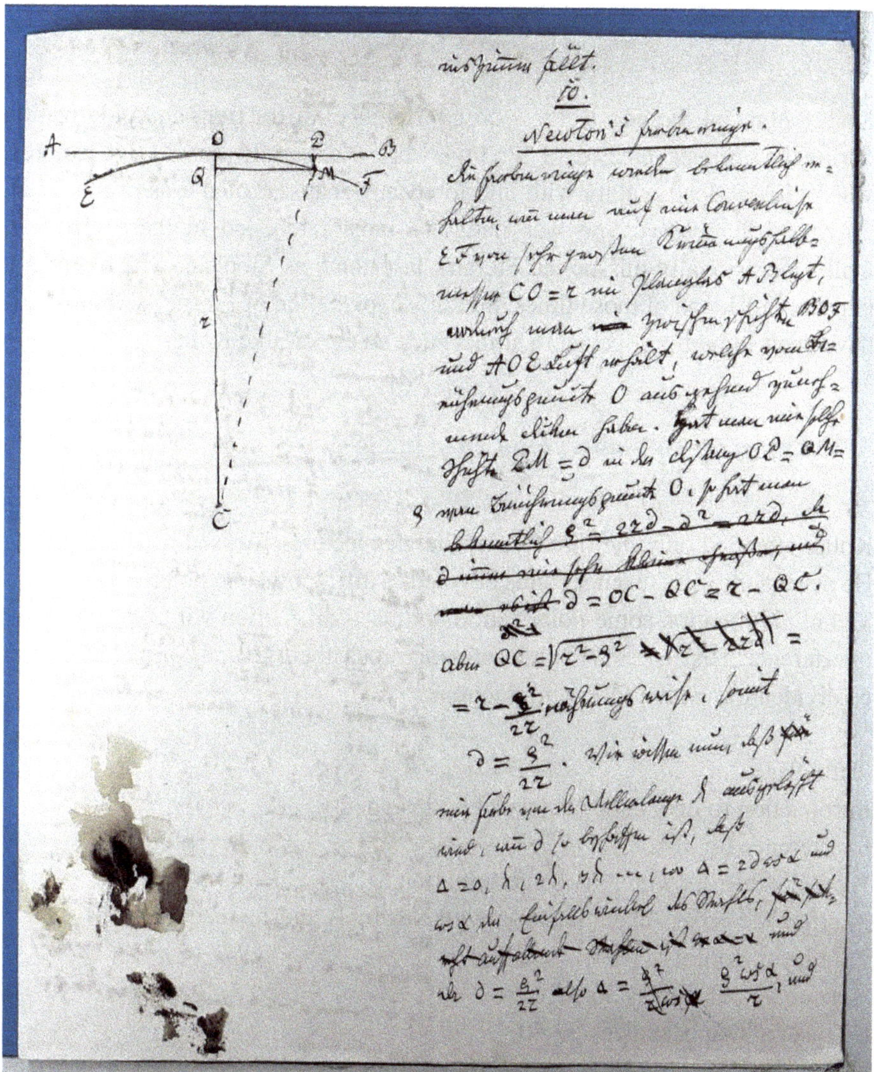

Figure 14: Koller's notes on Stefan's lectures on the optics of colored rings as a series of concentric light and dark colored bands (Koller-Manuskripte: Stefan, Über Interferenz, Beugung, Polarisation, summer semester 1863/64, folio 8c (p. 36), Directions-Archiv der Sternwarte Kremsmünster, Photo: Amand Kraml).

205

Koller attended Stefan's lectures on interference, diffraction, and polarization during the summer semester of 1863/64, while noting only the first five lectures, where Stefan dealt mainly with interference. Because Koller was primarily an astronomical observer, he was naturally most interested in these chapters. Koller occasionally misspelled Stefan's last name as Stephan, which was not unusual in Vienna of those times. Stefan's explanation of Fresnel's optics must have been something special to attract such an expert like Koller.

Structure of Stefan's interpretation of interference

Koller attended only the first five of Stefan's lectures focused on interference. He missed the next seven lectures devoted to diffraction and polarizing of light. Stefan did mention some polarization and used diffraction within his study of interference. His course with titles of paragraphs, theories and computations can be divided into seven following sections:

(Title Page)
Introduction to undulatory theory of optics pp. 2-6,
Interference pp. 6-8,
Fresnel's mirrors with Airy's interpretation 8-14,
Interference prism (in conjunction with Newton's interference rings) 14-21,
Thin layer's colors 21-36,
Newtonian color rings 36-39,
Diffuse light appearances 39-40.

1) 1st and 2nd subsection, inside 1st lecture on 12 April 1864 (4 pages 2 to 6), Introduction to the theory of wave optics.

2) 3rd subsection of 1st lecture (2.5 pages, pages 6-8): Interference.

3) Lecture 2 on April 19, 1864 (5.5 pages, pages 8–14), 4th Subsection: Fresnel's Mirror Experiment.[16]

---

[16] George Biddell Airy, *Mathematical tracts*, Cambridge: Macmillan and Co.

4) 5th and 6th Subsections: 2nd lecture of 5th subsection, followed by calculations at 6th subsection within 3rd lecture on 26 April 1864 (7 pages, 14–21): Interference Prism.

5) 7th - 9th Subsections: 3rd Lecture on 3 May 1864 and 4th lecture with calculations in the other sub-section (pages 21-36): Thin layers' colors.

6) 10th subsection within 4th lecture (3 pages 36-39): Newton color rings.

7) 11th subsection within 5th lecture on 10 May 1864 (one page, 39-40): Scattered images in diffuse light.

Stefan's Lectures on Fresnel's wave theory of Light

On 20 October 1863, with his 1st lecture, Stefan began his course of Fresnel's theory of light, or at last Koller titled it in that way, because for Stefan there was no other valuable theory but those Fresnel's undulations, while Koller as a student was still educated in frames of Newtonian emissions particle theory of light and had to accept Fresnel's reformulations as a grow up astronomer. Koller entitled his manuscript by Stefan's name in Slovenian alphabet as the "professor Dr J. Štefan" which is another link proving close relationships among both Slovenian erudite. Despite some efforts, even the main Slovenian scientific institute still uses the name Stefan and not Štefan.

Differently from his other courses, Stefan provided no historical introductory notes at the very beginnings of that course. He initially described the diffraction of light in an experiment involving rod and the area of medium. By the hypothesis of emissions' theory, the diffraction of vibrations of light might contradict the observations in nature.

In next part # 2 (Koller-Manuskripte, 26: page 2) Stefan stated that the experiments with refraction of light on the border between air and water does not support that emission theory. Jean Bernard Léon Foucault's (* 1819 Paris; † 1868 Paris) measurements proved otherwise: light travels more slowly through water than through air contrary to Newtonians. The breakdown of the second paradigm of optics between 1830 and 1853 has triggered the attempts of former friends and classmates Armand-Hippolyte-Louis Fizeau (1819 Paris-

1896 Venteuil in north-eastern France) and Léon Foucault (1819 Paris–1868 Paris). They wished to measure the velocity of light in water compared to the speed of light in air between 1849 and 1853, a decade before Stefan's lecture, while Stefan was still a high school student. Foucault and Fizeau were only the teenagers during Koller's visit of Paris in 1838, therefore Koller did not mention them in his memoires. Following suggestions by Koller's acquaintance François Arago, Léon Foucault and Fizeau collaborated in a series of investigations on the interference of light and heat and their speeds, although Foucault admired Napoleon III much more than the liberal Arago.

In his part # 2, Stefan detailly described Foucault's interference experiments by the wire of platinum with its square opening. He used Heliostat with a plane mirror which turns to keep reflecting (sun)light primarily focused to the predetermined target (Koller-Manuskripte, 26: page 2). Stefan also described and likely also demonstrated Foucault's experiments with prism, rotating mirror, and lenses (Koller-Manuskripte, 26: page 3). That was his pedagogic style: he loved to pick up one among famous experimentalists or theoretician, usually French or German, and praised him way too much to persuade his student audiences that there could be no doubt in physicists' truth which Stefan decided to support. It is interesting to note that Foucault was also the favorite of Stefan's high school class teacher Karel Robida, who published about Foucault's pendulum demonstrated in Paris in February 1851. Robida had his paper printed only four year later, on 5 April 1855, just few months after Stefan gradated in his class.

The differences of speed and amplitude of vibrations of light in air and water were decisive. As the supporters of undulatory theory, Stefan and Koller believed that light has greater velocity in air, which is less dense, than in water (Koller-Manuskripte, 26: page 6). Their ratio 4:3 is still used today.

The undulatory hypothesis is therefore professed by the differences of velocities of light in unequally dense media, as Stefan stated in the beginning of his part # 3 with no declared underlined title (Koller-Manuskripte, 26: page 7). The nature chose undulatory hypothesis as its tool. Fresnel used for it the (Pierre de Fermat's) principle of the least time of 1662 as the predecessor of Leibniz-Maupertuis-Euler's principle of least action of 1707-1744 (Koller-Manuskripte, 26: page 11). The principle of the least time used to be the connection step between ray theory of light in geometrical optics and the undulatory theory.

Stefan gave his second lecture on 27 October 1863 beginning with a part # 4 with no declared title (Koller-Manuskripte, 26: page 11).

The undulatory hypothesis needs the existence of a medium of aether which is filling the space. The vibrations of aether give the sensation of light as Stefan stated in the beginning of his part # 5 with no formal title (Koller-Manuskripte, 26: page 16). The concept of aether was in accordance with European Christian metaphysical thinking of those times which supposed a lot of invisible or at least hidden entities. But the Spring of Nations changed the opinions of the younger generations. Therefore, the young Stefan's students were somewhat confused with the untouchability of aether, even more than in the cases of invisible atoms-molecules, which forced Stefan to post two additional explanations for necessity of the existence of aether in the outer space:

1) For spreading of light through cosmic space you need a medium.

2) The speed of light is so great that no kind of ordinary matter could be its media.

Both ideas were Fresnel's ad hoc suppositions never endorsed without suspicions because of apparent absence of frictions along the paths of flying celestial bodies. Certainly, even Descartes and Newtonians needed aether.

Because the aether does not noticeably disturb the movement of bodies in space, it must be of considerable delicacy (Koller-Manuskripte, 26: page 17). It is known that Johann Franz Encke (1791 Hamburg-1865 Berlin) sought an influence of the aether on the movement of his comet; Encke predicted its return for 1822 and Koller observed its comeback in late 1838 as its period was 3.3 years. All the aether consists of the same matter, and the vibration theory does assume different types of aether, just as the emanation theory accepts different required kinds of light. We also think of the aether as consisting of the smallest particles like molecules, atoms. The atoms of the aether act on each other with forces. The nature of those forces is not known yet. All interactions between aether particles and the other material take place by collisions because the aether penetrates the body and changes density of the body. The aether is also set as an imponderable, which is only to be taken with a reserve. Namely, the interaction between ponderable substances and the aether, which takes place as caused by the aether, has also been stimulated by aether's natural elementary points, which must therefore be difficult (to imagine as an imponderable, Koller-Manuskripte, 26: page 18). So, if the particle could make an oscillating movement in aether, it has originally an unstable equilibrium, and the forces with which the aether particles work must enable the restoration of equilibrium. Therefore, an aetheric particle bombards (the matter), so the matter strives back to its rest position. The size of this striving towards the rest position is dependent on the size of the displacement and its direction. The displacement of an aetheric particle will be propagated in the medium precisely because of the interaction of the particles with one another.

Stefan therefore professed the aether hypothesis, as Stefan began his part # 6 with no declared title but with a lot of computations (Koller-Manuskripte, 26: page 18). For the beginning, he proposed the motion along the straight lines just like Lamé did two years earlier in 1860/61: first, you develop the differential equation for the simplest case which is not real in physics, then you get it complicated by the perturbation with further parts of Fourier's expansions into series to get as close as possible to physical reality. Lamé did it to connect the theory of elasticity with the theory of transport of heat which were both intuitively supposed to relay on inner structure of matter although it was not clear yet how that structure was geometrically related to the observed macroscopical symmetries of René Just Haüy's crystals (Lamé, 1861, XV). Lamé connected his new crystallography with conduction and elasticity where exponential power series were just modified into cosines power series, while a generation later Maxwell used their similar velocities to connect light and electromagnetism. Stefan preferred hydrodynamical analogies over analogies with electricity. He switched from solids to gases in his research while following Lamé's method of approximations with few simple first coefficients of Fourier series to begin with. He added additional ones as perturbations during his next step.

In their series of points along the straight line the particles of aether are in equilibrium. These points lie in the same straight line, mowing away perpendicular to that line, so we have the equation of the (Fourier) series.

In that simplified case, the original distances between three points are equal and denoted by ξ in Stefan's attempt to describe the force of aether as mechanical entity, which was not unusual in his times. So, he got the known formula of classical mechanics, using the mass multiplied by ds/dt^2 to get Brook Taylor's theorem (Koller-Manuskripte, 26: pages 22-23). Taylor's theorem published in 1715 gives an approximation of k-times differentiable function around a given point by a polynomial of degree k, called the kth-order Taylor polynomial. Brook

Taylor's (1685 Edmonton-1731 London) theorem is one of the central elementary tools in mathematical analysis. The distance between the points of Stefan's aether was ξ. Each of those points had the mase m, therefore Stefan's aether was not a real imponderable. The density of aether was therefore:

$$\rho = m/\xi$$

which enables the calculation of the force by Stefan's equation no. 4.

Stefan gave his third lecture on 3 November 1863 beginning with a part # 7 with no declared title (Koller-Manuskripte, 26: page 24). For the beginning, Stefan answered the question of his student Plaschek (Plašek, Kriček, Vinček) from Prague, probably one of Stefan's students in the audience, who later never distinguished himself with any publications. That Koller's remark might provide some light into Stefan's relations with his students, including later Boltzmann. Like previously on 17 June 1862, at the beginning of this lecture 3 November 1863, the professor Stefan announced that one of his listeners-students, Mr. Plaschek from Prague, noted the sentence deduced above (equation no. 4, valuable only for one refracting surface of revolution). He stated that for a refracting surface, whose rays all come from one point of a beam, all rays need the same time to get from the first to the second point (Koller-Manuskripte, 26: page 24; Koller-Manuskripte, 18: page 196). Any number of such refracting surfaces could get involved, while axes of rotation lie in a straight line. Stefan has proven that in a straight way, at the same time characterized by an elegance, which was praised by Koller. Then Stefan continued with his subject matter of mathematical deductions.

In the continuations of his mathematical derivations Stefan did not mention the disputed aether much more. He now only wanted to consider the displacement in the direction of the particles themselves (therefore in longitudinal direction). Let us note that direction by AB (Koller-Manuskripte, 26: page 31). We therefore have the equation for the interaction of the particles of aether with a shift in the direction of their alternating position (Koller-Manuskripte, 26: page 33).

In his part # 9 Stefan demonstrated the integration of that equation (Koller-Manuskripte, 26: page 34). He offered a knowledge of useful analogies, related to the other more classical parts of physics. In his later research, Stefan's favourite source of analogies and inspirations was the classical hydrodynamics.

Stefan gave his 4th lecture on 10 November 1863 beginning with a part # 10 with no declared title (Koller-Manuskripte, 26: page 36). His equations no. 4 and 6 described the displacement of the aether particles in the direction in which the particles lay (Koller-Manuskripte, 26: page 36):

$$A^2 = P/\rho$$

where ρ is density of aether, while P in one case measures the stresses (tensions) of the aether particles in the equilibrium position, and in the other case it is the coefficient of elasticity (Koller-Manuskripte, 26: pages 37, 40). From there Stefan got his second solution of that same problem.

In introductory note to part # 11 Stefan figured out that he got the integral of the same equation (named as formula no. 0) which has two parts (Koller-Manuskripte, 26: page 39).

In another introductory note to his part # 12 Stefan dealt once more with the second integral of his equation of part # 10 where he got the equation for P which in one case measured the stresses (tensions), of the aetheric particles in their equilibrium position and in the second case it represented the coefficient of elasticity (Koller-Manuskripte, 26: pages 37, 40, 41). Essentially, it was Lamé's idea again as Stefan proposed the analogy of the elasticity and other properties of matter like Lamé's thermal conductivity and Stefan's stresses (tensions) of the aetheric particles because Fresnel's concept of aether might have been somewhat too new to be incorporated into Lamé's still hidden atomism before the Spring of Nations of 1848.

The quantity ξ was in both parts of Stefan's formula the result of the integral (summation of the series) where Stefan or maybe just Koller reminded his audience about the same integral treated in the second edition of Oscar Xaver Schlömilch's (Oskar, * 1823 Weimar; † 1901 Dresden) *Compendium der höheren Analysis* (Braunschweig, 1862-1866, 1: 299). The Dresden professor Schlömilch was Koller's friend while Stefan published in Schlömilch's magazine already in 1859 and in 1862, probably by Koller's recommendation. The equation $F(x) = u$ was used to describe the particle of aether in time $t = 0$ while the other formula $f(x) = du/dt$ described the equilibrium of those particles in time $t=0$ (Koller-Manuskripte, 26: pages pages 45-46).

Stefan presented his 5th lecture on 17 November 1863 beginning with a part # 13 with no declared title (Koller-Manuskripte, 26: page 47). Stefan said: Let us make the series expansion in a definite (=asymptotic or Poincaré expansion published in 1886). There, truncating the series after a finite number of terms provides an approximation to a given function, like the velocities of the displaced particles as a function of time by the previous equation no. 10 (Koller-Manuskripte, 26: page 47).

In his part # 14 Stefan offered the solutions of his formulas number 9 and 11 (Koller-Manuskripte, 26: page 52). Stefan next discussed his integral equation (Koller-Manuskripte, 26: page 53). He began the calculations for the positive values of the variable x (Koller-Manuskripte, 26: page 54), and continued for the negative values of the variable x in his part # 15 (Koller-Manuskripte, 26: page 57).

Stefan gave his 6th lecture on 24 November 1863 beginning with a part # 16 with no declared title except for the underlined term "instant as a point of time" (Koller-Manuskripte, 26: page 61). He stated that we will now examine the state of our instant point only at a certain position for a given x at another time (Koller-Manuskripte, 26: page 61). Stefan's discussions could be valuable even for the modern students except for his aether mysteriously involved: he certainly proved his great trainings in mathematical physics. His course is still useful today if we change his notion of the particles of aether by some modernized notion of almost equally mysterious vacuum as the media of spreading of light.

Stefan discussed the motion of the wave with the (wave)length 2λ and defined the concept of period of oscillation T (Koller-Manuskripte, 26: page 63). If the length of a wave is L, and the velocity is a, we have:

$$T = L/a$$

Stefan continued by part # 17 with his discussion of the definite integral equation for u as defined between the smallest interval of time from t-x/a to t+x/a (Koller-Manuskripte, 26: page 64). On the end, Koller or Stefan himself provided two prolonged appendix notes as annotations, both on two pages, with a question mark in the first note which included point 3 of his discussion of the cases of variable velocity along the constant amplitude (Koller-Manuskripte, 26: page 66). The question mark in that mathematical derivation was obliviously

Koller's as Stefan hardly could provide questioning his own formulas as a lecturer.

Stefan gave his 7th lecture on 1 December 1863. He began by his No. 17 which was exceptionally not intended to be a new part of his course although all his other lectures began by new numbered parts. Stefan spoke about his underlined title: Interference. It was his first lecture with a clearly defined underlined title in that course (Koller-Manuskripte, 26: page 69), and several of that kind of titles followed. Stefan discussed the interference of continuous wave trains (Koller-Manuskripte, 26: page 70).

Stefan advanced his course by part # 18 of his course with a second point of his discussion of parallelogram (of forces) began in his previous page by his point 1 (Koller-Manuskripte, 26: page 74).

Stefan continued his course by part # 19 of his course with the (interference as) composed vibrations, therefore, vibrations as composed of other components. Every vibration can be divided down into two oppositely turning circular vibrations (Koller-Manuskripte, 26: page 75-76). That idea was certainly a clear product of Fourier's mathematical thinking, also connected with then popular ideas of pairs of forces. Then Stefan turned to the third point of his discussion of parallelogram (of forces).
If we have among the interfering vibrations those of unequal forms or of unequal speeds of propagation, and we look at each other's whereabouts, then the first oscillation curve interferes on the higher degree with the curve of the second, also by interference between its own wave movements.

So, we get the curve of fourth order (Koller-Manuskripte, 26: page 77). Stefan had in mind the leminiscate, as any of several figure-eight or ∞-shaped curves. Then Stefan turned to the last fourth point of his discussion of parallelogram (of forces) to explain his fourth type of vibrations by the law of parallelogram of forces, to get the congruence, involving two or more geometrical figures, contacting each other at all points when one is put on top of another.

In his part # 20 Stefan discussed the nature of vibrations in air. He explained his phenomena of interference and diffraction separately as an air jet consisting of a series of oscillating particles (in fact, the particles of aether). We assume that the oscillations were longitudinal, lying in the direction of the jet. Young and Fresnel examined the diffraction of vibrations and Fresnel performed his mirror experiment. In 1816 Fresnel and Arago discovered their novelty, showing that in fact interference from oppositely polarized rays appears to be smaller. Fresnel directed a view that the rays of light (radiations) are transverse. Because also the vibrations of a particle in new polarized rays vibrate, we could divide them into three components, one in the direction of the spreading and the other two perpendicular to it (Koller-Manuskripte, 26: page 77-78). The first is longitudinal, the other two are transversal. Therefore, Stefan still supported the possibility of longitudinal vibrations of light together with Helmholtz and Hertz, which was disputed by many other Stefan's contemporaries and poses a kind of challenge even today including John Henry Poynting's radiation pressure calculated from the strength of the longitudinal elastic waves generated by the pressure from an electromagnetic plane wave in a dielectric in 1911.

Stefan dealt with the theories of Young, Fresnel and Arago but did not mention that Arago initially opposed Fresnel's revolutionary transversal novelties (Koller-Manuskripte, 26: page 78). Such approach was pedagogically a great success: Stefan developed a huge prolonged mathematical theory of vibrations in aether, and than, suddenly, returned half of a century backwards, when no one among attendants except Koller was even born, to describe still valid theory behind all his sophisticated mathematics: namely, the ideas of Fresnel.

There was no Koller's note about Stefan's lesson on 8 December 1863, probably because of the St. Nicolas' fest. Stefan gave his next 8th lecture on 15 December 1863 beginning with a part # 21 by the underlined title Dispersion of Light. The polarized light falls on the obstacle situated in the direction of its spreading. The reflected light retained just its transversal component of vibrating particles of aether (at the polarized rays of double-slit experiment). The velocity of spread of light depends only on the density of the medium. All light has the same refraction index (a) in the same medium. It also explains the refraction of light in prism and the coloured rays thereof. After the white light is dispersed at the prism, the violet light has the greatest and red light the smallest (index of refraction). Also, considering the velocity: violet light has the smallest and red light the greatest speed; that Stefan's idea was refuted by Maxwell few years after! According to Stefan, in the moving bodies there are also other consequences which are not verified yet. But if we disperse light through a prism and determine its directionality, therefore the angle of refraction of the different colors of rays, we will find it greatest for violet light and smallest for red light. So, the speed for violet light had to be the smallest, and the greatest for red light. Therefore, the change must happen in the refracting body, when refracting, after making other influences, which our previous investigation did not take into its account yet (Koller-Manuskripte, 26: pages 81-82). Stefan's equation with the velocity equal to frequency multiplied by wavelength must not be challenged, but his speediest violet light was clearly the possibility which nature did not chose to follow, as Weber and Maxwell proved already in those Stefan's and Koller's times, which makes the speediest violet light look somehow strange. Certainly, different types of wave of light will travel at different speeds as the speed at which the individual crests and troughs of a plane wave, filling the whole space, with only one frequency propagated, which is called the phase velocity. A physical signal with a finite extent (a pulse of light) travels at a different speed. The largest part of the pulse travels at the group velocity, and its earliest part travels at the front velocity. The idea of a group velocity distinct from a wave's phase velocity was first proposed by W.R. Hamilton in 1839, when Stefan was a little kid, and the first full treatment was provided by Rayleigh in his "Theory of Sound" in 1877. So, group velocity and phase velocity might be on Stefan's mind when he seemingly departed from Weber's and Maxwell's idea of uniform velocities of all colours of light as the extension of Faraday's earlier proof that all kinds of electricity are the same, no matter of their origin.

Stefan stated that there are just the aether particles which vibrate according to his equation no. 6. The force is proportional just to the distance between both particles (ξ) and to the masses (m') of the particles. Stefan also considered the number of particles of aether which influenced each other. The greatness of force f(ξ) designed as A depends on the constitution of bodies. a=$\sqrt{A}$ is the velocity of propagation of light. It depends on the gravitation and on the colour of light. Therefore, (it depends) on the obstacles which particles provide to the spreading of the light. The other equation depends just on the constitution of (particles of material) body and not on nature of light (Koller-Manuskripte, 26: pages 82, 86-88).

Stefan noted his velocity of light in air as "a_{oE}" and its wavelength as λ, while Stefan's velocity of light in medium was a. So, we have the formula for dispersion (Koller-Manuskripte, 26: pages 89, 91). Considering the phenomena of diffraction, we have λ_o as the wavelength of coloured rays of light. We could determine the diffraction according to the ideas of Joseph Fraunhofer (* 1787; † 1826) developed in Munich.

Stefan used the method of least squares as a statistical procedure to find the best fit for a set of data points by minimizing the sum of the offsets or residuals of points from the plotted curve. We could use it for rays whose wavelength λ is known, where for the solid bodies it is determined to the fifth decimal place. The further experimental verifications are not provided as far, but they are needed to prove those decimal places.

Those facts do not hold for liquids because there the fourth decimal place is still uncertain (Koller-Manuskripte, 26: page 91). Stefan here suggested something similar as Kelvin and Helmholtz did later: the whole physics looks like finished science resembling (classical) mechanics or Gaussian optics, while task of the future researchers is mostly to provide further decimal places of the measured constants. Certainly, not very much attractive for future physicists among Stefan's students! The approaching quantum mechanics soon turned all that kind of logics upside down! Certainly, few years too late for Kelvin, Stefan, and Helmholtz.

Also, Stefan used the method of least squares to describe the liquids, especially the water (Koller-Manuskripte, 26: page 92). Stefan developed his equation by the series of cosines. For conclusion of that part # 21, Stefan discussed the wavelength of light.

Stefan gave his 9th lecture on 12 January 1864 beginning with a part # 22 without underlined title as he further researched the dispersion of light. Stefan expanded series of cosines to the fourth coefficient. The value ξ was used for the particles of aether. The particles of aether influence the wavelength of rays of light according to Fraunhofer's formula for diffraction D needed to calculate the wavelength (Koller-Manuskripte, 26: pages 92, 94-95). By measuring the other substances, we get the different results than for the air.

In his part # 23, Stefan discussed the propagation of the movement of the aether in linear direction. Then he extended his approach to the (three dimensions of) space. He discussed the particles of aether with their mass m moving from the coordinates in space x, y, z to the position in space x', y', z'. For the mass m he calculated the real kinetic energy (Koller-Manuskripte, 26: pages 96, 98).

Stefan finished his part # 23 with discussion about the equation of 3rd degree involving the particles of aether in rest. He wished to get his results valid for the chosen direction (Koller-Manuskripte, 26: page 105). So, Stefan, Cauchy, Tesla, and many others tried with huge efforts to calculate the behavior and characteristics of the aether. That fact explains their reluctance to accept the negative results of Michelson-Morley experiments in 1880/81 and 1887, until the new Einstein's generation preferred to quietly put that infamous aether aside, much more silently than Lavoisier cancelled the phlogiston in his times. Clausius and Boltzmann put much additional efforts to find the mechanical analogies of entropy before they discovered that they are making the job of Sisyphus. Why did such exact scientists ignore Ernst Mach's warning that a clever guy must deal just with the entities he/she could see? Hard to explain for the era before the technology enabled tools much more powerful than the human eye, four centuries after the similar aim of Galilean telescope. Most of Mach's opponents were devoted Christians who deeply believed in unseen miracles, and even in atoms where they finally defeated Mach even if they almost forever lost their Don Quixotic battle for equally invisible aether, even if N. Tesla never abandoned aether. The socialist Mach had his doubts, and his peers respected his experimental abilities like the modern respects of Andrej Detela's talents, but the mainstreamers were too scared to share their doubts in fundamentals of their own prevailing mainstream research. No wonder that Boltzmann, after hearing those Stefan's courses about aether, never really abandoned its use, although he or Stefan never published anything really focused on the theory of aether like Cauchy or the Benedictine Karl Puschl (baptized as Joseph, * 1825, † 1912) did in their era.

Stefan gave his 10th lecture on 19 January 1864 beginning with his part # 24 without any underlined title. He discussed the dislocation σ of particles of aether in direction of the resultant R. He also computed the angle of the dislocation σ and the coordinates of axes. Again, Stefan connected his deductions with the direction of axis of elasticity. That enabled him to discuss the force of elasticity (Koller-Manuskripte, 26: pages 105, 109, 110).

Stefan lectured on his next to last part # 25 without underlined title (Koller-Manuskripte, 26: pages 111-114). Stefan discussed his part # 25 to check the validity of his previous research considering the congruent T connected with the equation of curved area. The research of area gave the vector of elasticity in the point. That is Fresnel's surface of elasticity. After calculations, Stefan obtained the equation of elasticity and finally the equation of the ellipsoid for the surface of elasticity (Koller-Manuskripte, 26: pages 110-112, 114-115).

Stefan again lectured on his last part # 26 without underlined title (Koller-Manuskripte, 26: pages 115-119). He wished to derive his previous results again by another way. He described the working force as the sum of components. He got the equation which included the fourth power of ξ (Koller-Manuskripte, 26: page 116).

Therefore, that Stefan's course about optics had just ten lections although we would expect fourteen of them in any ordinary winter semester of those Habsburgian times.

Stefan on interference: overview

After Stefan's aether theory of light which followed Fresnel's approach, Koller attended Stefan's course focused on interference of light. According to Stefan's wave theory, the point source of light causes a swinging motion. It spreads equally to all sides. The distance from the surface during constant motion is the path determined by the oscillations at time (T), wavelength (λ) and translational velocity (v). The wave surface has a different shape depending on the speed of light in different directions. If this speed is the same in all directions, the surface forms a ball when the distance from the illuminating points is large enough. As in the case of Fourier's heat theory, Stefan (and Koller with him) certainly adopted the wave theory of light (and heat) of Fresnel instead of William Robert Grove-Robida's intermediate vibrational hypothesis about electricity, which was criticized by Stefan's senior collaborator Wilhelm Joseph Grailich. Fresnel's essay "About light" was published simultaneously with Fourier's final theory of heat. As novelties in their times, they both determined the viewpoints of Koller who was then in his middle twenties. Koller's four decades younger protegee Stefan was already initially influenced by the experts who were only born then in the era of great changes in physics in 1821-1824, like Helmholtz, Clausius and Kelvin. Fresnel-Fourier's approach of Koller's youth promoted the basic French inventions of mathematical series expansion pf trigonometric functions based on smooth shape of most curves involved, the luminiferous aether and transversality combined with Oersted-Faraday-Ampère's electromagnetism. The new leading mainstream generation of Stefan's youth shifted from revolutionary Parisians into Germanic areas with a clear Ettingshausen's concept of conservation of energy and less clear German approach to the French Carnot's entropy. Except for Cauchy's aether, which both Koller and Stefan got wrong from our present standpoints, the generational gap determined several huge differences between Koller and Stefan's approaches, especially considering the kinetic atomism never quite accepted by the Benedictine elite including Robida and Puschl, and the statistical entropy of later Boltzmann's ergodic hypothesis which remained unacceptable even to Simon Šubic. As an observer of astronomical, geomagnetic, and meteorologic data full of statistical calculations of mean squared errors' averaging and probabilities, Koller might have been more responsive to the new modern statistical physics of Stefan and his student Boltzmann.

Stefan used to think in the frames of granular aether; with Δ he marked the distances between the aether and the particles of matter in the waves of light. He followed the theories of Koller's friend George Biddell Airy (1801 Alnwick, Northumberland-1892) from the Greenwich Observatory, or at least Koller interpreted Stefan's lecturing by Airy's data.

On April 19, 1864, in his 2nd lecture, Stefan described two Fresnel's mirrors in a pair forming an angle of few minutes less than 180°. He used them to observe the disturbances of the coherent light beam, as proposed by Fresnel in 1816. That small angle between the normal planes of the mirrors was adjustable. The superposition of waves of light causes the interference, which is observed through a magnifying glass. In white light, we can see a white light strip, then two black ones, then a colored ribbon.

After his flat prism and the colors of thin layers, Stefan discussed Newton's colors of the rings as series of concentric dark rings, bordered by light rings, and colored strips next to them. They could be observed between two glasses, when one's convex side lies on the flat surfaces of the other: there is a layer of air between them.

In the last Stefan's lecture about optics which Koller wrote down, Stefan described scattered images in diffuse light on 10 May 1864.

Wave theory for Stefan's General Optics

Stefan included some optics in his other lectures. Without any visible warning in the last pages of Stefan's lectures about mechanics, Koller switched from acoustic to optics, while Stefan was certainly still discussing waves. The dated note on 27 May 1862 was next, entitled as "Strange effects", with the illustration of refraction (Koller-Manuskripte, 18: page 189= folio no 4). Koller took no notes of Stefan's lectures on 3 June and on 10 June 1862, at least none of them is preserved today.

Then Koller dated his note on 17 June 1862 entitled Birefringence (Koller-Manuskripte, 18: page 193). To his calculations on page 196 Koller added a note about somebody named Plaschek (Plašek, Kriček, Vinček), probably one of Stefan's students in the audience mentioned again in Koller's note on 3 November 1863 (Koller-Manuskripte, 18: page 196; Koller-Manuskripte, 26: page 24).

Then followed Koller's dated note on 24 June 1862 (Koller-Manuskripte, 18: page 196) this time exceptionally in the middle of a page, not distinguished by any underlined title. Koller recorded no Stefan's lection on 1 July 1862. Then he dated note on 8 July probably in 1862 (Koller-Manuskripte, 18: page 208) with no underlined title. The final additional paper photographed with a kind of matrix was added to the end of that booklet by Amand Kraml. It concluded Koller's inserted notes about wave motions of optics added on the end of Stefan's course of mechanics (Koller-Manuskripte, 18: 189-215).

Stefan's Lectures about Electrodynamics and Induction

Koller attended Stefan's lectures about electrodynamics and induction in the summer semester of 1863. He noted them on 26 handwritten folios, therefore on 102 pages containing texts with a separate title page added later. The course began on 21 April 1863. Stefan stated that the connection between electricity and magnetism was known for a long time. Everybody suspected that both have a mutual context but only Hans Christian Oersted detected and proved the effect of electrical current on magnet in 1820. In autumn, Arago learned about the discovery and presented Oersted's invention to Parisian Academy in September 1820. Soon the theory of that phenomena naturally developed with extreme speed, mostly due to Ampère's efforts at the Parisian Academy by the end of that year. Everybody was interested, just like during the productions of nuclear weapons or the vaccines against pandemic viruses two centuries later. Electromagnetism of those days might not promise such big businesses to inexpert observers including Koller's own students, but the profitability soon became oblivious in all three cases of accelerated state-supported research.

Ampère formulated the law by which the electric currents influence each other. On the left side of his notebook Koller's sketched Ampère's movable and stationary wires producing magnetism, termed as Ampère's model.

The electricity is measured by Ampère's method or by Charles-Augustin de Coulomb's method of torsion balance by determining the elasticity of a suspended wire, resembling the measurement of gravitation and similar approaches. Stefan and Koller did not use that term, but today we might call the first method electrodynamical, and the second electro-statical. The first way of determining the electricity by Ampère's invention is based on four points focused on interactions between the currents. In his first evaluation of Ampère's point, Stefan noted that the mutual influence of the currents on one another does not change if we reverse the direction of currents. After he stated those four points, in second part of his first lecture Stefan discussed the direction of electrical currents by dividing it into the smallest components, just like Ampère did in his molecular theory of magnetism, although Stefan in Koller's interpretation did not mention molecules as the smallest magnetic entities (Koller-Manuskripte, 32: bottom of page 3). The same goes for third part of Stefan's starting lecture with introduction of trigonometrical functions (Koller-Manuskripte, 32: page 6). The second lecture followed on 28 April 1863 stating that the interactions between two currents do not change if we make the distance between their midpoints equally small or big (Koller-Manuskripte, 32: page 8).

Stefan's third lecture was on 5 May 1863 (Koller-Manuskripte, 32: page 11). Stefan obtained his results by the integration of partial differential equations of trigonometric functions. The next section of that lecture included fourth part of discussion as those parts did not end with the finished lecture (Koller-Manuskripte, 32: page 14 = folio 4a). There, Stefan introduced an integration per partes. Fourth part ended with appendix note (annotation) which might have been Koller's own mathematical comment and not a part of Stefan's original explanations (Koller-Manuskripte, 32: page 15). The fifth part dealt with motions of points which caused the vibrations of force (Koller-Manuskripte, 32: page 16). Stefan integrated his trigonometric formula from the negative to positive infinite number (∞), which was also extended to the sixth part of Stefan's course included in his fourth lection on 19 May 1863 (Koller-

Manuskripte, 32: pages 17-18). Stefan introduced Ampère's differential trigonometric formula, as well as Ampère's equation for mutually influenced induction (Koller-Manuskripte, 32: pages 23, 26-27). Stefan gave no lecture on 12 May 1863 or on 26 May 1863, at least not by Koller's notes. None of those lectures had any formal title. Koller exceptionally numbered his illustrations on the left sides of his pages with the Roman numerals.

Stefan's fifth lection followed on 2 June 1863 by part nine because Stefan gave no lecture on 26 May 1863, at least not by Koller's narrations, which might indicate that Stefan lectured in electrodynamics and induction only every second week for a while (Koller-Manuskripte, 32: page 28). The sixth lection followed on 7 June 1863 by part 12 and Koller's figure XX with description of mutual induction of two closed circuits (Koller-Manuskripte, 32: page 39).

The seventh lection followed on 16 June 1863 by part 15 without any main initial figure as illustration as the focus went to mathematical integrations (Koller-Manuskripte, 32: page 51). Stefan began with the introduction to Weber's electrodynamics. From the year 1839 onwards Koller published in the magazine *Resultate aus den Beobachtungen des Magnetischen Vereins* which was edited by Carl Friedrich Gauss and Wilhelm Weber in 1836-1841. On 31 October 1857 Stefan reported about his own undergraduate studies of Weber's theory of undulations (Šubic, 1902, 79). As twenty years old student Weber and his almost a decade older brother, the professor of Anatomy at Leipzig Ernst Heinrich Weber, published a book on the *Wave Theory and Fluidity* (*Wellenlehre auf Experimente gegründet oder über die Wellen tropfbarer Flüssigkeiten mit Anwendung auf die Schall- und Lichtwellen*) in Leipzig in 1825. That book made considerable reputation and developed acoustics as Weber's favourite science, although that work was not translated into English language in Weber's lifetime. Therefore, Koller and Stefan both knew all about the merits of Wilhelm Eduard Weber (1804-1891). Weber again tried to answer question of Ampère's law: how do masses of electricity influence one another? He discussed the force of electrical current in conductor as a function of time. Stefan used a Greek letter ε for interaction among the elements of current, which Koller especially underlined. As usual, the time dependence was described by differential equations to solve the big questions of position of elements of current of electricity in the moment of time, which was ultimately reduced to Ampère's law (Koller-Manuskripte, 32: pages 52, 55, 60).

Stefan's eighth lection followed on 23 June 1863 by part 18 and no figure involved at the beginning. Just some sporadic illustrations were included in later discussions. Ampère's law of electrodynamics was interconnected with Coulomb's electrostatic $1/r^2$ law as its special condition although Stefan did not yet explicitly use that today oblivious division of theory of electricity, although he noted that the first one corresponds to the statical rest and the other corresponds with the dynamical motion (Koller-Manuskripte, 32: pages 61-62). Ampere's law was bound to fail for unsteady current until Maxwell fixed it by purely theoretical consideration in 1861, to later to be included in Stefan's lectures few months later. Maxwell's paper "A dynamical theory of the electromagnetic field" published in 1865 was finally confirmed by Hertz in 1888.

In his part 19, Stefan checked Weber's theory of ε by then available experiments. The verifications involved both Coulomb's and Ampère's laws. By error, Koller repeated his number 19 of that part again (Koller-Manuskripte, 32: page 67). Stefan checked the general validity of an extended Ampère's hypothesis which was proved only for a special case so far.

Stefan's ninth lection followed on 30 June 1863 by part 20 and no figure involved on the beginning and just sporadic illustrations included in following text. Stefan noted Weber's ideas about the time dependence of electrical influences. Koller concluded that Stefan's 20th part with an unusually long comment waging three points of similarities of Weber's and Ampère's ideas by Koller's annotation (Koller-Manuskripte, 32: pages 71, 76-80).

Stefan's 21st part about the theory of induction was almost the single one with a formal title. Koller's acquaintance Faraday discovered basic facts of it while he examined the closed circuits approaching or distancing from another circuit in 1830. The current in electrical conducting wires became stronger or weaker. Because of those changes, there appears the electromotive force produced by the induction of current. Could we explain Faraday's induction by Weber's theory, Stefan asked (Koller-Manuskripte, 32: pages 79-81)? The very posing of a question like that revealed Stefan's deep doubts and Koller eventually shared them. That was fundamental question of those times when British experimental-industrial influences overpowered the European mainland for a while. Soon even Stefan preferred Maxwellian approach as mathematization of Faraday's ideas. Shortly afterwards, Stefan's switched his physics from French towards British and Berliner influences as the Berliner Helmholtz preferred

Maxwell over the Gottingen based Gauss and especially Weber who taught there in 1831-1837 and again after 1849. As the imperial center, Helmholtz's Berlin suddenly became more important, politically, and scientifically, compared to the traditional university towns like Gottingen. The English Cambridge and Oxford somehow avoided that evil fate also because no major political change troubled the Victorian England of those times, while Bismarck's unification of Germany additionally shifted to Berlin the center of the new German state's sciences and its great money involved.

Stefan's tenth lection followed on 7 July 1863 by part 22 and no figure involved on the beginning and just sporadic illustrations included in later discussions. Stefan formulated Weber's law for the mutual exchange of electric influences alias induction, also for negative electricity. Stefan took induction current as function of time variables in his differential equations. Next, he took the velocity as a variable. He examined the velocity as he discussed equation for the electromotive power of electrical current noted as Q, which enabled Stefan's formula no. 22 noted as i = a·e·u. Somewhat mockingly, it resembled the good old Habsburgian motto A.E.I.O.U. In his part # 23 Stefan calculated the special case involving interaction among the elements of current with induction represented by the letter ε (Koller-Manuskripte, 32: pages 81, 82, 84, 85, 87-88).

In the part # 24 of his course Stefan checked the usefulness of Ampère's law in the case of induction with current and distance dS as dependent variables in differential equation. Stefan examined the square of velocity by his formula no. 22 developed earlier in his part # 22 (Koller-Manuskripte, 32: page 88). Seven years earlier in 1856 in 5th volume of *Acts of Academy of sciences of Saxony* under the title Elektrodynamische Maass-Bestimmungen with a shorter version published in *Pogg.Ann.* as Ueber die Elektricitätsmenge, welche bei galvanischen Strömen durch den Querschnitt der Kette fliesst, Weber and his student Rudolf Kohlrausch (1809–1858) demonstrated that the ratio of electrostatic to electromagnetic units is close to the value of the speed of light, which they denoted as c. They were the first worldwide to embrace such step which today seems to be oblivious. That idea enabled Maxwell's new concept of light as an electromagnetic wave, and his competing against Weber's theory of electrodynamics.

In his part # 25 Stefan described the excitation of electrical currents from one wire to another, propelled by induction (Koller-Manuskripte, 32: pages 90-94). Stefan denoted the velocity of light with a letter marked v as a special case of ordinary velocity u as he did not use Weber's sign c yet. The sign c became a customary denotation for the speed of light only gradually, peaking with Einstein's papers of 1905, also because Einstein was educated in German scientific milieu. Stefan noted his element of conductor as ∂S to calculate the induction ε (Koller-Manuskripte, 32: pages 91, 94).

The last eleventh Stefan's lection followed on 14 July 1863 by part # 26 and no figure involved on the beginning with just sporadic illustrations included in later discussions (Koller-Manuskripte, 32: pages 94-102). Stefan still discussed the possibilities of Weber's induction with the speed denoted u as a function of time noted by the usual letter t. It figured as a variable in differential equation of intensity of induced current ε. Stefan examined Weber's fundamental proposition. Stefan concluded that Weber's fundamental proposition also gives Ampère's law when the changing (variable) dielectric current and its conductor are moving (Koller-Manuskripte, 32: pages 94, 96, 98).

In part # 27 Stefan used again his formula # 22 for the electromotive force noted as Q. The inductive effects were noted in the first part of the equation number 28 for Q, while the second part of that same equation represented the changing velocity of the intensity of the current. Stefan summed both to get the total indictive effects (Koller-Manuskripte, 32: pages 99, 100).

In his final part # 29 Stefan examined the interactions between the element of conductor (parts of circuit) ∂S and $\partial S'$ illustrated by the next to last sketched figure XXXIII to calculate the effect of electromotive force Q on conductor. As his example, Stefan discussed the cause of two parallel elements of conductor ∂S and $\partial S'$. Then he examined the induction current in closed circuit, resembling Ampère's approach with the time involved. Introduction of variable time was special novelty of Stefan's era, which enabled Stefan's problem and Stefan's number in the era of somewhat postponed Habsburgian primitive accumulation of capital when time became money. Stefan discussed the induction currents in closed circuits introduced by Ampère's function of time (Koller-Manuskripte, 32: pages 100-102). On that point Stefan concluded his course whose final part was focused on checking of the validity of Weber's theory as the possible explanation of Faraday's induction. The final proof was the reducibility of Weber's ideas to Ampère's law which was considered as proved by decades of previous works. Just by examining Weber's formulas, Stefan proved that he had some deep although up to date hidden doubts in their truth. So, the stage was cleared for Maxwell's theory as the mathematization of

Faraday's ideas, although Stefan did not mention then still young Maxwell by name as Maxwell was just four years older than Stefan.

Therefore, Stefan held just eleven and not all twelve lectures which were usual in summer semesters of those days, at least by Koller's notes. Again, like in his lectures about the heat, Stefan began his course with deep and even historical introduction, but provided no equivalent general conclusions on the end although somewhere in the middle of his course Stefan mentioned Faraday's discovery of induction in 1830 (Koller-Manuskripte, 32: page 79). In his electrodynamics Stefan again initially relied mostly on French authors like Arago or Ampère with the only exception of the Dane Oersted and the Londoner Faraday, just like Stefan's discussions of heat focused on the theories of the Frenchman Fourier. The modern reader would expect more about the experimental induction of the Englishman Faraday, whom Koller personally met in London in 1838. In the second part of his course, Stefan mainly discussed Weber's electrodynamic, while successfully comparing and reducing it to Ampère's law. In December 1837, the Hannover government dismissed Weber, brothers Grimm and the other liberal Göttingen Seven who protested the royal abolition of the Hannover constitution as an early seed of an incoming Spring of nation of 1848. Those brave fellows refused to give their oath to the king. During next six years Weber widely traveled also to England, but it is not certain that Koller met him during his own journey in autumn of 1838.

Not many new ideas about electrical light, telegraphy, dynamos and similar applications of electrodynamics and inductions were included in Stefan's lectures as Stefan offered just a theoretical course while his interests in applications followed only two decades later when he became a technical director of Viennese electric exposition in 1883. Likewise, Stefan provided no clear applications inside his theoretical lectures about heat, where his applicative pursuits involving Stefan Problem of formation of ice followed very soon.

Concluding remarks on Koller-Stefan's connections

After Easter Sunday on 27 March 1864, from 12 April 1864 to 10 May 1864 Koller listened only to five Stefan's lectures focused on interference. He noted much smaller part of that course compared to the previous 12 out of 14 lectures about heat which he had attended a year and a half earlier from 21 October 1862 until 10 February 1863, just before Stefan became the youngest professor of mathematical physics at the University of Vienna and co-director of the Physics Institute on 9 March 1863. During that time, Koller also published a related optical-astronomical discussion Über das Passage-Instrument in the *Annually of Brno Natural Sciences Society* in 1863. Koller began attending Stefan's lectures on interference together with the student Boltzmann two weeks after Easter 1864, but Koller stopped taking notes just before the middle of the course, which was likely to consist of twelve Stefan's lectures in the summer semester.

Stefan probably taught two hours in a row about all his themes, except in his longer experimental courses which involved several hours of measurements of his students. We usually follow the same timetable in modern universities. Normally, in the same semester he taught two different topics, mostly about heat and optics.

In the summer semester of 1863/64, Boltzmann attended Stefan's lectures on heat theory, as well as Kunzek's lectures on heat, both without Koller in the audience. Stefan's second theoretical course about heat in the winter semester of 1865/66 may have been more filled with ideas involving the entropy. That one was attended by Boltzmann but not by Koller, who already finished his task of promoting Stefan. Petzval also taught optics and sciences of heat with Boltzmann in his classroom during the winter semester of 1863/64, but Koller was not there, as Petzval was already an established professor, and Koller usually listened to the younger promising stars of Viennese university. Koller visited the lectures of young Petzval long before he promoted Stefan. Koller attended five Petzval's lectures between 1850/51 and 1855/56 to promote the talent of young Petzval, much like Stefan's abilities were exposed a dozen years later by Koller's notes. Loschmidt himself lectured about molecular physics during the winter semester of 1866/67 for Boltzmann and his classmates. However, Koller was not in his class: molecular physics might not be Koller's favorite topics or the newly promoted Loschmidt was not young enough anymore for Koller's promotions. Most probably, overburdened by administrative tasks Koller ceased to attend any university lectures and even the Viennese academic sessions in his last two years.

Stefan lectured on Introduction to Physical Experiments ten hours per week during the summer semester of 1866, probably in his laboratories, where he recently became a boss. Among his audience was Boltzmann, but not Koller.

The groundbreaking Loschmidt's course about the molecules followed only in Boltzmann's fourth year in the winter semester of 1866/67. Boltzmann attended Loschmidt's class and befriended him, just like Stefan. Koller has already passed away, but his strong followers remained in charge.

Third Part

Stefan published in London, Paris, Geneve, Berlin

Introduction

Koller was the main designer of Stefan's initial success, even if Stefan concentrated his research on other narrowly defined parts of physics and not on Koller's experimental-observational geomagnetism, meteorology, and astronomy. Both remained fans of mathematical physics. Stefan obsession became heat also because his technical leadership enabled the success of Viennese electric exhibition focused on bulbs in 1883. Stefan Problem and Stefan Number are among the most widely studied topics of modern (multi) phase transitions involving movable borders between phases. Despite of great attention paid worldwide, nobody until now noticed that many Stefan's articles about those ideas were almost simultaneously published in London, Paris, Berlin, and Geneve. Stefan mostly reported in German language to his domestic Viennese academy, but his fans carried his achievements worldwide. In that aspects, he resembled Cauchy who used his father-in-law's enterprises to publish quickly enough in his own journal, or Faraday, who simply published his carefully edited laboratory notebooks.

Stefan was exactly thirteen years old when the March Revolution brought his Spring of Nation to Vienna and to lesser extent even to Stefan's native Klagenfurt areas. Koller was exactly seventeen years old when the Napoleonic Illyrian provinces brought the imperialized ideas of French Revolution of 1789 to his Carniolan homeland. Both teenagers felt the same on those occasions: the world, at least its European areas, had suddenly deeply changed, just like during the modern pandemics. Hand in hand with the social-political changes came the revolution in physics. The concept of heat as the new branch of physics was specially affected as it was basically changed at least twice during Koller's

231

lifespan: first Lavoisier changed phlogiston for caloric in the time of Koller's birth, which Jacquin soon promoted in Vienna. While Stefan was about to finish his studies and Koller replaced his monastic observatory with high political function, the kinetic theory of atoms dismissed any kind of substances resembling caloric to complement new ideas of conservation of energy and entropy combined with the spread of steam engines' industrial revolution. Another profound change dethroned Newtonian optics for Fresnel's transversal waves in imagined aether, also to promote the short-lived wave theory of heat. The newly proved electromagnetism might have even deeper impact crowned by Maxwellian unification soon after Koller's death. In the middle of those profound changes, the physical ideas promoted by Stefan and Koller were necessary somewhat different because of generation gap of more than four decades between them, but they shared basic belief in atomistic kinetic and fundamental Galilean faith in borderless ability of mathematics to explain all physics and maybe even the branches of knowledge beyond physics.

Stefan was also the Commissioner for Physics and Physical Textbook Consultant at the department of state. In that capacity, he might have attended the third Parisian World's Fair (*Exposition Universelle*) from 1 May through 10 November 1878. Among the many inventions on display was Alexander Graham Bell's telephone. Electric arc lighting had been installed all along the Avenue de l'Opera and the Place de l'Opera. In June, a switch was thrown, and the area was lit by electric Yablochkov's arc lamps, powered by Zénobe Gramme's dynamos. Thomas Edison had on display a megaphone and phonograph. International juries judged the various exhibits, awarding medals of gold, silver, and bronze. At Parisian International Exposition (*Exposition universelle d'art et d'industrie*) from 1 April to 3 November 1867 Tegetthoff also frequented until July 1867. There, or more probably at Parisian World's Fair (*Exposition Universelle*) from 1 May to 10 November 1878 Stefan also participated. That French spectacle might have been one of Stefan's rare journeys abroad. The second Parisian exposition with Carniolan Agricultural company participating with cereals was in 1867, the third under general commissioner navigational-construction engineer Jean-Baptiste Krantz (1817–1899 Paris) in 1878. Otherwise, Stefan avoided personal travels, but everybody interested learned about him from his published works. He was forty-eight years old in 1883, when he as the technical boss of the Viennese exposition finally personally met most of his fans, including Siemens and W. Thomson. Similarly, Koller was nearly forty-six years old during his European scientific tour visiting Berlin, Hamburg, London, Paris, and Freiburg im Breisgau. The leadership of the Viennese exposition made Stefan a leading Habsburgian public figure in 1883, just like a job in state ministry gave Koller the same opportunity in 1847. Both enjoyed that political limelight just for few months less than two decades, which made their lifepaths somewhat similar.

Koller published a lot abroad. Stefan's works were mostly anonymously translated into English by himself, his fellow Viennese academicians, or by the Brits. At least one of his British translators could be identified. *Phil.Mag.* also published several articles written by Boltzmann's best student the Carniolan Ignac Klemenčič (Klemenčič, 1885, 393-395; Klemenčič, 1890, 284; Klemenčič, 1892, 396; Klemenčič, 1893, 537-538), and by the grandson of Stefan's teacher Andreas baron Ettingshausen, Albert Ettingshausen, also in collaboration with Walther Herman Nernst. Stefan's students Boltzmann and Albert von Obermayer, as well as Stefan's antagonist Ernst Mach also widely published in *Phil.Mag.*

Stefan's publications in *Phil.Mag.*

At least twenty English summaries of Stefan's treatises printed in *Philosophical Magazine* were mostly translated from Viennese academical publication *Wien. Anz.* (*Anzeiger der Kaiserlichen Österreichischen Akademie der Wissenschaften, Mathematisch-Naturwissenschaftliche Klasse*). There, they issued a kind of stenographic records of the Viennese academic sessions from 8 January 1864 onwards, regularly appearing after their first volume printed in 1864. The approved among presented discussions, mostly belonging to the mainstream of then sciences, were subsequently published in the main Viennese academic newsletter *Wiener Sitzungsberichte der kaiserlichen Akademie der Wissenschaften in Wien* (*Wien.Ber.*). The longer works used to be printed in *Denkschriften der Kaiserlichen Akademie der Wissenschaften. Mathematisch-naturwissenschaftliche Klasse* (*Mathematical-natural science class*). Less likable contributions were announced only in *Anzeiger*, such as the papers of Nikola Tesla's professor the Catholic Croatian Martin Sekulić formulated against H. Helmholtz' ideas,[17] or some works of the Carniolan Simon Šubic (*

---

[17] Sekulić, *Anzeiger* 1878, 129.

1830), who criticized the ergodic hypothesis of his colleague at the University of Graz, Boltzmann.[18]

The English translator of Stefan's works might have been John Tyndall as one of the five or occasionally four editors of the *Philosophical Magazine* from 1856 to 1862, while the active managing editor was William Francis. After 1871 William Thomson (later lord Kelvin) was Francis' co-editor. The only signed translator of Stefan's work were James Alfred Wanklyn in English language and August Friedrich count Marschall in French language. The translations mostly acquired from the German original of *Wien. Anz.* were:

Stefan, J. (14 April 1863). Über die Fortpflanzungsgeschwindigkeit des Schales in gasförmigen Körpern, *Ann.Phys.* 4th series, 118 (28, 194)/3: pp. 494-496 (noted in index as J. Stephan but in text as J. Stefan, sent from Vienna on 13 March 1863). Translated report quoted data from *Ann.Phys.* as that paper was Stefan's exception and never appeared in that abridged form in *Wien.Berichte*: LXVI. Intelligence and miscellaneous articles. On the velocity of the propagation of sound in gaseous bodies. *Philosophical Magazine*, 4th series, 25th Volume, number 170 (June 1863) pp. 490-491. There, Stefan supported August Karl Krönig (* 1822; † 1879) and Clausius's kinetic theory which were widely translated and published in *Phil.Mag.* as thew were needed to correct Newtonian and Laplacian formulas for the velocity of sound. That was first Stefan's work published in *Phil.Mag.* after some six years of his other publications, echoing his new advanced status at the Viennese university and academy, but also beginning new trend of publishing woks of Habsburgian researchers in *Phil.Mag.* In the same volume published Stefan's colleague Edmund Reitlinger (* 1830 Pest; † 1882 Vienna) who was professor of physics at the Viennese High School of Technology where he taught Rudolf Steiner. Reitlinger discussed the stratification of electric light in Geissler's cathode ray tubes on pages 317-318.[19]

Stefan, J. (1863). *Bemerkungen zur Theorie der Gase, Wien.Ber.* II, pp. 47: 81–97 (session on 22 January 1863). Reprint 1863: Kleinere Mittheilungen XII. Bemerkungen zur Theorie der Gase (I. Wärmleitung), *Schlömilchs Zeitschrift für Mathematik und Physik,* volume 8: 355-368; Abbreviated reprint: *Ann.Phys.*

---

[18] Šubic, 1864, 22-25, 134-136; Šubic, 1872, 26.

[19] Edmund Reitlinger, On the stratification of the electric light, *Philosophical Magazine*, 4th series, 1863, 25th Volume, number 168 pp. 317-318.

119 (29): pp. 492-496 (30 July 1863). Abbreviated translated report noting data from *Wien.Berichte* volume 47: XIV. Intelligence and miscellaneous articles. Remarks on the theory of gases. *Philosophical Magazine*, 4th series, 27th Volume, number 179 (February 1864) pp. 75-77. Stefan supported Clausius's kinetic theory which was widely translated and published in *Phil.Mag*. Stefan was against Puschl, Hoppe, and Jochmann as he supported Clausius, Maxwell, J. Fourier, and Dulong-Petit. On page 94 of *Wien.Ber.* (*Zeitschrift für Mathematik und Physik* p. 366) Stefan criticized the error in Puschl's calculation of kinetic energy (Lebendige Kraft) during collisions of molecules and their flight upwards against the gravitation. Karl Puschl (* 1825 Wolfsbach in Lower Austria; † 1912 Seitenstetten in Lower Austria 49 km east of Kremsmünster) was a Benedictine. Stefan's former class-teacher the Benedictine Robida had the same problem with gravitational determination of the height of atmosphere as Robida directly criticized Stefan later in the next volume of *Zeitschrift für Mathematik und Physik* 1864 9: p. 218 at the first page of Robida's article, sent from Klagenfurt on 10 November 1863, a decade after Stefan received Matura in Robida's classroom in 1853. The other Benedictine, Stefan's mentor Koller, was probably not really upset.

Stefan, J. (1864). Über die Dispersion des Lichtes durch Drehung der Polarisationsebene im Quarz, *Wien.Ber.* 50, II, no. 15, pp. 88–124. Abbreviated report with somewhat different title: Ein Versuch über die Natur des unpolarisirten Lichtes und die Doppelbrechung der Polarisationsebene im Quarz in den Richtung der optischen Axe. *Wien. Anzeiger* 1/23 (3 November 1864): 175-177. Abbreviated translation from *Wien.Ber.* provided from *Wien. Anzeiger* by the professor of Chemistry at the London Institution from 1863 and FRS James Alfred Wanklyn (1834 Ashton-under-Lyne by Manchester–1906 New Malden suburb of south-west London), an analytical chemist who is remembered today chiefly for his "ammonia method" of determining water quality. His translation was not published in *Intelligence and Miscellaneous Articles* like other Stefan's papers, but in the main part of magazine under the title: On the Dispersion of Light by Quartz, owing to the rotation of the plane of polarisation. *Philosophical Magazine*, 4th series, 28th Volume, number 187 (August 1864) pp. 137-140. There Stefan supported Cauchy's law while criticizing Jean-Baptiste Biot's (* 1774; † 1862) law, also because the baron Augustin-Louis Cauchy (* 1789 Paris; † 1857) used to be a friend of Stefan's mentors Ettingshausen and Koller. George Gabriel Stokes (* 1819 County Sligo in Ireland; † 1903) signed his footnotes to the translated Stefan's paper as G.G.S. on pages 139 and 140. Stokes stated that Stefan committed an error.

According to Stokes, Stefan erroneously omitted his arbitrary constant after integration and obtained negative constant in dispersion formula, which indicated that Biot's law is still maintained in Stefan's experiments. Besides Brewster, Stokes was the greatest British expert in optics of those days and probably the initiator of that translation but Stokes himself supposedly lacked time or his knowledge of German language was insufficient for the relevant translation. Stokes' criticism might have contributed to the fact that Stefan's abandoned further research in optics after 1866, except for his two papers (Stefan, 1871, 223–245; Stefan, 1872, 325–354) which were not translated in *Phil.Mag.* Maybe Stokes' friend W. Thomson opposed the translation as the associate editor after July-December 1871. Stokes was close to W. Thomson who later befriended Stefan in Vienna in 1883, but Stokes' criticism may have had even some political motives as Cauchy used to be the most Catholic conservative of all top scientists, while his Parisian antagonist Biot remained the last valuable supporter of Newtonian corpuscular theory of light and furthermore Biot disliked aether. Cauchy widely travelled with the exiled French court, but mostly through Catholic states including Gorizia and Prague.

Stefan, J. (1865). Über einige Thermoelemente von grosser elektromotorischer Kraft, *Wien.Ber.* 51, pp. 260–262. Read at 9th session on 23 March 1865. Also in reprint: VIII. Über einige Thermoelemente von grosser elektromotorischer Kraft, *Ann.Phys.* 124 (4), pp. 632-635 (20 April 1865) with two bibliographic footnotes of the editor Johann Christian Poggendorff (1796-1877), signed as P. on pages 633 and 634. Abbreviated translated report from *Ann.Phys.*: XI. Intelligence and miscellaneous articles. On some thermo-elements of great electromotive force. *Philosophical Magazine*, 4th series, 30th Volume, number 201 (July 1865) pp. 77-78. As the vivid reader of *Phil.Mag.*, Stefan commented the measurements of Robert Wilhelm Bunsen (* 1811; † 1899) published in *Phil.Mag.* in January-June 1865. Stefan also discussed the success of the former Stefan's mentor Carl Friedrich Wilhelm Ludwig's (Karl, * 1816 Witzenhausen 20 km south of Gottingen; † 1895 Leipzig) Viennese mechanician the Jew Siegfried Samuel Marcus (Markus, 1831 Malchin, in the Grand Duchy of Mecklenburg-Schwerin-1898 Vienna). Marcus invented his petrol-powered vehicle in Vienna in the previous year 1864 (Marcus, Eine neue und sehr kräftige thermo-elektrische Säule (Kupfer-Zink-Nickellegierung/Antimon-Zink-Wismutlegierung). *Wien.Ber.* 1865; Marcus, Eine neue Thermosäule, *Schlömilchs Zeitschrift für Mathematik und Physik,* 10 (1865): 333-336, Reprint: *Philosophical Magazine*, 29/197 (1865 05): 406-407). Stefan also discussed thermo-piles of John Frederic Daniell (Frederic, * 1790; † 1845

London). In the same volume at pages 319-320, Stefan's antagonist Ernst Mach of Graz wrote on physiology of retina while focusing on the incoming light.[20]

Stefan, J. (1866). Über (Ueber) eine neue Methode die Lange der Lichtwellen zu messen, *Wien.Ber.* 53, II, pp. 521–528 (Viennese academic session on 26 April 1866). Abbreviated reprint: XXXVIII. *Schlömilchs Zeitschrift für Mathematik und Physik,* 11 (1866): 549-550; Abbreviated translated report: LXXIX. Intelligence and miscellaneous articles. On a New Method of Measuring the Lengths of Luminous Waves. *Philosophical Magazine*, 4th series, 31st Volume, number 212-Supplement (June 1866) pp. 550-551. Stefan continued the measurements published by Fizeau, Foucault, Fraunhofer, Jacob Fredrik Emanuel Rudberg (1800 Stockholm-1839 Uppsala), his doctoral student Anders Jonas Ångström (1814-1874) and Stefan's student Leander Ditscheiner (1839 Vienna-1905 Vienna) who later taught at the Viennese Technical University. Stefan used theory of diffraction to determine the wavelength of the spectral lines of quartz.[21]

Stefan, J. (1866). Über den Einfluß der inneren Reibung in der Luft auf die Schallbewegung, *Wien.Ber.* 53, II, pp. 529–537 (Viennese academic session on 16 April 1866). Abbreviated translated report: LXXIX. Intelligence and miscellaneous articles. On the Influence of Internal Friction un the air on the motion of sound. *Philosophical Magazine*, 4th series, 31st Volume, number 212-Supplement (June 1866) p. 551-551. Stefan adapted the differential equations of the description of propagation od sound for his favorite hydrodynamics according to the measurements of Oscar Emil Meyer (1834 Varel-1909

---

[20] R.W. Bunsen, On some thermo-electric piles of great activity. *Phil. Mag.,* Januar-Junij **1865**, *29/194*, 159-162; R.W. Bunsen, Simple method of preparing thallium. *Phil. Mag.* **1865**, *29/194*, 168-168; E. Mach, Über die Wirkung der räumlichen Vertheilung des Lichtreizes auf die Netzhaut : vorgelegt in der Sitzung am 3. October 1865, *Wien.Ber.* **1865**, *52/2*, 303-322. Povzetek: On the Visual Sensations Produced by Intermittent Excitations of the Retina, *Phil. Mag.*, Series 4, **1865**, *30*, 319-320.

[21] (a) J. Stefan, Über eine neue Methode, die Lange der Lichtwellen zu messen, *Wien. Ber.* II, **1866**, *53*, pp. 521–528, read at Viennese academic session on 26 April 1866. Abstract: *Wien.Anz.*, **1866**, *3/11*, pp. 95–96. Abbreviated translated report: LXXIX. Intelligence and miscellaneous articles. On a New Method of Measuring the Lengths of Luminous Waves. *Philosophical Magazine*, series 4, junij **1866**, *31*st volume, no. 212 (Supplement), pp. 550–551; (b) A. J. Angström, *Neue* Bestimmung der Länge der *Lichtwellen*, nebst einer *Methode*, auf optischem Wege die fortschreitende Bewegung des Sonnensystems zu bestimmen, *Ann.Phys.*, **1864**, *199/11*, pp. 489–505; (c) L. Ditscheiner, Theorie der Beugungserscheinungen in doppeltbrechenden Medien. *Wien.Anz.*, **1866**, *3/21*, pp. 193–195.

Wroclaw), who after a doctorate with Franz Ernst Neumann in Kaliningrad became a leading researcher of the friction in fluids and the kinetic theory of gases.

Stefan, J. (1868). Anwendung der Schwingungen zusammengesetzer Stäbe zur Bestimmung der Schallgeschwindigkeit, *Wien.Ber.* 57, II, pp. 697–708 (Viennese academic session on 30 April 1868). Abbreviated translated report: XIII. Intelligence and miscellaneous articles. (On the) Application of the vibrations of compound bars to determining the velocity of sound. *Philosophical Magazine*, 4th series, 36th Volume, number 240 (July 1868) p. 80-80. Stefan improved the experimental methods of Ernst Florens Friedrich Chladni (1756-1827 Wroclaw)[22] for his measurements of the speed of sound in (white) beeswax) dissolved in a glass tube at a constant temperature, in a wooden stick, in a glass tube, cork. sealing wax, chalk, (white and vulcanized) rubber. He compared his results with measurements of the speed of nerve stimuli obtained by Herman Helmholtz (1821-1894), with whom they jointly paved the way for the kinetic theory of gases and for (Maxwell's) theory of the electromagnetic field.[23]

Stefan, J. (1871). Über das Gleichgewicht und die Bewegung, insbesondere die Diffusion von Gasgemengen, *Wien.Ber.* 63, II a, pp. 63–124. Not translated at *Phil.Mag.* although Stefan discussed Maxwellian formulas for Maxwell–Stefan diffusion's model for multicomponent systems. The equations describing these transport processes have been developed by James Clerk Maxwell for dilute gases in 1866 and by Stefan for fluids in that paper (Crepeau, 2008, 32).

Stefan, J. (1872). Untersuchungen über die Wärmeleitung in Gasen I, *Wien.Ber.* 65, II, pp. 45–69 (6th session on 22 February 1872). Not in *Phil.Mag.* A year and half later Maxwell cited that Stefan's paper from German language original (Maxwell, 1873, 299; Maxwell, 1873, 439). It was therefore not published in *Phil.Mag.* as Maxwell would otherwise prefer to note English language version for reader of the magazine *Nature* who mostly did not master German language. Maxwell there also noted Loschmidt's German language paper as basis of Maxwell's own method of statistical mechanics borrowed from the social

---

[22] E.F.F. Chladni, *Kurze Übersicht der Schall und Klanglehre, nebst einem Anhange die Entwickelung und Anordnung der Tonverhältnisse betreffend*, Mainz, **1827**.

[23] H. Helmholtz, Messungen über Fortpflanzungsgeschwindigkeit der Reizung in den Nerven. *Archiv für Anatomie, Physiologie und wissenschaftliche Medicin*, **1852**, *19*, 199-216.

sciences (Maxwell, 1873, 299-300; Maxwell, 1873, 438-400; Loschmidt 1865; Crepeau, 2008, 33-34; Crepeau, 2013). That Loschmidt's work was therefore also not published in *Phil.Mag.* Those more theoretical papers were probably not welcomed in *Phil.Mag.* (or in *Ann.Phys.*) because Maxwell was not the person in charge there, although Maxwell's own papers published by Royal Society were regularly abstracted in *Phil.Mag.*, also in that year 1872 (Maxwell, 1872, 529-538).

Stefan, J. (1872). Über Schichtungen in schwingenden Flüßigkeiten, *Wien.Ber.* II, 65 (April-May 1872): pp. 424–427. Reprints: *Sitzungsberichte der Königlich Preussischen Akademie der Wissenschaften zu Berlin (Berlin.Ber.*, sessions of Prussian Academy of Sciences, April and May 1872); *Naturforscher,* 1873, volume 6, no. 7: p. 67-67, *Der Naturforscher Wochenblatt zur Verbreitung der Fortschritte in den Naturwissenschaften* was published in Berlin by the heirs of Ferdinand Dümmler (1777-1846) in 1868-1885 and edited by Wilhelm Sklarek (1836-1915) in 1868-1885. At least two French language summaries were published. Stefan was noted as J. Stephan in the index on the end, but his name was spelled Stefan in the title of his article issued by the formal editor De la Rive at the journal of the Genevan publisher Alfred Cherbuliez (1838-1895), who inherited the company A. Cherbuliez et c. with branches in Paris and Lausanne. The title of that article was: Stratification dans un liquide animé d'un mouvement oscillatoire, *Archives des sciences physiques et naturelles (Bibliothèque Universelle et Revue Suisse,* Geneve: A. Cherbuliez (Paris & Lausanne), Nouvelles Période (2), 1873, 46: 270-271; Stratification dans un liquide animé d'un mouvement oscillatoire, *Kosmos. Les mondes; revue hebdomadaires des sciences et leurs applications aux arts et à l'industrie* (Paris, ed. Moigno) 1873, 32: 707-708, 1874, 33: 564-565. Abbreviated translated report of the Swiss French language summary: XL. Intelligence and miscellaneous articles. (On) Stratification in a liquid in oscillatory motion. *Philosophical Magazine,* 4th series, 45th Volume, number 300 (April 1873) pp. 320-320. That Genevan magazine was published under title cited here after 1846 while edited by Auguste de La Rive (1801-1873) from 1846 to 1857. Rive's student the Swiss chemist Jean-Charles Galissart de Marignac (1817-1894) was a joint editor from 1846 to 1857, while their co-editor was also the zoologist Jules Francois Pictet de la Rive (1809 Geneve–1872). Their names were used even after 1879 in editorial title pages. In the time of Stefan's publication, Auguste de La Rive was still the leading physicist in Geneva as president of the Helvetic Society of Natural Science from 1845. He was a researcher of voltaic piles and electric discharges in rarefied gases which led him to form a new theory of the aurora borealis. But in the times of Stefan's publication in 1873, Auguste de La Rive's health declined, affected by paralysis. Stefan continued the Graz experiments of his student Boltzmann and his collaborator August Töpler (Toepler, 1836-1912), as well as August Kundt (1839-1894) and Mach's assistant at the University of Prague Clemens Neumann (Klemens, Mírumil, 1846–1873). Stefan linked their acoustic research to layers in a Geissler cathode ray tube. Stefan reported on his experiments with Geissler's cathode ray tubes based on August Kundt's data. The molecules of the gases illuminated by the passage of the current in Geissler's tube would take on the same role as the grains of filings in the acoustic experiment.

Stefan, J. (1873). Versuche über die Verdampfung, *Wien.Ber.*, II 68 (preliminary report at 24th session on 23 October 1873, full publication read at 26th session on 13 November 1873), pp. 313, 385–423. Stenographic notes: Versuche über die Verdampfung, Experimenteller Theil, *Wien.Anzeiger.* 10/24 (23 October 1873): 161-164; Versuche über die Verdampfung, Theoretischer Theil, *Wien.Anzeiger.* 10/26 (13 November 1873): 173-174. *Naturforscher* 3 January 1874 7/1: pp. 5-6 (from *Wien.Anzeiger.* 10/24 (23 October 1873): 161-164) & 1874 7/30: 282-284 (from *Wien.Ber.*). Translations of abbreviated report containing the experimental part from that notes of session on 23 October 1873: LXIV. Note: Intelligence and miscellaneous articles. Experiments on evaporation. *Philosophical Magazine*, 4th series, 46th Volume, number 308 (December 1873) pp. 483-484; Résultats des expériences sur l'évaporation, *Journal d'Institut (Mémoires de l'Académie des sciences de l'Institut de France Académie des sciences; Mémoires présentés par divers savans à l'Académie royale des sciences de l'Institut de France, et imprimés par son ordre: sciences mathématiques et physiques, des savants étrangers)* 1874 tome 39 (tome (volume) 20 was published in 1872, tome 21 in 1875 and none appeared in 1874); 10. Sur l'évaporation, *L'institut: journal universel des sciences et des sociétés savantes en France et à l'étranger,* Paris: E. Donnaud (Paris: L'Institut), 17 December 1873, 41 (nouvelle série 1re année no 50): 400-400, 31 December 1873, 41 (nouvelle série 1re année no 52): 414-414. The reports on sessions of Viennese academy on 22 October 1873 (sic!, in fact 23 October) and on 13 November 1873 were published by Cte Marschall (zoologist and paleozoologist August Friedrich count Marschall von Sudoc auf Burgholzhausen und Tromsdorf (1804 Vienna-1887 Oberneidlingen)). Reprint: Résultats des expériences sur l'évaporation, *Kosmos. Les mondes; revue hebdomadaires des sciences et leurs applications aux arts et à l'industrie* (Paris, ed. by Cauchy's friend former Jesuit abbé François-Napoléon-Marie Moigno (* 1804 Guéméné in Brittany; † 1884 Saint-Denis (Seine)) 1874, 33: 272-272. The article cited Stephan and *Journal l'Institut* (not published in *L'institut: journal universel des sciences et des sociétés savantes en France et à l'étranger,* nor in the *Revue des sociétés savantes de la France et de l'étranger*). That used to be the first paper on Stefan Problem and Stefan Number, certainly called by that name not sooner than a century later.

Stefan, J. (1874). Zur Theorie der magnetischen Kräfte, *Wien.Ber.* II, 69 (12 February 1874), pp. 165–210. Abbreviated translated reports: Contribution to the theory of magnetic forces. *Philosophical Magazine*, 4th series, 47th Volume, number 312 (April 1874) pp. 318-319; Magnétisme, théorie des forces Magnétique, divisé en trois parties, *Kosmos. Les mondes; revue hebdomadaires des sciences et leurs applications aux arts et à l'industrie* (Paris, edited by Moigno) 1874, 33: 691-692; Forces Magnétiques, *L'institut: journal universel des sciences et des sociétés savantes en France et à l'étranger,* Paris: E. Donnaud) 1874, 42 (nouvelle série 2e année no. 82): 254-255 (notes on session of Viennese academy by Cte Marschall, in fact the zoologist and paleozoologist August Friedrich count Marschall von Sudoc auf Burgholzhausen und Tromsdorf (1804 Vienna-1887 Oberneidlingen).

Stefan, J. (1874). Versuche über die scheinbare Adhäsion, *Wien.Ber.* II, 69 (session on 30 April 1874), pp. 713–728; Abbreviated: *Ann.Phys.* 154, (230/2 (1875)): pp. 316-318; *Naturforscher* 1874 7/36: p. 339-339 (from *Wien.Anzeiger.* 1874 volume 12). Abbreviated reports: Experiments on apparent adhesion. *Philosophical Magazine*, 4th series, 47th Volume, number 314 (June 1874) pp. 465-466; Recherches sur la cause de l'adhésion de deux plaques superposes (Adhésion Apparente). *L'institut: journal universel des sciences et des sociétés savantes en France et à l'étranger,* Paris: E. Donnaud (Paris: L'Institut), 1874, 42 (nouvelle série 2e année no 90): 318-319 (notes on session of Viennese academy by Cte Marschall). J. Stefan proved that the adhesion of two marble plates does not involve forces between the invisibly small constituents which were proposed earlier during two centuries to prove the existence of vacuum. Stefan refused the old ideas about the vacuum between two flat pressed planes as he had in his disposal many other more relevant tools to prove his kinetic atomism. Stefan experimented with the plates of glass immersed in water, salted water, alcohol, and air to find that their mutual force diminished with the fourth power of distance. He connected the problem with the hydrodynamics of capillarity and no more to statics. Stefan determined the coefficients of inner friction (viscosity) by Arago's and Ampère's student at École polytechnique Jean Léonard Marie Poiseuille (1797 Paris–1869 Paris), Oscar Emil Meyer and Maxwell. The similar "Huygens effect" was explained by Huygens' Parisian confidant, the poet-critic Jean Chapelain (1595-1674), with the assumption of the pyramidically shaped atoms that supposedly take the leading role in water under exhausted air. Of course, Huygens refused that idea too. He himself saw in the phenomenon the pressure of a fine substance, like Hooke. Later, Laplace attempted to explain Huygens' experiment with capillarity. In the 19th century, the phenomena interested the Belgian Jean-Jacques-Daniel Dony (abbé Dony, 1759 Liège-1819), the English lecturer librarian of the Royal Institution Charles Frederick Partington (Wartington, † 1857), and after them the German Helmholtz. In the 20th century, the research was continued by Huygens' fellow Dutchmen Kamerlingh-Onnes, his student Willem Hendrik Keesom (1876 Texel-1956 Leiden) and Casimir at Philips Research Laboratories in 1948. The electromagnetic causes of the Casimir effect have shown that Newton may have been amongst the closest to the truth, but the mechanical forces of adhesions are today supplemented by intermolecular forces to cause the adhesion of dissimilar particles and the cohesion of similar particles. Therefore, despite Stefan's attempts, the vacuum is still involved in modern explanations.

Stefan, J. (1878). Über die Diffusion der Flüssigkeiten. I. Abhandlung. *Wien.Anzeiger*, volume 15, number 26 (5 December 1878), pp. 222-223. The entire article: Über die Diffusion der Flüssigkeiten (1: Über die optischen Beobachtungsmethoden), *Wien.Ber.* II. Volume 78 (26th session on 5 December 1878) pp. 953, 957-975. Translation: On the Diffusion of Liquids, *Philosophical Magazine*, 5th series, 7th Volume, number 40 (January 1879) pp. 74-75. Stefan relied on the concept which in 1855 Adolf Eugen Fick (1829 Kassel–1901 Blankenberge in Belgian Flanders) introduced as Fick's laws of diffusion, which govern the diffusion of a gas across a fluid membrane and influenced Stefan's Viennese mentor Carl Ludwig (1816-1895) who taught Fick in Marburg. In 1870 Fick measured the cardiac output, using what is now called the Fick principle.

Stefan, J. (1879). Über die Diffusion der Flüssigkeiten, 2 Abhandlung (2: Berechnung der Graham'schen Versuche, *Wien.Anzeiger*, volume 16, number 3 (1879), pp. 24-27. The entire article: Über die Diffusion der Flüssigkeiten, 2. Abhandlung, *Wien.Ber.* II. Volume 79, number 1, 3rd Session, 3 January 1879) pp. 161-214. Abbreviated translation, in fact taken from *Wien.Anzeiger*: On the Diffusion of Liquids, *Philosophical Magazine*, 5th series, 7th Volume, number 43 (April 1879) pp. 295-297.

Stefan, J. (1879). Über die Beziehung zwischen der Wärmestrahlung und der Temperatur, *Wien.Ber.* II. Volume 79, number 1 (8th Session, 20 March 1879): pp. 391-428. Abbreviated report: *Wien.Anzeiger*, volume 16, number 8 (1879), pp. 87-89. English translation of that paper: "On the relation between thermal radiation and temperature" was not published in *Philosophical Magazine* where they preferred more experimental research. Tyndall was certainly glad that Stefan used his data there, but he was no more in charge for *Phil.Mag.* The other journal *L'Institut, journal universel des sciences et des sociétés savantes en France et à l'étranger* ceased its publication three years earlier with its volume 44 in 1876. French critique was published by Jules Louis Gabriel Violle (1841 Langres in Haute-Marne-1923 Fixin in Bourgogne-Franche-Comté). Stefan. - Ueber die Beziehung zwischen der Wärmestrahlung und der Temperatur (Sur la relation entre le rayonnement calorifique et la température); Sitzungs berichte d. K. Akademie d. Wissenschaften in Wien, p. 84. *J. Phys. Theor. Appl.*, 1881, 10 (1), pp. 317-319. DOI: 10.1051/jphystap:0188100100031700. jpa-00237795

Stefan, J. (1879). Über die Abweichungen der Ampèreschen Theorie des Magnetismus von der Theorie der elektromagnetischen Kräfte, *Wien.Anzeiger*, volume 16, number 10 (1879), pp. 110-111. The entire article: Über die Abweichungen der Ampèreschen Theorie des Magnetismus von der Theorie der elektromagnetischen Kräfte, *Wien.Ber.* II. Volume 79, number 2 (10th Session on 17 April 1879) pp. 659-679. German language reprint: *Ann.Phys. (Wiedmann's Annalen der Physik und Chemie)* 12: pp. 620-638 (1 March 1881). Abbreviated translation, in fact taken from *Wien.Anzeiger*: On the deviations of Ampère's theory of Magnetism from the Theory of the Electromagnetic Forces, *Philosophical Magazine*, 5th series, 8th Volume, number 46 (July 1879) pp. 83-84.

Stefan, J. (1880). Über die Tragkraft der Magnete, *Wien.Ber.* 81, pp. 89–116. Abbreviated publication: *Wien.Anzeiger* (15 January 1880) pp. 14-15. English translation: On the carrying-power of magnets, *Philosophical Magazine*, 5th series, 9th Volume, number 55 (March 1880) pp. 232-233.

Stefan, J. (1881). Über die Verdampfung aus einem kreisförmig oder elliptisch begrenzten Becken, *Wien.Ber.* II, 83 (read in May 1881, printed in June 1881), pp. 943–954; German language reprint communicated (mitgetheilt) by Stefan: *Ann.Phys. (Wiedmann's Annalen der Physik und Chemie)* 17: pp. 550-560 (1 September 1882). Not printed in *Phil. Mag.* Modified reprint: Extension of Stefan's treatment of a tube was published in his: Versuche über die Verdampfung, *Wien.Ber.* II, 68 (1873), pp. 385–423. It was needed for other forms of geometry of evaporated media, in analogy with Maxwell's electrostatics noted at *Ann.Phys.* 17: pages 553-555 and connected to Stefan's Theorie des Psychrometers nach Maxwell und Stefan, *Zeitschrift der österr. Ges. für Meteorologie* Wien 1881 16: 177-182 whish was noted at *Ann.Phys.* 17: page 558.

Stefan, J. (1883). Über die Berechnung der Inductionscoeffizienten von Drahtrollen, *Wien.Ber.* 88, II, pp. 1201–1211; German language summary: Über die Berechnung der Inductionscoeffizienten von Drahtrollen, *Naturforscher (Der Naturforscher Wochenblatt zur Verbreitung der Fortschritte in den Naturwissenschaften*, Berlin, published by Ferdinand Dümmler in 1868-1885, edited in 1868-1885 by Wilhelm Sklarek (1836-1915)) 1879 (in fact published in 1884), volume 17, pp. 231-232. Not printed in *Phil. Mag.*

Stefan, J. (1888). Über die Herstellung intensiver magnetischer Felder, *Wien.Ber.* II a, 97, pp. 176–183 (9 February 1888). English translation from *Wien.Anzeiger*: On the production of intense magnetic fields, *Philosophical Magazine*, 5th series, 25th Volume, number 155 (April 1888) pp. 322-323. Stefan updated the research of Carl Fromme (1852 Kassel-1945 Gießen). In 1880, Carl Fromme received the extraordinary chair for theoretical physics and geodesy at the suggestion of the physicist Wilhelm Conrad Röntgen, who was then working at the Justus Liebig University in Giessen, and Fromme was also appointed director of the institute there. For his own experimentation, Stefan used the big inductor of Heinrich Daniel Ruhmkorff (1803-1877). In 1880, a Viennese university private docent for physics, Ivan Pavlovich Pului (Johann Puluj, Іван Павлович Пулюй, Iwan Pawlowytsch, 1845 Hrymailiv in Galicia, now in Ukraine-1918 Prague), borrowed J. Stefan's Viennese inductor of Ruhmkorff for Puluj's experiments. Right after Stefan's paper, they printed the paper of his student A. von Obermayer about the flames of St. Elmo on pp. 323-324.[24]

Stefan, J. (1889). Über die Verdampfung und die Auflösung als Vorgänge der Diffusion, *Wien.Ber.* II a, 98 (21 November 1889), pp. 1418–1442; German language reprint: *Ann.Phys. (WA)* 41/12 (1890), pp. 725-747. English translation from *Wien.Anzeiger*: On evaporation and solution as processes of diffusion, *Philosophical Magazine*, 5th series, 29th volume, number 176 (January 1890) pp. 139-140. Stefan updated his 16 years earlier experiments with evaporation in ether and Carbon disulfide (CS_2) published at *Phil.Mag.* as Experiments on evaporation. The toxicity of carbon disulfide was not so obvious in those days, so experiments with it certainly contributed to Stefan's Bright's disease (a form of nephritis named after Richard Bright (1789 Bristol-1858 London)) and finally to stroke three and a half years after the start of Stefan's carbon disulfide and ether experiments. Like H. Davy before him, Stefan was far too unaware of the dangerous fumes in his laboratory next to his apartment where he had stayed most of his time before his wedding. A year and a half before his dangerous experiments Stefan's long-time roomie the sixty-six years old Josef Loschmidt moved away after his marriage on 10 December 1887 which might have made Stefan less careful. Stefan carelessly observed the evaporation of the ether and the carbon disulfide from narrow glass tubes, 2 to 6 mm in diameter. He filled them to their brims with the (dangerous) liquid and placed in his airy room. As the liquid evaporates, so does its level lower in the tube. The speed of this sinking is a measure of the speed of evaporation, which was already Stefan Problem of the movable border of phase transition. Two simple laws could be derived directly from the observations. According to the first, the rate of evaporation is inversely proportional to the distance between the surface of the liquid and the open end of the tube (brim); according to the second, it is independent of the cross-section of the tube. Therefore, already in 1873, Stefan concluded that the speed of evaporation is inversely proportional to the distance between the level o fluid and the open end of tube above it and does not depend on the cross-section of the tube. A decade and half later in 1889, Stefan wished to apply the new (Graham's and Maxwell's 1868) theory

---

[24] J. Stefan, Über die Herstellung intensiver magnetischer Felder, *Wien. Ber.* II a, **1888**, *97*, pp. 176–183 (9 February 1888). Translation: On the production of intense magnetic fields, *Philosophical Magazine*, Series 5, April **1888**, *25*th volume, no. 155, pp. 322–323; C. Fromme, Zur Frage der anomalen Magnetisirung, Ann.Phys., **1887**, *269/1*, pp. 236-237; Obermayer, Albert von. Fire of St. Elm. *Philosophical Magazine*, 5th series, April **1888**, Volume *25*, number 155, pp. 323-324.

of diffusion of gases to those two experimental facts. Stefan continued Adolph August Winkelmann's (1848 Dorsten-1910 Jena) experiments from the noon of 23 June 1889 to the noon of 9 July 1889 by observing the evaporations through the telescope with the aid of a slightly magnifying telescope, since the boundary between the still undissolved rock salt and the liquid above it is presented as a sharp line parallel to the lines of the division. In 1875, Winkelmann from the Technische Hochschule at Aachen proved the validity of Maxwell's prediction that the thermal conductivity of air is independent of the pressure down to a pressure of 1 mm Hg. Winkelmann measured with brass diathermometer of Stefan, in which the distance between the walls is reduced from 1.5 to 2 mm. He recommended observing in a completely glassy container and proved that the vacuum does not translate heat. Maxwell described the gas as the system of (Bošković's) points mutually acting on one another by the inverse ratio of fifth power of distance. Stefan never fully accepted Maxwell's "ad hoc" model of molecular force proportional to inverse-fifth power of the distance, even if Maxwell's model provided the acceptable diffusion coefficient proportional to the temperature (Mitrović, 2012, 12-15; A. Winkelmann, 1875. Ueber die Wärmeleitung der Gase. *Ann.Phys.* 156: 497-531). However, Maxwell's model did not allow the determination of the internal properties of gas. That is why Stefan published his dynamic model of gas in which the radius of the molecule decreases with some power of the absolute temperature between 1 and ½. For some gases it is proportional to the absolute temperature; but if one calculates the interaction of the gas molecules on the assumption that they behave like elastic spheres, which calculation, however, cannot be carried out exactly, one finds the proportionality to the square root of absolute temperature. According to the first hypothesis, it would be proportional to the first power and according to the second hypothesis to the square root of absolute temperature. Both theories behave in the same way in relation to the dependence of the coefficients of internal friction and heat conduction on temperature. As for the latter of the two, experiments have also shown for the coefficients of diffusion that their dependence on temperature is not the same for all gases, and that the exponent which determines this dependence has values between 1/2 and 1 (J. Stefan, Sitzungsber. LXV, 2. Abh., 323—363, 1872; J. Stefan 1889, 1428). Stefan also applied Maxwell's idea to hydrodynamics of diffusion of fluids combined with Dalton's principle. In Stefan's model, the radius of faster molecules diminished due to increased possibilities of interpenetration during collisions. The repulsive force is then proportional to the velocity, just as French Claude-Louis-Marie-Henri Navier (1785 Dijon-1836 Paris) argued in his lectures on mechanics at Parisian École Royale des Ponts et Chaussées already in 1826. The theoretical assumptions of those times were often seen on rather shaky legs because of the supposed suspicious background of the German Philosophy of Nature; therefore, that kinds of theories were not easily published in Poggendorff's *Ann.Phys.* or in *Phil.Mag.* of Koller's and Stefan's era. (J.C. Maxwell, On the dynamical theory of gases. *Phil.Mag.*, 4. Series, **1868**, *35*, 129-145, 185-217. Reprinted from the *Phil.Trans.*, **1868**, *157*, 49-88. Stefan quoted *Phil.Mag.*, p. 199 on his page 1424; J. Stefan, *Wien.Ber.* LXIII, 2. Abhan., 63—124, 1871). Stefan also relied again on Frick's theory of diffusion (J. Stefan 1889, 1436). That was one of main papers about Stefan Problem and Stefan Number, therefore available to English language readers just few weeks after its Viennese publication.

Stefan, J. (1890). Über die elektrische Schwingungen in geraden Leitern, *Wien.Ber.* 99, II a (9 January 1890 & 16 January 1890), pp. 319–339; German language reprint: *Ann.Phys. (WA)* 41/11 (3rd series volume 177, 1890), pp. 400-420. English translation: On electrical vibrations in straight conductors, *Philosophical Magazine*, 5th series, 29th Volume, number 179 (April 1890) pp. 373-374 & number 180 (May 1890) pp. 450-452. Stefan used the measurements of Kirchhoff's Jewish doctoral student Gabriel Lippmann (1845 Luxemburg–1921) (G. Lippmann, Sur une loi générale de l'induction, dans les circuits dénués de résistance, *Comptes rendus*, **1889**, *109*, 251–255). Among the first, Stefan studies new Hertz's discoveries. Stefan supported new oscillators of H. Hertz who used to be a student of Stefan's main supporter Helmholtz (H. Hertz, Ueber die Fortleitung electrischer Wellen durch Dräthe, *Ann.Phys.*, 273/7 (1889): 395-408). Stefan followed Maxwell and lord Rayleigh but did not derive his equitation of skin effect from Maxwell's theory of field, but from Neumann and W. Weber's electrodynamic potential of two elements of electrical current, as Stefan already published two years earlier even if Stefan had some doubts in Weber's electrodynamics already in early 1860-s according to Koller's notes (J. Stefan, *Über veränderliche elektrische Ströme in dicken Leitungsdrähten*, Wien.Ber., **1887**, *95*, 917-934). Stefan also used the measurements of conductibility of sulfuric acid compared to mercury as published by Rudolf Kohlraus h (* 1809 Göttingen; † 1858 Erlangen). The shielding effect of the electrolyte weas already researched by J.J. Thomson who surprisingly replaced Maxwell and Rayleigh in Cavendish on 22 December 1884 (J.J. Thomson, The resistance of electrolytes to the passage of very rapidly alternating currents, with some investigations on the times of vibration of electrical systems (Elektrolitski upor prehajanju visoko frekvenčnih izmeničnih tokov, z nekaterimi preiskavami o periodah vibracij električnih sistemov), *Proceedings of the Royal Society*, vol. 45, 1889, pages 269-290). Therefore, Stefan used brand new research published just few months before his own paper.

Stefan, J. (1890). Über die Theorie der oscilatorischen Entladung, *Wien.Ber.* 99 (2 June 1890), II a, pp. 534–548; German language reprint: *Ann.Phys. (WA)* 41/11 (3rd series volume 177, 1890), pp. 421-434. English translation: On the theory of oscillation discharge. *Philosophical Magazine*, 5th series, 30th Volume, number 184 (September 1890) pp. 373-374 & number 180 (May 1890) 282-283. In his continuation of his studied of Hertzian waves, Stefan relied on Kirchhoff's and Stefan's new friend William Thomson's theory of oscillatory discharge of Leyden jar which differed from the ideas of pendulum. Stefan remembered an independent Genevan scientist Édouard Sarasin's (1843-1917) and Auguste's son Lucien de la Rive's (* 1834 Choulex; † 1924 Geneve) measurements (Sur les oscillations électriques rapides de M. Hertz (communiqué à la Société de physique et d'histoire naturelle de Genève dans sa séance du 5 septembre 1889), *Archives des sciences physiques et naturelles* (edited by Sarasin), 3rd series, **1889**, *22*, pp. 283-288). Félix Savary (1797 Paris-1841 Estagel) was the first to notice the oscillatory discharge of a Leyden jar connected to an inductor in 1827. In 1847, Hermann Helmholtz derived the oscillation of electric discharge currents from his own law of conservation of energy to provide the origin of the modern arc plasma science. The easiest way to describe what happens during an oscillatory discharge is to imagine an LC circuit. The high frequency oscillations present in ESD waveforms can have unidirectional or oscillatory discharge waveforms.

Stefan, J. (1891). Über Wheatstone's Bestimmung der Geschwindigkeit der Elektrizität, *Wien.Anzeiger (Akademischer)*, 28/10: pp. 106-107 (10th session on 23 April 1891). Announcement which just noted a title: The chairman, Prof. J. Stefan, presented a communication: About Wheatstone's determination of the speed of electricity (Über Wheatstone's Bestimmung der Geschwindigkeit der Elektrizität). *Wien.Ber.* 2a 100/1 (23 April 1891): 468-468. English translation: On Wheatstone's determination of velocity of electricity. *Philosophical Magazine*, 5th series, 31st Volume, number 193 (June 1891) pp. 519-520 & 32nd Volume, number 198 (November 1891) pp. 480-480. Stefan's last paper published in *Phil.Mag.*, based on Kirchhoff's (1857), Wheatstone's and Herz' measurements.

Discussion

In July-December 1842, the editors of Koller's astronomical republication in *Phil.Mag.* were Brewster, Robert Kane, Richard Taylor (1781–1858) and the quaker chemist Faraday's friend Richard Phillips (1778 London-1851 Camberwell) who was also a joint editor of *Annals of Philosophy*, the founding member of Chemical Society in 1841 and a prominent member of British Association. With his albeit short publications in *Phil.Mag.*, Koller paved the way for the publication of his protegee J. Stefan there. In Stefan's times, the main editors (conductors) of *Phil.Mag.* were Robert Kane and William Francis (1817 London-1904 Richmond). The Irish chemist Robert John Kane (1809 Dublin–1890) was still the editor in 1887 and he certainly did not publish many works of females as he officially opposed their admissions to the universities in his final decade. William Francis edited *Phil.Mag.* from 1851 until his death. He also published his translations with *Taylor's Scientific Memoirs magazine, Selected from the Transactions of Foreign Academies of science and Learned Societies and from Foreign Journals*. That magazine offered a series of books edited and published by Richard Taylor in London between 1837 and 1852. Francis translated for Taylor's magazine H. Helmholtz's work *Über die Erhaltung der Kraft* and Ohm's masterpiece *Die Galvanishe Kette*. Francis's frequent travels and residences abroad enabled his fluent uses of French and German languages.

Richard Taylor was co-editor of *Phlil.Mag.* with Tyndall, Brewster, Kane, and Francis while they produced the volumes 12-16 of *Phil.Mag.* from July-December 1856 up until Taylor's death in December 1858. Taylor was replaced by David Brewster and Tyndall with no new co-editor included.

Tyndall, Brewster, Kane, and Francis were co-editors of volumes 17-24 from January 1859 until December 1862. In that time Tyndall widely published in *Phil.Mag.* as he used to be among the rare Brits who recognized many priorities of the continental Europeans including those of Robert Mayer and Rudolf Clausius. For *Phil.Mag.* Tyndall translated the works of Rudolf Clausius about the mechanical theory of heat, Hermann Helmholtz's correlation of forces, and Angström's work of 1853/54, in English translation of 1855 published as

Optical Researches in *Phil. Mag.*[25] Tyndall also translated publications of J. Plücker and many others.

After Tyndall left the editorship of *Phil. Mag.*, Brewster still coedited volumes 25-34 in collaboration with Kane and Francis from January 1859 until Brewster's death in February 1868.

During next three years Kane and Francis edited *Phil.Mag.* alone. In July-December 1871 William Thomson began co-editing 42nd volume of 4th series of *Phil.Mag.* together with the old tandem Kane and Francis. Therefore, in 1871-1890, the editors (conductors) of *Phil.Mag.* were Robert Kane, William Francis, and William Thomson, after 1892 Lord Kelvin. Stefan personally met William Thomson as the technical director of the Viennese international electrical exposition in 1883 where Brits attracted visitors with W. Thomson's personal measuring devices, original Wheatstone's bridge, and a piece of submarine cable. In mid-October 1883, with the help of the Polish-Lithuanian electro-engineer Bruno Abdank-Abakanowicz (* 1852 Ukmergė in east Lithuania; † 1900 Parc Saint-Maur in Île-de-France), W. Thomson personally measured currents of new Lalande's voltaic element at Stefan's Viennese exposition as Thomson reported to the Glasgow philosophical society proceedings in volume 15: p. 378 on 9 January 1884. Bruno Abdank-Abakanowicz was the best friend of Nobel prize winner Henryk Sienkiewicz. The Frenchmen Félix Paul Ernest de Lalande (1845 Albi in Tarn of Southern France-1919 Paris) from Parisian Seine department and Georges Chaperon as a director of Mines d'Alosno at Puebla Guzman of Province de Huelva in Spain designed their first alkaline storage battery. They simultaneously employed oxide of copper as the depolarizing-body and of caustic potash or caustic soda in solution as the exciting-liquid. They obtained French patent on 25 June 1881 and USA patent on 20 March 1883. Edison soon developed it further as his type S battery set, needed for the powering of fans, phonographs, and telephones. As W. Thomson used to say: No one else is Edison. Stefan's newly befriended Thomson left Vienna with Sir William Siemens to visit Helmholtz in Berlin, but W. Siemens suddenly died few days after on 19 November 1883.

---

[25] A. J. Angström, Optiska undersökningar, *Svenska vetenskapsakademiens Handlingar*, received 16 February 1853, **1852**, *40*, str. 333-360. Translation: Optische Untersuchungen, *Ann.Phys.*, **1855**, *170/1 (94)*, str. 141-165. Translation from German language: Optical Researches. *Phil. Mag.*, Series 4, **1855**, *9*, pp. 327-342.

British editors of *Phil.Mag.* constantly tried to keep a balance between the Irish, the Scots, and the English editorships. After the Irishman Kane illness and death on 16 February 1890, in January 1890 Thomson and Francis's co-editor of 29th volume of 5th series became another Irishman George Francis FitzGerald (* 1851 Dublin; † 22 February 1901 Dublin) who finally introduced the term electron in 1897. That trio was still in charge through 1890s, until Irish physicist and professor of geology at the University of Dublin famous for his development of radiotherapy in the treatment of cancer John Joly (1857-1933) replaced his fellow Irishman the deceased Fitzgerald as the co-editor of second volume of sixth series in June 1901. That means that Thomson used to be the editor of last two decades of Stefan's publications in *Phil.Mag.*, therefore through all fifty volumes of 5th series and more. He might have been the most important Stefan's English language editor, especially because after they met each other in Vienna in 1883.

Koller, Andreas Ettingshausen and Andreas Baumgartner of older Viennese generations never published in *Phil.Mag.* except rare Koller's republications, as in their era the motto Publish or Perish was still not evident (Južnič, 2020, 700, 1124). Certainly, like Stefan and Boltzmann, Andreas Ettingshausen communicated his articles published in *Wien.Ber.* for reprint in *Ann.Phys.*, for example in 1882 at 17: pp. 272f. That nice habit of republications of relevant works ceased in late 20th century except for the reviews of books, just like the bombastic dedications to powerful wealthy Maecenas ceased even somewhat earlier.

Stefan(-Boltzmann) law among Stefan's contemporaries

To understand the development of treatment of Stefan Problem in could be interesting to learn more about the simultaneous acceptance of Stefan's law. On 1 July 1876, Bartoli of the University of Bologna announced that thermal radiation in a vacuum of radiometer opposed the second law of thermodynamics. With his supposed contrary evidence Bartoli announced a potentially fatal blow to all Boltzmann's networks. His ideas emerged from his research of vacuum in a radiometer, which Crookes presented before the Royal Society in London three years earlier. Unlike Crookes, Bartoli was immediately

251

confident and convinced that the radiometer did not measure the absolute vacuum[26] because Bartoli relied on myriads of his Florentine academic ancestors' research of vacuum including Torricelli.

The native Florentine Bartoli studied physics and mathematics in Pisa until 1874. In 1876 he became an assistant at the University of Bologna and a physicist in Arezzo. In 1878 he was in Sasari, in 1879 at the Technical Institute in Florence, in 1886 at the University of Catania, and at the University of Pavia in 1893 when Pavia of Bošković and Volta was no longer ruled by Boltzmann's Habsburgians. Bartoli has published seventy-nine papers in the leading Italian newspaper *Il Nuovo Cimento* in Pisa. He mostly researched electrical and thermal phenomena. Bartoli's theory of the pressure of light resembled Maxwellian ideas and enabled the first measurements. Lebedev published his experiments on the foundations of Bartoli's suppositions in Moscow in August 1901; the Russian Lebedev used G. Kahlbaum's Berlin vacuum pump and measured his pressures with McLeod's manometer.

Thus, the experimental data about the radiation through vacuum began to influence the research before the publication of Stefan's law. In 1883, Bartoli's idea was supported by Henry Turner Eddy (1844-1921), a professor of mathematics and astronomy at Cincinnati University who studied in Yale, Paris, and Berlin. Boltzmann criticized his peer Eddy. Like Stefan, Boltzmann read about Bartoli's research only in Eddy's work after Stefan gave Boltzmann the English vocabulary to learn enough for his readings of Maxwell's works. The associate professor at Leipzig E. Wiedemann later edited the central physical magazine entitled *Ann.Phys.* with his father G. H. Wiedeman in Leipzig and had a much better overview of Italian publications; so, he read Bartoli's work among the first Germans.

Boltzmann dealt with Bartoli's work in 1884, just before Boltzmann's theoretical derivation of Stefan's law.[27] Even Stefan probably did not know about Bartoli's reflections of radiations in vacuum of a radiometer from 1876 until Eddy supported Bartoli's ideas in his publications. Therefore, in scientific networks, the geographically nearby Pavia was much more distant from Vienna than the faraway USA, which became even more true in following century when the USA ownership of scientific and other media became so prominent that only

---

[26] Bartoli, 1879, 274.
[27] Höflechner, 1994, 1:80

the more democratic web provided some slight chances to others to challenge USA monopoly of media. That is why Bartoli's radiation studies did not decisively influence Stefan's path to the radiation law in 1879.

Naturally, like all other physicists of his era, Stefan was interested in vacuum experiments with a radiometer. Stefan had even close contacts with Italian researchers, especially with his former colleague at the Viennese Physical Institute Blaserna, who was the presidents of the Roman Academy dei Lincei since 1904.

From 11 August 1883 to 3 November 1883 the electrical exposition in Vienna could have influenced Boltzmann's interest in radiation theory. On 25 October 1883 Stefan as a scientific leader of the exposition met with the Honorary Chairman of the Technical-Scientific Commission William Thomson, the later Lord Kelvin. Both scientists, together with William Siemens, were solemnly accepted at the English embassy in Vienna. In Vienna William Siemens as a German-born British industrialist even lectured about the radiation inside vacuum.[28] Unfortunately, Siemens did not mention Stefan's law, although he knew our hero Stefan well. From his short resentment on radiations in vacuum we will never know all Siemens' whereabouts and ideas about Stefan's law, since Siemens suddenly died in London only fifteen days after the end of Viennese exposition.

Just four years after Edison's invention, Stefan presented the facts of Edison's light bulb to the Viennese mayor Eduard Uhl (1813-1892) with his municipal representatives included and even the Habsburg Crown Prince Rudolf (21 August 1858-30 January 1889) present just after the birth of his only legitimate daughter Elisabeth on 2 September 1883.[29] Rudolf sadly passed away five years later together with his other girlfriend, which was another fatal blow to the dying Habsburgian monarchy. Stefan explored the electrical incandescent light bulb as an example proving his own theory of radiation. At the same time, he inherited the tendency towards his useful application of knowhow from his teacher Ettingshausen and emphasized the advantages of the light bulb against other sources of light. In that capacity, the Scientist Stefan also developed a great talent for lobbying and even businesses, after he successfully installed fans of his kinetic atomism at all relevant Habsburgian academic posts except for E.

---

[28] Sitar, 1993, 87, 92-93
[29] Sitar, 1993, 93

Mach's Prague. As the members and co-workers of Stefan's Viennese physics institute, Friedrich Wächter and Puluj also presented their own designs of vacuum light bulbs at the Viennese exposition under Stefan's leadership. Ivan Pavlovich Pului (Johann Puluj, Іван Павлович Пулюй, Iwan Pawlowytsch, * 1845 Hrymailiv in Galiтia, now in Ukraine; † 1918 Prague) had already taken over his new chair in Prague, but he kept his vivid contact with the Viennese researchers of vacuum techniques. Stefan kept evaluating the capacity and usefulness of light bulbs during several years after the Viennese exposition. He expected that the very experiments with radiation in vacuum light bulbs will help to promote his theory of radiation. Indeed, the modern experiments fully justified his hopes.[30]

The Viennese exposition and subsequent important papers published about light bulbs have greatly contributed to the popularity of Stefan's radiation law. By overseeing the Viennese exposition in 1883, Stefan became authority for the manufacturers, producers, and users of the light bulbs; thereby Stefan paved the success of his own law. Nevertheless, during the first five years after its publication, Stefan's equation was supported mainly by physicists writing in German language.[31]

---

[30] Prasad, Mascarenhas, 1976.
[31] Lummer, 1900, 61-63; Brush, 1976, 511, 517.

Table 1: Important papers on radiation in the first five years following the publication of Stefan's Law.

Name	Nationality	Place of publication	Year of publication	Business Place	Endorsed Stefan
Leo Graetz (* 1856 Wro law; † 1941 Munich)	German Jew	Leipzig	1880	Assistant Professor in Munich	Yes, pp. 913-914, 929-930
Paul Quentin Desains (1817-1885), Curie	French	Paris	1880	Sorbonne, Paris	
Ludwig Valentin Lorenz (* 1829; † 1891)	Dane	Leipzig	1881	Professor in Copenhagen	Yes, p. 587
Charles Augustus Young	USA	New York	1881	Princeton	Did not mention Stefan
Jules Violle	French	Paris	1881		Against
Ernst Lecher	German Habsburgian	Vienna, Leipzig	1882	Professor in Innsbruck, Vienna	Yes, p. 498
Charles-Alcide Riviére (1856 Grenoble-1939 Paris)	French	Paris	1882, dissertation 1884	Professor at Lycée Saint-Louis	Did not mention Stefan
William Siemens	German, Naturalized Englishman	London	1883	Inventor in London	Did not mention Stefan
Abney, Festing	Englishman	London	1883		Did not mention Stefan
Christian Christiansen (1843-1917)	Dane	Leipzig	1883	Professor Polytechnic in Copenhagen	Yes, pp. 268, 280, 283
Bottomley	Englishman	London	1884	Glasgow	No
Peter Tait	Scotsman	London	1884	Professor	Advocates Dulong-Petit's law
Boltzmann	German Habsburgian	Leipzig	1884	Professor in Graz	Yes
Schneebeli	German Swiss	Leipzig	1884	Neuchâtel in east Switzerland after 1879 Zürich	Yes, citing Stefan on pp. 433-434

The table contains only works published in books, printed in the central German language scientific journal *Ann. Phys.*, or announced by main contributions of Habsburgians (*Wien. Ber.*), French (*Comptes Rendus*), or English (*Proc. Roy. Soc. London*, *Philosophical Transactions*, and *Philosophical Mag.*). Researchers have largely agreed that the shape of energy curves and energy distribution is not dependent on radiating substance and its temperature. The elderly Desains and barely twenty years old French experts, the brothers Paul-Jacques Curie and later Nobel Prize laureate Desains's preparator Pierre Curie, were using platinum and copper plates glowed up to their white radiations in the faculty of Science in Paris in 1880; they had identified the point of maximum energy. In 1882 in Innsbruck and Vienna, the son of chief editor of Viennese daily *Neue Freie Presse* where Karl Marx also reported form England, Ernst Lecher (* 1856 Vienna; † 1926 Vienna), noted that the shape of energy curves and the distribution of energy does not depend on the substance and temperature, which was in accordance with Kirchhoff's Law. In Stefan's Viennese Physics institute under supervision of Victor von Lang the former Stefan's student Ernst Lecher discussed the experiments with thermo-multiplier of Macedonio Melloni (1798-1854). Lecher also addressed the research of André-Prosper-Paul Crova (1833 Perpignan-1907 Montpellier) in Montpellier, Crookes' fellow spiritualist Leipzig professor of astrophysics Johann Karl Friedrich Zöllner (* 1834; † 1882), and Edison's bulb (Lecher 1882, 515). Lecher also mentioned Violle and the father of Edison's friend Henry Draper (1837-1882), John William Draper (* 1811 St. Helens Lanlashire, England; † 1882 Hastings-on-Hudson 32 km north of Manhattan). Lecher cited the professor at Sorbonne Paul Quentin Desains (* 1817; † 1885), Belquerel, and Tyndall's measurements. Lecher explained their results by Stefan's law, by the data of Wüllner, by Kirchhoff's law, as well as by the research of Fourier, Fresnel, and Cauchy (Violle, 1879, 171; Lecher, 1882, 498, 515).

Graetz in Bavaria and Strasbourg (Graetz, 1880, 913-914, 929-930) and Kelvin's nephew James Thomson Bottomley (1845 Fort Breda in Ireland-1926 Glasgow) in Glasgow carefully measured thermal radiation in the vacuum. By doing so, they got rid of the possible additional translation of heat, which probably bothered the experimental attempts of their Londoner colleague the German William Siemens a year earlier. Experiments in the vacuum were becoming necessary for accurate results as the researchers gradually understood the influences of environment. However, in 1893 Dewar defended the law of the third power of the temperature instead of Stefan's law which professed forth power.[32] Only in 1920 Dewar used Stefan's law, but it was to late for any Stefan's personal victorious triumph as by then he was eventually buried long ago.

The early French and English physicists did not mention Stefan's equation. Some of them denied his formula or even insisted on Dulong-Petit's obsolete law like professor at the Faculty of Sciences in Lyon Jules Louis Gabriel Violle (probably the secret alchemist Fulcanelli, 1841 Langres in Haute-Marne-1923 Fixin in Bourgogne-Franche-Comté) (Violle, 1881, 317-319). Like Violle, Charles Augustus Young (1834 Hanover in New Hampshire-1908 Hanover) was also the solar physicist and spectroscopist, but Young failed to mention otherwise specialized Stefan's work in his monograph *The Sun*, which was translated into German and French languages in 1883.

The English manufacturer of photographic and spectroscopic devices William Abney de Wiveleslie (* 1844; † 1920) and major-general turned chemist Edward Robert Festing had not yet known Stefan's law in 1883. Abney was a member of the London Royal Society, but he was not interested in achievements developed on the European continent. Festing was elected FRS on 4 June 1886. More friendly Stefan's fans were of course researchers who wrote in German language; among them was a teacher of physics in Neuchâtel in Switzerland Heinrich Schneebeli (1849 Ottenbach by Zürich-1890 Zürich), who learned about the vacuum experiments as Kundt's student in Zurich and discussed radiations of small bulb of Joseph Wilson Swan (* 1828 Pallion in Sunderland; † 1914 Warlingham) (Schneebeli, 1884, 433-434, 437).

Three quarters of year after Stefan's death, the Association of German naturalists and physicians had a meeting in Vienna on 24 September 1894. The professor at Erlangen E. Wiedemann lectured on radiation there during his invited talk. Towards the end of 19th century, the centre of research of radiation and its measurement with vacuum techniques passed from Habsburgian countries to Berlin, where it was crowned by Planck's theory of quanta and with Einstein's achievements in Einstein's wonderful year 1905; the anniversary of these events is repeatedly celebrated worldwide as those theories still form the basis of modern physics.

---

[32] Lummer, 1900, 63, 65; Dewar, 1927, 353-355

Stefan Problem and Stefan Number

Besides Stefan Law, Stefan Problem is the most popular field connected with his name. We follow Tarzia's approach of Stefan Problem's bibliography (Tarzia, 2000). Tarzia's data intended to be useful for the active and incoming researchers of Stefan Problem, therefore his bibliography was arranged by the topics and by the sorts of publications in proceedings, journals, and books. However, our principal aim is historical, so we chronologically arranged the updated Tarzia's bibliography. Our goal is to find how the similar but gradually sophisticated mathematical treatment switched from early case studies to the other including aviation and rocketry which were beyond the imaginations of early 18th and 19th century pioneers of studies of the puzzle later called Stefan Problem. Our second goal is to follow the changing of geographical distributions of the academic groups studying Stefan Problem as well as their mutual interactions connected with the modernizing of (Western) political structures from the Napoleonic wars up to the fall of Berliner Wall Iron Curtain. Except for the Japanese, the early studies of Stefan Problem outside our Western scope are hardly included as they were not published in modern scholars' ways, although it is oblivious that all civilizations and cultures always had some interesting and useful ideas about puzzles now called Stefan Problem. Our third goal is to explain how the sophistication of involved mathematics including teamwork, computerizations, and Web changed the approaches to Stefan Problem. Our fourth and most complicated goal is focused on predicting how all those three aspects of research of Stefan Problem might develop in 21st century.

Early Beginning of Stefan Problem Before Stefan

Stefan Problem began as a puzzle of cooling of planets until Stefan formulated it for freezing of water.

The earliest modern mathematical formulation of Stefan Problem belonged to Gabriel Lamé (* 1795 Tours; † 1870 Paris) and his friend Benoît Paul Émile Clapeyron (1799 Paris-1864 Paris). Both studied at École Polytechnique and at the École des Mines in Paris in the era when both Napoleonic networks, of

Laplacians and fans of Fourier, already challenged each other. After graduating, Lamé and Clapeyron went to Petersburg in 1820 to serve the generous Russians for the next twelve years. The victorious Russians invaded Paris of defeated Napoleon and somewhat attracted several Frenchmen.

In cold Petersburg climate both friends attacked the problems of freezing. They were also ignited by the increasing discrepancy of demands of their era, propelled by the estimated geological ages of Earth and Sun versus their supposedly much quicker physical cooling.

Those problems bothered even Kelvin before Ernst Rutherford's failed attempt to introduce the supposed solar radioactivity into this field, focused on solar radiations. Lamé and Clapeyron supported Cauchy's novelties in Russia but returned to Paris just as Cauchy left with the exiled Bourbon court.

A few weeks before the July revolution of 1830, with the genial rebellious mathematician Évariste Galois participating in it, Lamé and Clapeyron's paper was read at the Parisian academy under the title "Memoir on the Cooling Solidification of a Liquid Globe." Despite their title, a spherical geometry was not of their primary interest.

Their problem, factually addressed in their final two-page note, was that of the formation of a flat solid crust by cooling the upper surface to a constant temperature, while the undefined liquid below remained at its constant and uniform solidifying temperature. The temperature of their liquid was initially everywhere equal to the crystallization temperature. Lamé and Clapeyron found that the thickness of the crust is proportional to the square root of the time but did not determine the coefficient of proportionality (Jaime Wisniak, 2006, 193). Four years later in 1834, Clapeyron popularized Carnot's work soon called by the Clausius' term entropy, which belonged to the same broader branch of research of heat which included Stefan Problem. Besides mechanics and optics, heat became new fashionable topics of research in physics, while the electromagnetism was also knocking on the door of new industrializations.

Eight years after Lamé and Clapeyron's memoir, on 27 August 1838, Stefan's future mentors Marian Koller, A. von Ettingshausen and August Kunzek von Li⬚hton (* 1795; † 1865) attended a meeting of that same Parisian Academy of Sciences, where they met scholars such as François Arago, Bouvard, Félix Savary, the baron Siméon-Denis Poisson, the mathematician Jacques Charles

François Sturm (1803 Geneve-1855 Paris), the chemist Michel Eugène Chevreul (1786–1889), the zoologist André Marie Constant Dumeril (Duméril, 1774–1860), the geologist Alexandre Brongniart (1770–1847), the volcanologist-botanist Jean-Baptiste Geneviève Marcellin Bory de Saint-Vincent (1778 Agen-1846 Paris), Jacques Babinet (1794-1872), the pioneer of research of phase transitions baron Charles Cagniard de la Tour (1777–1859), and Alexander von Humboldt.

Clapeyron and Lamé were not members of the Parisian academy yet, although Lamé and Clapeyron were already corresponding members of the Russian academy and Lamé was a newly appointed member of Prussian academy. Lamé became a member of Parisian academy in 1843 while Clapeyron was elected there only in 1858. Therefore, Koller did not feel obliged to note his eventual meeting with those two in 1838. However, the seven years old idea of Stefan Number was there circulating through Paris of Koller's touristic days and the mathematically minded Koller could not fail to recognize its potential, and to suggest it later to Koller's best discovery named Stefan.

Half a century after Koller's journey, Koller's protégé Josef Stefan upgraded Lamé-Clapeyron's ideas as well as Charles Cagniard de la Tour's research of phase transitions from solid to liquid in Stefan's seven papers published between 1889 and 1891.

Stefan widely lectured on Fourier's theory of thermal conduction with Koller in his Viennese audience in the winter semester of 1862/63. Boltzmann attended similar courses delivered by Stefan in the summer semester of 1863/64 and in the winter semester of 1865/66. All those courses offered 12 or 14 two-hour lectures which is still common today except that the most modern courses use both semesters allover the schoolyear.

Stefan told his audience all about the cooling of finite or infinite rods but did not particularly discuss spheres or movable borders between different states of aggregation, as that might be too complicated for Stefan's undergraduate students of those times.

Stefan's theory of thermal conduction addressed the broader context of his and Boltzmann's interest in transport phenomena, particularly the phase transitions from liquid to gaseous state and parallel chemical reactions which Stefan researched in 1873-1889.

Stefan's work was a transition to modern research from the first experimental and analytical attempts to describe the change in the solid-liquid phase, developed by the pioneers like Agricola, Kepler, and the 18th century Joseph Black's (* 1728; † 1799) exploring of the latent heat required for the phase transition.

Black's work became known after the posthumous publication of his lectures by Black's successor at his chair in Glasgow, the anti-Jacobin R. Bošković's fan John Robison entitled *Lectures on the Elements of Chemistry, Delivered in the University of Edinburgh by the Late Joseph Black* in 1803. Black's followers included Koller's acquaintance Charles Cagniard de la Tour, J. Fourier, Thomas Andrews, as well as Lamé and Clapeyron in 1831. Three decades later their ideas were used in unpublished Kaliningrad (Königsberg) lectures of Franz Ernst Neumann (* 1798; † 1895) with Neuman's comments of Riemann's data in 1860.

In the years up to 1901, Heinrich Martin Weber (1842 Heidelberg-1913 Strasbourg) published Riemann's legacy and Stefan used Heinrich Martin Weber's comments in 1890 as a leading Habsburgian researcher of specific heats of compounds.

Among the best arctic measurements at Stefan's disposal were those of his fellow native of then Slovenian areas Wilhelm von Tegetthoff (1827 Maribor-1871), a Habsburgian commandant of the fleet at the North Sea during the Second Schleswig War. The Habsburgian North Pole expedition was an Arctic voyage to find the North-East Passage that ran from 13 July 1872 to 1874 on the main ship Tegetthoff, named for the Admiral Wilhelm von Tegetthoff, who died 15 months before that Siberian expedition. It is hardly possible that Stefan and Tegetthoff met each other on some Viennese occasion as they belonged to the different social classes and occupations despite of their common Slovenian-German linguistics. Their only mutual interest happened to be Stefan Problem.

Stefan's work on thermal conduction was therefore not anymore propelled by the problems of cooling of Earth (or Sun), but by the difficulties with oceanic ice-faction which was causing problems to the Habsburgian imperialistic search for the northern passes to America and China at their second *Deutsche Nordpolarfahrt* in 1869-1870, according to the British meteorologist Richard Strachan who also reported to Londoner Geographical society on journeys by White Nile to Khartoum in 1874, after Konblechar's expeditions. Stefan's Artic ice model described the seawater initially at the freezing temperature and the air in contact with the water to be at a constant temperature below the freezing point. Thus, the circumstances triggered formation of ice at the air-water interface (Font, 2014, 5; Strachan, 1873; Strachan, 1879-1888; Vuik, 1993, 93-107). The resulting growing ice layer was found to be proportional to the square root of time.

Stefan suddenly died before his students began to research that new economically useful problem en masse, also because Stefan's best student Boltzmann preferred more abstract problems of thermodynamics. Soon followed the fatal disease of the whole Habsburgian monarchy. Therefore, at least for a while, Stefan Problem ceased to be a relevant prestigious mainstream area of research and the northeastern passage never became very usable except later for the modern transport of Russian gas and oil into EU. That was the reason behind the next four decades of neglect of Stefan Problem which became popular again for the modern researchers serving Soviet oil industry and metallurgy. Their Western and Japanese contemporaries were not lagging much behind even if Stefan Problem of arctic ice and permafrost was more Russian than anybody else's.

In honor of Stefan's work, which includes the boundary between solid and liquid that moves freely over time at polar ice, the concepts of Stefan Problem and dimensionless Stefan Number are often used in multiphase systems research today.

If the Stefan Number is large, the phase transition does not have a significant effect; if Stefan number is small, the phase transition has a significant effect on the movable border between the multiphase system (Šarler, 1995, 83; Stefan, 1889, 965; Šarler, 2011, 10, 21, 87). Stefan in Carinthia and Koller in his native Alpine Bohinj, of course, learned all about the ice caps already during their local juvenile wanderings.

In 1889-1891 Stefan introduced his ideas, which are now referred to as the Stefan Problem and the Stefan Number. He announced his ideas in seven of his publications printed by the Viennese academy, some of them reprinted in *Ann.Phys.* and the first of them translated under the title Evaporation and solution as processes of diffusion in *Philosophical Magazine* in January 1890. Unlike Stefan law (Stefan–Boltzmann law), at last three of Stefan's papers about Stefan Problem were quickly translated in *Phil.Mag.* and in Paris for immediate uses of the Anglo-Saxon and French researchers in Europe and Americas beginning with Experiments on evaporation (*Philosophical Magazine* December 1873, Résultats des expériences sur l'évaporation, *Journal d'Institut* 1874), On the Diffusion of Liquids (*Philosophical Magazine* January and April 1879), and Evaporation and solution in January 1890. The simultaneous French translations might have been the reason why the naturalized Parisian L.M. Brillouin first used the keyword Stefan Problem 55 years after its first translation in Parisian *Journal d'Institut* which Brillouin certainly read as then nearly twenty years old freshmen of École normale supérieure in 1874, after his family returned to Paris following Bismarck's and Communard's retreats.

In 1831, Lamé and Clapeyron's publication proved to be the first European attempt to reach the general solution of the puzzle later called Stefan Problem (Rubinstein, 1971, 4-5). It took almost a century after Lamé before Stefan Problem became a part of Brillouin's mainstream again.

Early 20th Century Stefan Problem Before the End of WW1

The renaissance of attention of Stefan Problem in 20th century began with the development of parabolic equations needed for is solving. The early 20th century breakthrough belonged to the unhappy Isonzo (Soča) front WW1 Italian soldier Eugenio Elia Levi whose elder brother was also a mathematician. There were also the Swedish academicians whose iron used to be the best for centuries also in competition with Koller's boss Zois. Traditionally successful French academic descendants of Lamé included Maurice-Joseph Gevrey (1884 Fauverney, Côte d'or, France-1957 Fauverney) who obtained his PhD at Faculté des sciences de Paris in 1913 needed for his teaching at the University of Burgundy after 1919. The works of those years included:

Levi, Eugenio Elia (1883 Torino-28 October 1917 Cormons 12 km west of Gorizia in WW1 action), "Sull'equazione del calore", Atti Accad.Naz.Lincei Rend., 5, N.16 (1907), pp.450-456.

LEVI Eugenio Elia, "Sull'equazione del calore", Ann.Mat.Pura Appl., (Milano: Tip. Rebeschini di Turati e C.) 3rd series, N.14 (1908), pp.187-264.

HOLMGREN Erik Albert (1872 Stockholm-1943 Uppsala, docent in Uppsala in 1898, professor there in 1909-1937), "Sur l'équation de la propagation de la chaleur. Deuxième note", Arkiv Matem.Astron.Fysik (Uppsala: Almqvist & Wiksell), 4, N.18 (1908), pp.1-28.

HOLMGREN Erik Albert, "Sur l'équation de la propagation de la chaleur", Arkiv Matem.Astron.Fysik, 4, N.14 (1908), pp.1-11.

BLOCK Henrik (Gabriel, PhD ar University of Lund in the southern Sweden in 1909), "Sur les équation s aux dérivées partielles du type parabolique", Arkiv Matem.Astron.Fysik., 6, N.31 (1911), pp.1-42.

GEVREY Maurice-Joseph, "Sur les équations aux dérivées partielles du type parabolique", J.Math.Pures Appl., 9 (1913), pp.305-471 (abridged PhD at Faculté des sciences de Paris in 1913, Paris: Gauthier-Villars).

GEVREY Maurice-Joseph, "Sur les équations aux dérivées partielles du type parabolique (suite)", J.Math.Pures Appl., 10 (1914), pp.105-148.

GEVREY Maurice-Joseph, "Sur la nature analytique des solutions des équations aux dérivées partielles", Ann.Sci.Ec.Norm.Sup., 35 (1918), pp.129-190.

20th Century Stefan Problem: from WW1 to the Great Depression

The practical but less mathematical Japanese metallurgists soon entered the research of Stefan Problem. The iron and steel technology of modern Japan has basically evolved from the transplanted western technology after the 1850s. Masatomo Sumitomo (住友 政 友, * 1585 Fakui north of Kyoto; † 1652) designed his shop of medicines and books in Kyoto in 1630. His adopted son Tomomochi Sumitomo (Riemon, * 1607; † 1662) studied mining with the Dutch merchants in Deshima to help with copper mining in the company of his father Riemon Soga (* 1572; † 1636). Riemon founded the Company Izumiya in 1590; he successfully married Masamoto's older sister and learned Western procedures for obtaining copper and silver from minerals. Under the influence of Sumitomo, every sixth of the 300,000 inhabitants of Osaka dealt with copper reselling in 1685. Nearly two-thirds of the annual production of 6,000 tons of Sumitomo's copper was exported to China in 1697 for the manufacturing of coins. Of course, the entrepreneur Sumitomo did not avoid even the direct exchange of goods, and he also acquired a lot of Chinese books or the books of Chinese Jesuits, although such reading was officially not allowed in Japan before 1720 as the prohibition of ordinary folks does not apply to the wealthy. In any case, Osaka became the center of modern Japanese metallurgy, electric and vacuum experiments. The inauguration in 1901 of the state-run Yawata Steel Works (officially named the Imperial Japanese Government Steel Works under the competence of the Ministry of Agriculture and Commerce) started new page of the history of iron and steel technology in Japan. In 1904, Kichizaemon VI Sumitomo Tomoito (住友 吉 左衛 門, * 1865; † 1926), the fifteenth heir of the Sumitomo family, founded the library in his hometown Osaka, and two decades later, he donated to it his own valuable collection including his hundred and fifty rare old books. The library is now named Nakanoshima in the Prefecture of Osaka. The Iron and Steel institute of Japan as the first engineering society specializing in iron and steel was established at the initiative of Noro Kageyoshi, Imaizumi Kaichiro, Tawara Kuniichi and others in February 1915. It was an academic body specializing in research on iron and steel. The Research Institute for Iron and Steel (the present-day Research Institute for Iron, Steel, and other metals) of the Tohoku Imperial University was founded in 1919. After those administrative successes, the Japanese iron and steel technology began to have bridgeheads at which it could contact the pertinent sciences. After the WW1, among the distinguished Sumitomo's employee was Seizo Saito (Seizō Saitō, 斎藤, 省三, * 1889) from the influential Saito family. In 1921 he was the first to research distribution of temperature in steel ingots for Tōhoku (Tohoku) Imperial University as quoted on 1 January 1930 by N. M. H. Lightfoot of Heriot-Watt College in Edinburgh, later resettled to the South England (Lightfoot 1930, 97). In 1933 Seizo Saito lived at Toyonaka-shi(Cho) residential area of Osaka Prefecture while he patented with Nobutaka Yamamoto from Osaka a Brake Apparatus for Wheels in Japan and USA in 1933-1934. Seizo Saito and Nobutaka Yamamoto jointly published their articles about the steel while they worked together at the Research Laboratory, Copper Works (Later at Steel works in 1943) of the Sumitomo Metal Industries during the WW2 in 1939-1945. The works of those years included:

Saito, Seizo (Seizō Saitō). 1921. On the distribution of temperature in steel ingots (Wear of Metals = 磨耗試驗 = Mamō shiken). *The Science reports of the Tōhoku Imperial University. First series, Mathematics, physics, chemistry.* Sendai, Japan: Maruzen Co. 10: 305-329, 496.

LIGHTFOOT, Nicholas Morpeth Hutchinson, The effect of latent heat on the solidification of steel ingots, Third Report of the committee on heterogeneity of Steel Ingots, Section V, J.Iron Steel Inst. (Journal Iron and Steel Institute, London: Iron and steel Institute), 109/1 (1929), pp.364-376.

Lightfoot, Nicholas Morpeth Hutchinson, The Solidification of Molten Steels, *Proceed, London Math. Soc.* (*Proceedings of the London Mathematical Society*), series 2, Vol. 31/1, 1930 (received 16 April 1929), pp. 97-116.

The practical metallurgy was not the only success of those times as the needed mathematical tools also advanced with the textbook of Édouard Goursat (1858 Lanzac in Lot-1936 Paris) who studied and taught at the Parisian École normale supérieure.

GOURSAT Édouard, "Cours d'analyse mathématique", Gauthiers-Villars, Paris (1927).

20th century Stefan Problem: from the Great Depression to WW2

The study of parabolic partial differential equations used to describe a wide variety of time-dependent phenomena continued even during the Great Depression. Its main media was Richard Edler von Mises' (1883 Lvov-1953 Boston) *Zeitschrift für Angewandte Mathematik und Mechanik*, began in Berlin in 1921 as a German-language journal published by Verein Deutscher Ingenieure (VDI) established already in Stefan's times in 1856. The Jew Mises obtained his PhD in Viennese university in Stefan-Boltzmann's applicative traditions in 1908. Slowly, the practical application of metallurgy of Stefan Problem embraced again Stefan's initial puzzle of freezing at the geological department of MIT. Stefan Problem also served the needs of emerging oil industry of the USA Gulf Research & Development Company after the fossil oil as transport fuel began to grow in Stefan's days. Morris Muskat (1906/1907 Riga in Latvia-1998 Pasadena, California) emigrated to the USA as a kid to work for the Gulf Research & Development Company in 1929-1950. He published about the Stefan Problem related to oil production following the massive overproduction of oil in the 1930s and the subsequent low prices that resulted in the virtual collapse of the U.S. oil industry. Stefan original Problem of freezing and ice formation was researched at the Freiburg university and the MIT which soon became the most influential.

The Moscow Steklov Institute of Mathematics (Математический институт имени В.А.Стеклова)) also joined the research of Stefan Problem headed by Ivan Petrowsky (Petrovskii, Petrovsky, Petrovskij, Иван Георгиевич Петро́вский, 1901 Sevsk in Russia-1973 Moscow). The Orthodox dissident Dmitri Fyodorovich Egorov (Его́ров, 1869-1931) supervised Petrowsky who finally became even the member of the Presidium of Supreme Soviet of USSR in 1966 under the chairman Nikolai Viktorovich Podgorny. Petrowsky taught at Steklov Institute of Mathematics and at Moscow Lomonosov university. In 1935 while he was awarded his PhD in Moscow, Petrowsky published his probability calculus focused on the first boundary value problem of the heat conduction equation at the newly established journal *Compositio Mathematica*, edited by Luitzen Egbertus Jan Brouwer (1881 Overschie, Netherlands-1966 Blaricum, Netherlands) and published by Noordhoff in Holland. That new journal was an outcome of frog and mice battle (Batrachomyomachia) between the Dutchman Brouwer and his much stronger antagonist fellow editor of *Mathematische Annalen* the Göttingen German David Hilbert described by Einstein as the prelude to the Nazis' promotion of German Arian science. As backed by the Berliner opponents of Hilbert, Brouwer objected to Ostjuden (German Jews of Eastern European descent) writing for the journal Mathematische Annalen, therefore Hilbert removed Brouwer from his position as editor. Next in 1937, Petrowsky extended his probabilistic ideas to biology in collaboration with the genial Kolmogorov who sadly testified against his supervisor Egorov. Petrowsky's PhD student was Olga Arsenievna Oleinik (Oleĭnik, О́льга Арсе́ньевна Оле́йник, 1925 Matusiv in central Ukraine–2001 Moscow) who in turn supervised the Russian-Jewess researcher of the tree-dimensional Stefan Problem Shoshana Kamin (Шошана Камин, Susanna L'vovna Kamenomostskaya, Сусанна Львовна Каменомостская, שושנה קמין, * 1930 Moscow) to complete Russian-Jewish contributions also by the female side. The works of those years included:

Schwarz, Carl (PhD engineer in Hamborn by Duisburg in North Rhine-Westphalia in 1933, in Aachen in 1937, in Salzburg in 1957, also Karl Ernst Schwarz), Die rechnerische Behandlung der Abkühlungs- und Erstarrungsvorgänge bei flüssigem Metall. I & II. *Archiv für das Eisenhüttenwesen* (Düsseldorf: Stahleisen), v5 n3 (September 1931): 139-148 & v5 n4 (193110 October): 177-191. Reprint: "Zur rechnerischen behandlung der erstarrungsvorgänge beim geiben metallen", Z.Angew.Math.Mech. (ZAMM), 13 (1933), pp. 202-223.

Muskat, Morris, "Two fluid systems in porous media. The encroachment of water into an oil sand", Physics, 5 (1934), pp.250-264.

Petrowsky, Ivan (Ива́н Гео́ргиевич Петро́вский), "Zur ersten Randwertaufgabe der Wärmeleitungsgleichung", *Compositio Mathematica*, 1 (1935), pp. 383-419.

Kolmogorov, Andrey Nikolaevich (Андре́й Никола́евич Колмогоров); Petrowsky, Ivan; Piskunov, N.S (Николай Семёнович Пискунов), Исследование уравнения диффузии, соединенной с возрастанием количества вещества, и его применение к одной биологической проблеме (Investigation of the diffusion equation combined with an increase in the amount of a substance and its application to a biological problem), *М., Бюлл. ун-та, А (Moscow Bulletin of the Institute of Academy)*, 1:7 (1937), 1-72.

Doetsch, Gustav (1892 Köln-1977, in Freiburg after 1931 where he initially supported Nazis), "Les équations aux dérivées partielles du type parabolique", *Enseig.Math.*, 35 (1936), pp.43-87.

Lachmann, Kurt (Berlin), "Zum Problem des erstarrens für den durch zwei parallele Ebenen begrenzten Korper", Z.Angew.Math.Mech., 15 (1935), pp.345-358 & 17 (1937), pp.379-380.

Huber, A., "Zum problem des erstarrens fur den durch zwei parallele ebenen begrensten korper", Z.Angew.Math.Mech., 17 (1937), pp.379-380.

Huber. A. (in Freiburg (Switzerland), later in Vienna after 1939), "Uber das Fortschreiten der Schmelzgrenze in einem linearen Leiter", Z.Angew.Math.Mech., 19 (1939), pp.1-21.

Pekeris, Chaim Leib (1908 Alytus in Lithuania-1993 Rehovot Israel. Worked at MIT Department of Geology in 1934-1941); Slichter, Louis Byrne (* 1896 Madison-1978 Los Angeles, professor of geophysics at MIT in 1932–1945), "Problem of ice formation", J.Appl.Phys., 10 (1939), pp.135-137.

Schwartz, H.A. (in 1925 at National Malleable Castings Co. Cleveland, Ohio), "Solidification of metals",
Trans.Amer.Foundrymen's Assoc.,
53 (1945), pp.1-35.

Brillouin and the Renaissance of Stefan Problem

The mainstream of researchers of Stefan Problem spread worldwide, but a century after Lamé's pioneering work Parisian were still at the forefront of then science and technology. In 1929 in Paris the mathematician Louis Marcel Brillouin (1854 Saint-Martin-lès-Melle-1948 Pariz) discussed Stefan's merits within Brillouin's notes about Clapeyron (Brillouin, 1931, 287, 280, 294, 296, 301). There, Brillouin also first used the term Stefan Problem (Brillouin, 1931, 300) which escaped the attention of later modern scholars. L.M. Brillouin's interests might have been purely mathematical even if his son used to be a leading quantum mechanician Léon Brillouin (1889 Sèvres by Paris-1969 New York). Therefore, the academician of Collège de France Louis Marcel Brillouin announced in 1929 but only published in 1930/31 the first serious focused work on Stefan Problem after Stefan's own contributions of 1889-1891. A long delay of the research of an interesting topic is certainly not unusual in the history of physics, resembling Einstein's prediction of the Stimulated Emission of laser in his paper The Quantum Theory of Radiation in March 1917. Only in the year of Einstein's death in 1955 Charles Hard Townes of Columbia University in New York named his reinvented Einstein's maser (microwave amplified stimulated emission of radiation) and three years later with Arthur Schawlow explained how to extend the idea to visible and infrared frequencies of the laser while the Soviet experts Nikolay Basov and Aleksandr Prokhorov did the same in those funny days of Cold War. The first working laser was built only in 1960, therefore the long path of Stefan Problem toward success in nothing unusual. The modern technological uses of scientific innovation might be accelerating by modern businesses' founding of scientists.

Vorkuta permafrost institute opened 1936 for the Soviet dissident as Stefan Problem followers: Leibenzon, Rubinstein and Redozubov in 1930s and 1940s

Our now popular usage of the Stefan Problem was introduced again only in the Soviet Union ravaged by the approaching WW2 and the early Cold War. The early contributor was Lev Isakovich Rubinstein (Rubinštein, Rubinštejn, Лев Исакович Рубинштейн, * 1914 Berdychiv (Бердичев, Бердичив, Berdičev, באָרדיטשעוו) now in Ukraine; † 2009 Jerusalem) after WW2. In 1947, on behalf of Soviet Turkmenia oil industry, Rubinstein solved Leonid Samuilovich Leibenzon's (Леонид Самуилович Лейбензон, Leĭbenzon, 1879 Kharkov-1951 Moscow) modification of Stefan Problem.

Leibenzon worked for the Soviet oil industry in Moscow in 1931 as the organizer of the first oil field laboratory in the USSR in Moscow in 1925 and professor at the Moscow Mining Academy in 1922-1930. Like Rubinstein few years later, Leibenzon was arrested and deported to Alma-Alta in Kazakhstan in 1936/37-1939, but Leibenzon was soon fully pardoned and elected academician in the middle of the WW2 in 1943.

On 4 April 1953 Rubinstein sent his research to Moscow from his Turkmenian branch of the All-Union (Soviet) Research institute in west Turkmenian center of oil industry in Balkanabat (Балканабат, Neftedag, Nebit-Dag). On 24 April 1953, his contribution was presented to the Moscow academy by one of the leading designers of Soviet nuclear bomb project the academician head of Moscow computer department Sergei Lvovich Sobolev (Сергéй Львóвич Сóболев, 1908 Petersburg-1989 Moscow).

American translation followed in December of the same year 1953 under the title *on the dynamics of evaporation of ideal multi-component liquid mixtures* in Oak Ridge. It was a modification of Stefan Problem to fit a composite boundary problem of the equation of heat conduction in a region of fixed boundaries. As excellently trained Soviet literati, Rubinstein mentioned all three Stefan's papers of 1889 (Rubinstein, 1953).

The scientific success did not save Rubinstein from the troubles of his turbulent times. The Russian-Ukrainian-Polish Soviet Jewish chemist Rubinstein was born in the city of Berdychiv, which had the largest Jewish population in what was then a Russian empire. Soon enough, it was not the best to be a cosmopolitan Jew under Stalinism or Nazism.

On 15 March 1941, the NKVD sentenced him to five years in prison in the coal-forestry gulag of the easternmost European city of Vorkuta (Воркута, Воркутинск) within the Arctic Circle, after he was accused as a socially dangerous element (СОЭ, социально-опасный элемент). That was in fact his great luck because by October 1941 the Germans had killed all tens of thousands of Jews in his native Berdychiv.

In the fourth largest city within the northern Arctic circle, Vorkuta, Rubinstein naturally advanced and developed further his interest in the Stefan Problem of arctic ice freezing, which he researched during the next quarter of a century even after moving to Hebrew University of Jerusalem. The artic ice permafrost problem of Stefan's Habsburgians benefited Soviet and subsequently also Israeli Middle East oil industry. Certainly, Stefan would be pleased to hear about that.

In Vorkuta, L. Rubinstein met his fellow prisoner Dmitry Vasilyevich Redozubov (Дмитрий Васильевич Редозубов, 1904 village Osmarinsk in Governorate of Semipalatinsk by Pavlodar (Семипалатинская губерния, Павлодарский уезд, provinca Semipalatinsk, okrožje Pavlodar, vas Osmarinsk) now in Kazakhstan-1978). He lived in Siberia in areas of Tomsk (Томская обл.). Redozubov received higher pedagogical education as a teacher of mathematics in Leningrad as a member of Communist party. On June 14, 1936, while still a student, he was repressed and condemned by a special meeting for five years in the camps under Article 58-10 (anti-Soviet activity). He was sent to the gulag Ukhtpechlag (Ухтпечлаг, Ухтинско-Печо́рский исправи́тельно-трудово́й ла́герь, УПИТЛа́г), from where he was transferred to Vorkuta in July 1936. Years of imprisonment made Redozubov a staunch anti-communist and a wonderful scientist. In the gulag zone, he was the eternal barrack watchman, read Gogol to the convicts at night, and then he stoked the stove and studied science. Dmitry Vasilyevich Redozubov spent about ten years in the camp. He became an employee of Vorkuta research permafrost station of the Obruchev Permafrost Institute of the USSR Academy of Sciences, opened in April 1936 (Воркутинская научно-исследовательская мерзлотная станция института мерзлотоведения имени Обручева Академии Наук СССР). It was necessary to build it, as the permafrost destroys everything. Redozubov was finally free on June 14, 1941. He was engaged in research of the temperature regime of permafrost strata, proposed new methods of thermal exploration. He headed the permafrost laboratory of the Central Research Base. He was one of the opponents of the theory of degradation of Sumgin (Михаил Иванович Сумгин, 1873-1942 Tashkent). In 1951, during a new wave of repression, Redozubov was removed from office, but continued to informally manage the laboratory. He taught at the Mining College, and in the 1960s at Vorkuta. Redozubov was the initiator of the organization of the Training and Consulting Center in Vorkuta. After complete rehabilitation he left for Leningrad. Calculation of the configuration of frozen strata by Redozubov's method is still included in the program of specialty "Engineering Geology, Permafrost and Soil Science" of Moscow State University. Respectively, many modern dissertations on permafrost refer to his work. Redozubov's work was one of the first works on the geothermal method in USSR, which began to be actively used to search for mineral deposits in the mid-60s.

From the gulag of Vorkuta, Lev Rubinstein moved to the city of Ufa of over million inhabitants in Russian Bashkiria between the Volga River and the Ural Mountains. There Isaak Rubinstein (* 16 September 1949 Ufa in Bashkiria, Russia) was born as the son of Lev and Batya Haskin married Rubinstein. In following years, their family moved to the west Turkmenian center of oil industry in Balkanabat where Lev researched in 1953. Next, the family moved to Latvia where already in 1967 Lev worked together with H.M. Geiman at Latvijskij Gosudarstvennyj universitet imeni Petra Stučki, Vyčislitel'nyj centr (Computing Centre named by Pyotr Ivanovič Stuchka (Pēteris Stučka) Latvian State University, Riga, USSR, Computer Science of the University of Latvia, Latvian: Matemātikas un informātikas institūts, IMCS). That Stuchka university of Riga computer research centre was founded in 1959. Isaak became Bachelor of Science, Master of Arts at Latvian State University in Riga in 1971 where he married Luba Gourevitch (Люба Гуревич) on July 27, 1971. H.M. Geiman and Rubinstein family finally moved to Israel in 1974 where Lev lived at first at Merkaz-Klita Alef, Dimona (דִּימוֹנָה) in the Negev desert, 30 kilometres to the south-east of Beersheba, 35 kilometres west of the Dead Sea and 100 km south of Jerusalem. Dimona Nuclear Reactor is located 13 kilometres southeast of the city where the heavy-water nuclear reactor went active sometime between 1962–1964, therefore a decade before Lev's arrival. The first nuclear weapons were supposedly produced there in 1967. Lev soon began working at School of Applied Science and Technology, The Hebrew University of Jerusalem. In 1978, Isaak became a Doctor of philosophy at Weizmann Institute of Science,

Rehovot, Israel and made his postdoc as MIT Cambridge research associate in 1978-1979.

In Israel, Lev Rubinstein published together with his son Isaak Rubinstein (Rubinstein, 1947, 1958, 1967, 1995). Isaak joined Ben-Gurion University of the Negev's Desert Research Department of Mathematics and Department of Solar Energy and Environmental Physics. There, Jaime Wisniak with Purdue PhD also joined their efforts at Negev's Department of Chemical Engineering. Lev worked in School of Applied Science and Technology, The Hebrew University of Jerusalem while still collaborating with the Riga based Bruno Janovičs Martuzāns (Martuzans). Already in early 1970s in Riga, Lev and his collaborators switched to the applications of Stefan Problem to the biophysics on semi-transparent membranes related to the later and still modern research of the viruses. Lev's theory of transfer of low-molecular nonelectrolytes across deformable semipermeable membranes of large curvature developed of 1974 was used to describe the dynamics of swelling and shrinking of a muscle fiber at the influx and efflux of low-molecular nonelectrolytes. Lev posed an abstract problem, the solution of which resembles the characteristic morphological feature of the initial stage of the macropinocytosis. It is the free boundary diffusion problem, which models the disturbance of osmotic equilibrium due to the interaction of lysosomal hydrolase with the large liquid drop, adsorbed at the external side of the pinocyte plasmalemma and treats this as the unique cause of appearance of the force triggering the motion of the drop toward the cell interior. Its solution imitates the initial stage of macropinocytosis in the case of generation of pinocytosic vacuoles (storage bubbles found in cells) inside the intracellulare tubular systems, that are either the endoplasmatic net, microtube system or system of cytoskeletal tubes.[33]

The others continued Rubinstein's research of Stefan Problem in Soviet Russia for the metallurgy and oil industry. In 1938/1947, the mathematical geophysicist Andrey Nikolayevich Tikhonov (Tychonoff, Tihonov, Андре́й Никола́евич Ти́хонов, 1906 Gagarin (Гага́рин, until 1968 Gzhatsk (Гжатск)) in Russia-1993 Moscow) used Leibenzon's solution (1939) of Stefan Problem for metallurgy of continuous ingots, as the continuous casting methods for ingot

---

[33] Emails of Isaak Rubinstein sent to the author from Israel on 26 March 2021 at 13:05 and at 19:26;
https://www.geni.com/people/%D0%9B%D0%B5%D0%B2-%D0%A0%D1%83%D0%B1%D0%B8%D0%BD%D1%88%D1%82%D0%B5%D0%B9%D0%BD/6000000051388989938.

processing (Rubinstein, 1971, 4; Tikhonov, 1938a; Tikhonov, 1938b; Leibenzon, 1931; Leibenzon, 1939). That was a third step: from artic ice freezing, through oil industry, all the way to metallurgy. Like a kind of dialectic of thesis-antithesis-synthesis.

Gulag Vorkuta Permafrost for Modernized Stefan Problem Applied to Research of Other Areas

Soviet success in solving Stefan Problems stimulated the research in Soviet satellite states including Bulgaria which by happy coincidences shared great pre-war ties with Parisians, just like their neighbouring equally Orthodox Serbians with Pavle Savić. Louis de Broglie's Sorbonne PhD student of Schrödinger's equation in June 1938 Asen Borisov Datsev (Dacev, Assène Datzeff, Асен Борисов Дацев, * 14 February 1911 Kamenar, Razgrad (Каменар, Разградско) in Bulgaria; † 1994 Sofia) researched the classical problem of thermal conductivity. That lead him to another problem in the theory of thermal conductivity, called the problem of freezing, or Stefan Problem. In many of his works besides the fundamental quantum mechanics, Datsev developed a method for solving Stefan Problem in different situations – different number of phases, different boundary conditions. He examined the appearance or disappearance of a phase. The main area in which he sought the solution was further divided into sub-areas with time-dependency. In 1963 with Broglie's preface and again in 1970 Datsev published about linear Stefan Problem in the Parisian *Mémoires de sciences physiques*. Datsev treated the temperature in the field of the space occupied by two phases of a given substance, usually in solid and liquid phase, for example water and ice. The functions, presenting the regions of different phases, comply with the corresponding heat equations, with the unknown separating surface of the phases resting at constant temperature, related with an equation called calorimetric condition. Stefan Problem leads to solution of systems of the (parabolic) partial differential equations or integrals with boundary conditions, some of which are changeable and had to be separately determined in each case. Stefan Problem has different important practical regions of application, for example hydrodynamics, aviation, freezing and ice thawing, internal motions inside Earth, crystal growth, metal melting, and rockets. Datsev's Parisian book of 1970 contains the results of his investigations on the linear problem of Stefan, published in Bulgaria and USSR mainly during 1947-1956 as exposed in several lectures in the Institute of Mathematics at the Florence University in May 1967 where Datsev met Giorgio Sestini's team (Datsev 1950; Datsev, 1970, 3, 4, 5).

In the first part of his book Datsev considered different cases of one-dimensional problem, beginning with two unlimited phases, and passing consecutively to the cases of limited phases, changeable phases, and different boundary conditions. In the second and third parts of the book, the results are generalized for two and three-dimensional cases, including anisotropic bodies. The key of Datsev's approach was the method of stitched solutions. Datsev published with the Soviet Academy of Science in Leningrad as well as in Paris and Firenze on both sides of the Iron Curtain, which means that he was fully aware of Rubinstein's merits (Datsev 1970, 2, 3) and of the work of Broglie's colleague L.M. Brillouin (Datsev 1970, 2), even if Datsev did not mention that Stefan was of a Slavic origin like himself. The prince de Broglie and Leon Brillouin participated in establishing wireless communications with submarines during WW1. Broglie was familiar with the works (1919–1922) of Louis Marcel Brillouin on Stefan's favourite hydrodynamic model of an atom and Broglie attempted to relate it to the results of N. Bohr's theory. In 1967, Datsev translated Albert Einstein and Leopold Infeld's work *Evolution of Physics* from French to Bulgarian language and stimulated research and conferences about vacuum, quantum mechanics, and Einstein's relativity in Bulgaria.

Datsev's Florentine collaborator researching Stefan Problem was the Florentine PhD student of Giovanni Sansone (1888 Porto Empedocle in Sicily-1979 Firenze), Giorgio Sestini (1908 Firenze-1991 Firenze). Sestini was the professor of rational mechanic in Parma in November 1949-1956 and later taught in Firenze. During his last period in Parma, Sestini approached the study of Stefan Problem as an effective representation of the solidification or liquefaction process. It was a field that Sestini successfully developed upon his return to his native Florence. In Parma, Sestini organized the mathematics institute with the collaboration of his friend Antonio Mambriani. He was secretary and administrator of the *Mathematics Journal of the University of Parma* (*Rivista di matematica dell'Università di Parma*) from the first issue, which appeared in 1950, to 1956. He also published about Stefan Problem there. Since the 1950s, he began to research the problems governed by parabolic equations. In 1957 he published his first work on phase change problems: About the uniqueness (aka unicity) theorem in one-dimensional problems analogous to Stefan's (Sestini,

1957, 516-519). In December 1960 in his paper dedicated to the 70th birthday of his supervisor Sansone, Sestini already quoted Lamé, Clapeyron, Neumann, Stefan, Brillouin, and Datsev's (Dacev) solving of bidimensional and tridimensional Stefan Problem (Sestini, 1960, 203-204; Datsev 1955).

Sestini approach to Stefan Problem solved the partial differential equations of domains whose boundary is partially unknown: for example, during the solidification of a liquid, the interface with the solidifying part changes its position which is therefore not known a priori. To be compatible with the description of the solidification process, some balance conditions between the discontinuities of the internal actions of the material considered during the liquid-solid phase transition must be verified through this interface. Precisely these conditions received the name of Josef (Jožef) Stefan. The fact that Sestini in Firenze might have acknowledged Stefan's Slovenian origin as mentioned in Sestini's biography is important in the light of Slovenian fight against Italians in WW2.

For example, in transport problems of phase transitions, such as that of ice-water, the speed of the growing of interface is required to depend on the discontinuity of the heat flow. Sestini was the first in Italy to deal with this large and fruitful class of problems. He had the merit of bringing it to the attention of the Italian and European mathematical community. Some of his results on Stefan-type problems opened-up a new field of research that is still active today, also for important industrial applications. Sestini's ideas were the basis for further developments that were studied by his pupils. The Florentine school developed in these directions with Demore Quilghini as the first, followed by Mario Primicerio and Antonio Fasano. That school established itself internationally in the field of mathematical physics.

Stefan Problem from Vorkuta to the New World: USA

The Bulgarian and Italian followers of Rubinstein's work soon got their competitors overseas. Stefan Problem was researched in New World immediately after the American publication of English language translation of Rubinstein's work. The early advancement was provided by George William

Evans II (1922-1972) and Eugene Isaacson (1919-2008) who followed their dying mentor James Keene Lorne MacDonald (* 1905 Cape Breton, Nova Scotia, Canada; † 1950) at New York University. George W. Evans II later joined Stanford Research Institute.

In 1928, MacDonald got his PhD from Montreal McGill University where Ernest Rutherford worked in 1898-1907. On 14 October 1949, Evans, Isaacson, and MacDonald sent for publication their study of Stefan-like problems from NYU. They quoted Rubinstein, the Bulgarian Asen Borisov Datsev and others as the alleged authors of the term Problem of Stefan (Evans, Isaacson, MacDonald, 1950, 312).

The works of Evans were praised by Jim Douglas jr. in his Theorem for the Solution of a Stefan Problem. Douglas thought that Datsev and Sestini demonstrated the existence of the solution of Stefan Problem; however, they gave no satisfactory uniqueness proof which Evans suggested but Douglas finally proved after the suggestion for simplification of his colleague at Duke university Thomas Muir Gallie Jr. (1925 New York City-2019 The Forest at Duke in Durham, North Carolina). Later, Gallie excelled as the professor of mathematics and computer science who wrote the grant to purchase the first computer at Duke University as a founder of the Computer Science Department. Douglas did not mention Louis Marcel Brillouin or Rubinstein. Douglas stated that Stefan Problem consists of solving a parabolic differential equation subject to a boundary condition on a moving boundary the location of which is not given in advance but is determined as a portion of the problem. Many physical problems give rise to such conditions, including the heat conduction involving phase change such as the melting or freezing of ice, evaporation or condensation, or recrystallization of metals. Others include various displacement processes of reservoir engineering as a branch of petroleum engineering, wherein one fluid partially displaces a second (and originally resident) fluid in flowing through a porous medium including oil which was extremely important for then Douglas' employer in Houston, Humble Oil and Refining Co. (Douglas, 1957, 402-403; Evans, 1951; Sestini, 1952; Datsev, 1950). That Company was founded in 1911 in Humble, Texas in the times when the oil industry mostly resembled hunting of the wild animals in the USA. In 1919, a 50% interest in Humble was acquired by John D. Rockefeller's Standard Oil of New Jersey which acquired the rest of the company in September 1959 and merged with its parent to become Exxon Company, USA in 1973 and ExxonMobil in 1999. Douglas simultaneously worked as a P.M. (project manager) at the Rice institute of Duke University United States Air Force Office of Scientific Research.

Evans, Isaacson, and MacDonald soon got their NYU followers in their research of Stefan Problem began in 1950. In late 1950s-1960s, the Jew Richard Courant's (1888 Lubliniec, Poland-1972 New Rochelle, New York) student Joseph B. Keller (1923 Paterson, New Jersey-2016 Stanford, California) organized the research of Stefan Problem at the Courant Institute of NYU. Among his postdocs was Walter Thomas Kyner (Tom, 1926-1999 Albuquerque NM) who obtained his Ph.D. from the University of California in Berkeley and spent two postdoc years at the Courant Institute at NYU. Next, Kyner worked for a year at Northwestern as he just could not live without his beloved West. In 1960-1970 Kyner was at the University of Southern California in Los Angeles, and in 1970-1997 Kyner was a mathematics professor at the University of New Mexico. In 1959, already residing in California, he published his postdoc research accomplished at Courant Institute at NYU about a nonlinear Stefan problem. His paper was received on 3 November 1958 as the research done while was a temporary postdoc member of the Institute of Mathematical Sciences, Courant Institute, New York University. Simultaneously, Keller's PhD student at Courant Institute at NYU was Willard L. Miranker (1932 Brooklyn-2011) who earned his Ph.D. in 1956, supervised by Keller. They published about Stefan Problem together in 1960. In 1961-1989 Miranker was at IBM Research Center, Yorktown Heights, N. Y., U.S.A.

A great part of research of Stefan Problem belonged to US businesses outside the universities. William F. Trench (1931-2016) was a Senior Engineer and Engineering Specialist at Philco Corporation, Philadelphia, PA in 1957-1959. In June 1959 he published about the explicit method for the solution of Stefan Problem. That work was his abridged doctoral thesis in Mathematics at University of Pennsylvania which he as a part-time student defended in 1958. He sent his article about Stefan Problem on 1 May 1958 by citing Evans, Isaacson, MacDonald, Rubinstein, Sestini, and work of Lawrence Bragg's Manchester student becoming the expert of diffusion at Brunel University at Acton John Crank (1916 Hindley, Lancashire, England-2006). Crank published his work in 1957. Trench also praised the Jew Landau's article published in 1950. Hyman Garshin Landau (1909 Chmielnik by Kielce with predominant Jewish population northeast of Krakow in then Russia, now in Poland with a Synagogue or less probably Khmelnik (Hmelnik, Хмільник, Khmilnyk, Хмельник) in Podolia region of the modern Ukraine where the Jews have settled in 16th century during the colonization of the Ukrainian stepprs-1966). Landau was born in Russian Empire, like his fellow Jewish researcher of Stefan Problem Lev Rubinstein five years later, but Landau left Soviet Union before Stalinist persecutions just to met similar trouble in the USA. In 1946, Landau got his PhD in statistics at Pittsburgh and joined Ballistic Research Laboratories, Aberdeen Proving Ground from where he sent his research of heat conduction in a melting solid related to Stefan problem, received on 19 November 1948 and published much later in 1950 (Trench 1959, 181-182). After 1950 Landau joined the Committee on Mathematical Biology of the University of Chicago which he had to leave after the heavy interrogations due to accusations by the House Un-American Activities Committee (HUAC). In 1952, as the victim of Joseph McCarthyism era lasting from the late 1940s through the 1950s similar to Rubinstein's troubles in the USSR, Landau was examined for alleged Subversive Influence in the Educational Process, but Landau was certainly never sent to gulag (Hearings Before the Subcommittee to Investigate the Administration of the Internal Security Act and Other Internal Security Laws to the Committee on the Judiciary, United States Senate, Eighty-second Congress, Second Session[-Eighty-fourth Congress, First Session], Volumes 4-13). Landau had luck as he subsequently found work at Columbia University. However, Landau's interest in what are now known as tournaments (graph theory) did not arise from his studying of sports competitions. As so often happens in mathematics, the source of Landau's interest was something rather different. He was interested in animal behaviour of pecking orders in chickens as previously researched by the Nazi Konrad Lorenz (1903 Vienna-1989 Altenberg) and Lorenz's friend SS Nazi prince Alfred Auersperg (1899-1968 Hamburg). Landau was an elder brother of the murdered television director Jack Landau (1922 Braddock, Pennsylvania-1967) which may have stimulated his political persecutions.

Stefan Problem soon became important for aviation and rocketry. Arthur Louis Ruoff (Art, * 17 September 1930 Fort Wayne, Indiana) obtained his PhD at Utah university in 1955. Then as an expert of high pressures he worked at Cornell University Department of Engineering Mechanics and Materials. In 1958 he also worked at Wright-Patterson Air Force Base, Ohio, Wright Air Development Center, Air Research and Development Command, U.S. Air Force. In 1958 he published an alternate solution of Stefan problem.

Stefan Problem at the New World: Canada

Besides USA, Canadians also joined the research of Stefan Problem. In July 1966, the Norwegian James R. Gunderson as Lock's graduate student used the term Stefan Problem and Neumann-Stefan problem at his master thesis in Alberta before Gunderson joined the Federal Department of Transport in Edmonton in Canada. Gunderson quoted Evans, Redozubov, and Leonard Rose Ingersoll (* 1880; † 1958). Ingersoll excelled as a professor of Physics at the University of Wisconsin-Madison where he established L.R. Ingersoll Physics Museum on the second floor of Chamberlin Hall as the first museum in the United States to focus solely on physics. Ingersoll collaborated with Otto Julius Zobell (* 1887) and Alfred Cajori Ingersoll (* 1920 Madison Wisconsin; † 1999). Gunderson also cited the research of steel solidification published in 1929 and 1930 by Nicholas Morpeth Hutchinson Lightfoot (* 1902 Jarrow in England; † 1962 London) of Heriot-Watt College in Edinburgh (Gunderson, 1966, 1, 3). The term Stefan Number was therefore coined much earlier on the other side of Iron Curtain than noted by many historians who cite Lock's pioneering merits. Gunderson's professor of mechanical engineering at the University of Alberta, Edmonton in Canada was Gerald Seymour Hunter Lock (* 30 June 1935 London) who published his research of Stefan Number in 1969, two decades after Rubinstein's work was translated in the USA. Lock and his students were probably unaware of Brillouin's work as they did not cite him. Lock also used the terms "problem of Stefan" and "classical Stefan solution", but not exactly the phrase Stefan Problem. After all that Soviet and American researchers of oil and metallurgy, Lock and his Canadian students went back to Stefan's original puzzle of one-dimensional ice formation (Lock, 1969, 285; Lock et all, 1969, 1343).

Stefan Problem at the New World: Bibliography of Stefan Problem in Latin America

The Latin America was not left behind in the research of Stefan Problem. After his PhD about Stefan Problem and habilitation in Paris in 1979 and 1991, Domingo Alberto Tarzia (* 20 December 1950 Rosario, Santa Fe, Argentina) joined the Universidad Austral in Buenos Aires, Argentina as director of research of Stefan's moving boundaries of phase transitions at the University of Nevada, Reno, USA. Since 1979, he excelled as one of the main bibliographers and researchers of Stefan Problem, but failed to note the Japanese Seizo Saito (1921), the Russian Leibenzon, the Viennese Franz Selig's publication of 1956, misspelled the American MacDonald family name into Donald, and overlooked the Norwegian turned Canadian Gunderson's dissertation of July 1966. Tarzia noted just the Bulgarian Asen Borisov Datsev (Assène Datzeff) publication of 1970, but not his earlier publications of 1950 and 1955 in Sofia and Leningrad in French and Russian languages (Tarzia, 2000).

Stefan Problem back in the areas of his Habsburgian Monarchy after WW2

Stefan's Vienna also trained domestic researchers of Stefan Problem including Franz Selig (Selling) of Vienna who also worked at Socony Mobil Oil Co. Inc., Field Research Laboratory in Dallas in 1966. Franz Selig published on Stefan Problem in Austrian Engineering Journal in 1956 and on similar topics in *Wien.Ber.* where Stefan used to publish his masterpieces a century earlier.

Tarzia's bibliographic work is continued today by the Slovenian physicist Šarler which makes Stefan Problem a kind of Slovenian ar at least Slavic topics in former areas of Habsburgian monarchy and Russian Empire. Among Šarler's collaborators is John C. Crepeau of Department of Mechanical Engineering at University of Idaho in Moscow.

Stefan Problem in the Modern Globalized World of Web

Stefan problem, also known as the moving boundary problem, continues to be a fruitful field of research. Besides the book by Rubinstein, other texts followed including *The Classical Stefan Problem* by Sushil Chandra Gupta (* 1937), and *Mathematical Modeling of Melting and Freezing Processes* by professor in the Mathematics department at University of Tennessee at Knoxville Vasilios Alexiades (* Xanthi in Greece) and Alan D. Solomon from Mathematics and

Statistics Research Department Union Carbide Corporation, Nuclear Division Oak Ridge, Tennessee. Solomon also translated L. Rubinstein's works from Russian to English language. The classic papers were provided by L.S. Yao at Department of Mechanical and Aerospace Engineering of Arizona State University at Tempe and PhD of University of Illinois Joseph Michael Prusa at Department of Mechanical Engineering at Iowa state University in Ames. W.F.M. Goss' professor of Engineering at of Argonne National Laboratory Raimondas Viskanta (Raymond, * 1931 Marijampolė, Lithuania), later professor at Purdue, published an excellent introduction into the study of melting and solidification of materials. Review articles about Stefan Problem were provided by James Caldwell (* 16 May 1943 Macosquin, Coleraine, County Londonderry, Northern Ireland) of Department of Mathematics, City University of Hong Kong in Kowloon, Hong Kong and Yuen-Yick Kwan of Center for Computational Science, Tulane University, New Orleans, L.A., USA. Luca Salvatori of Dipartimento di Ingegneria Civile e Ambientale (DICeA) Università di Firenze and Niccolò Tosi of School of Engineering Sciences, University of Southampton, Highfield, UK, gave an overview of numerical methods that have been used to solve the Stefan problem.

Future of Stefan Problem

We described how Stefan Problem of freezing switched to the similar but gradually sophisticated mathematical treatments al the way to aviation and rocketry, how the research output gradually affected its own changeable geographical distributions of the academic groups of researchers, and how the sophistication of involved mathematics including the internet changed the approaches to Stefan Problem. Now we are up to speculate about some prophecies of the future aspect of Stefan Problem in 3rd millennia.

The physics of post Stefan world was extremely successful in developing working mathematical tools based on experimentation. The deepest impulse used to be the WW2 Manhattan Project and its echoes in the USSR. That network soon flooded into sciences beyond physics and quantum mechanics while also exporting the local initially narrowly defined physical terms including the entropy, boundary conditions into an interdisciplinary theory of chaos, theory of cooperation on networks, non-equilibrium statistical physics in evolutionary game theory which modern grownup researchers used to play as kids, self-organization of major scientific ideas propagation across the expert literature in a simple mathematical regularity that is able to identify scientific memes.

There is a kind of imperialism of physics and genetical research in their best paid modern sciences which allows us to predict the expansive use of Stefan Problem beyond physics and technology in biology, economy, and humanities. They will be used in solving the slightly similar problems involving movable boundaries between the different phases, races, social categories, species. The modernized movable boundaries might be time-dependent or function of other more sophisticated parameters. Stefan could never have imagined, but his new dynamic world of angry Spring of Nations marching masses never really stopped despite of pandemics. And they need the mathematics of Stefan Problem to explain themselves.

Conclusions

The travel involving time-dependent changes is the basic of modern pre-pandemics' world based on Stefan problem. Curiously enough, Stefan's teachers and students Andreas Ettingshausen, Kunzek, Marian Koller and Boltzmann widely travelled, but not Stefan or Loschmidt. Despite of that failure which Boltzmann did not approve Stefan found his ways needed to spread his novelties around Europe and Americas. In 1838, Ettingshausen, Kunzek, and Marian Koller's journey to Hamburg, Berlin, London, Greenwich, Paris, and Freiburg am Breisgau also enabled Stefan's connections worldwide after he replaced the ill Ettingshausen. On the other hand, traveling could cheer up Stefan, so he would not die so young, scarcely older than his Slovenian predecessor among the Viennese professors of applied mathematics, Jurij Vega. His refusing to travel abroad might have dogged Stefan's international relations, but just until he became the technical leader of the Viennese electrical exposition in 1883. Stefan and Loschmidt's aversion of traveling or dancing was caused by their humble origins incomparable to the erudite raised in well-to-do Habsburgian families including Ettingshausen, Koller, Kunzek, Kunzek's Viennese replacement Viktor von Lang, Ernst Mach, or Lecher. Stefan never understood Humphry Davy's Consolations in travel, but Stefan's administrative organizational abilities nevertheless enabled him to rule over Habsburgian physics except for Mach's Prague, while heavy overwork without physical exercises in crowded metropolis finally damaged his ruined health.

Bunsen's postdoc student raised as a member of Protestant Moravian Church, Wanklyn, remains the only proved Stefan's English translator in 1864 while almost certainly also other experts did their similar job. Cauchy's friend the conservative Catholic Moigno and the Viennese informer of Parisians the paleo-zoologist August Friedrich count Marschall Sudoc auf Burgholzhausen in Saxony-Anhalt and the nearby Tromsdorf (1804 Vienna-1887 Oberneidlingen district of Esslingen, in Baden-Württemberg) provided abbreviated French translations of Stefan's publication including first Stefan's article about Stefan Problem and Stefan Number. As the son of honorary Viennese citizen Friedrich Ernst count Marschall (1748-1832 Vienna) and grandson of freemason-illuminati Ernst Dietrich imperial count Marschall of Burgholzhausen (1750-1824 Weimar), August Friedrich Marschall published his reports for French Academy month and a half after the relevant Viennese academic sessions where

Stefan spoke, while Moigno reprinted Marschall's data few weeks later. Stefan's French and English translators therefore held extremely different political views echoing the temporary depoliticization of sciences after Spring of Nations which included Stefan's own abandoning of nationalist Slovenian poeticizing which enabled him to became more suitable academic leader under the conservative emperor Franc Josef. Americans, Spaniards, and others were probably not so quick in commenting or republishing Stefan's works in those times.

We proved that Stefan Problem and Stefan Number were almost simultaneously known among the German, French, and English language readers. Stefan's research was welcomed in Paris also because Stefan himself, like his mentors Koller and Ettingshausen, initially mostly relied on French authors like Joseph Fourier, Gabriel Lamé (* 1795; † 1870), Benoît Paul Émile Clapeyron (1799 Paris-1864 Paris) or Cauchy.

Stefan Problem and Stefan Number got their names only much later compared to Stefan's T^4 law. Stefan Problem was easily provable theoretical and experimental puzzle which was quickly repeated, bettered, and accepted on the West once it was introduced in the mainstream mathematics and industrial research few decades after Stefan's death. On the other hand, Stefan's T^4 Radiation Law was a brave theoretical speculation comparable to Dewar's T^3 ideas or Maxwell's r^{-5} speculations. None of those was ever so effective as Newton-Coulomb's (r^{-2}) law of gravitational and electromagnetic forces. Stefan never accepted Maxwell's "ad hoc" model of molecular force proportional to inverse-fifth power of the distance, even if Maxwell's model provided the acceptable diffusion coefficient proportional to the temperature (Mitrović, 2012, 12-15). However, Maxwell's model did not allow the determination of the internal properties of gas. That is why Stefan published his dynamic model of gas in which the radius of the molecule decreases with some power of temperature. In Stefan's model, the radius of faster molecules diminished due to increased possibilities of interpenetration during collisions. The repulsive force is then proportional to the velocity, just as French Claude-Louis-Marie-Henri Navier (1785 Dijon-1836 Paris) argued in his lectures on mechanics at Parisian École Royale des Ponts et Chaussées already in 1826. The theoretical assumptions of those times were often seen on rather shaky legs because of the supposed suspicious background of the German Philosophy of Nature; therefore, that kinds of theories were not easily published in Poggendorff's *Ann.Phys.* or in *Phil.Mag.* of Koller's and Stefan's era. Hegel stated that it is too bad for the facts if they do not fit in his theories, and Petzval had similar idea while criticizing Doppler's effect, but the modernized experimental physicists just did not like those approaches anymore.

Our aim was to illustrate how the dying Habsburg Central-European monarchy topped the worldwide erudition by M. Koller, Franz Liszt, Stefan, E. Mach, Boltzmann, S. Freud's two months younger N. Tesla, Gustav Mahler, Gustav Klimt, both Johann Strauss, Stefan Zweig. After the Habsburg monarchy collapsed and following Einstein's Prague semesters from April 1911 through July 1912, the Viennese Erwin Schrödinger built his variant of quantum physics which mostly Budapest Jewish naturalized Americans developed into a mainstream that still reigns. In that context, the swansong of Habsburgian monarchy is not just the story of our past erudition, but also a tale about our future, especially if we have another collapsing empire(s) onboard today.

The generation of Stefan's and Boltzmann's heirs witnessed the collapse of the ideas of Danubian Monarchy during in two decades between World War I and Anschluβ. In their era, Vienna as a Central European heart in fact became a temporary cultural and scientific capital of the world with Boltzmann's heirs including Prague Professor Einstein's lecturing in 1911-1912, professor of Graz the Viennese Schrödinger (1936-1938), or the Viennese Circle of logical positivists of Verein Ernst Mach established in 1928 under the leadership of Moritz Schlick. Schlick and his Mach's fans gathered around the *Tractatus Logico-Philosophicus* of the wealthy Jew Ludwig Wittgenstein (1921), which attracted their fans including Gödel or even Karl Popper (1902-1994). One of the richest Jews of his times, the graduate of aeronautics in Berlin named Wittgenstein admired the logic of Gottlob Frege whose explicit syntax dominated over semantics as in subsequent linguistics of Noam Chomsky, although Gödel proved that syntax, however, cannot replace semantics. Additionally, Vienna and its antagonistic alter ego Prague hosted the Jew Sigmund Freud, the Jewish Austro-Marxists, Brahms' performances in Wittgenstein's house, B. Smetana's nationalistic Czech music, the painter Gustav Klimt and premium-class Jewish writers Stefan Zweig (* 1881, † 1942) and Franz Kafka (* 1883, † 1924), or Kafka peer-alter ego the Czech Jaroslav Hašek's (* 1883, † 1923) Good Soldier Švejk.

The Habsburgian Nikola Tesla as the best of them all finally brought the Balkan mixtures of fertile Bošković's novelties at the disposal of moneymaking US, although Tesla finally fell short with his wireless transmission station at

Wardenclyffe tower. In few dozen years between Stefan-Boltzmann's and Tesla's successes, the Habsburgian Bošković's heirs transformed sciences and technologies worldwide, even if Tesla hated Boltzmann's entropy law. Their success was an exported swansong of an once great Danube Monarchy, while the Viennese Karl Popper tried to provide the logical explanations for historical successes of sciences, albeit he was eventually not aware of his own connections with Habsburgians of Stefan-Boltzmann-Tesla's type. Tesla and Boltzmann knew more about their Habsburgian pedigree, therefore they publicly praised Bošković a lot.

The Danube monarchy's novelties were borrowed at the West after Habsburgians imported Netherlander enlightenment experts. The Habsburgians returned their loans with interest included in their feedbacks. The western Sciences enriched themselves via the Habsburgian statistical mechanics which became a mainstream. The scheme describing Western-Habsburgian mutual interactions will certainly help our notions of similar future Westerners' feedbacks borrowed from even more foreign scientific cultures, which might again perform onstage in resembling forms of swansongs of political declines of local empires like the Soviet Union, which is already dissolved but its impacts not yet researched in all its aspects. There is also a possible decline of the Chinese Belt-and-Road initiative under Trumps' and Dems' USA hammers.

The beginnings of modern physics were born among the Habsburg scholars educated in Viennese physics institute (Physikalischen Institutes) of Doppler (1850), and particularly in networks of his antagonistic successor Ettingshausen. Doppler was dismissed by political plot, which enabled the rise of Habsburgian kinetic atomism of his opponents even if Doppler supported similar atomistic ideas. Ettingshausen was a friend of Clemens von Metternich (* 1773; † 1859) who used to be almighty Chief until the spring of Nations. Ettingshausen was named and appointed on Doppler's post in 1852, Stefan succeeded him in 1866, and Boltzmann served as Stefan's heir in 1894-1900.

Those Viennese guys also serve as a reference example, the backfired feedback of the knowledge of peripheral universities of the Danube monarchy, focused on their returned impact into Berlin, Paris and mainly into Maxwellian Britain. Under Metternich's Habsburgian heirs with the Viennese-Graz based Schrödinger and Einstein's pre-WW1 Prague, they built their part of the relativistic quantum mechanics and "heretical" quantum wave mechanics which opposed the Copenhagen School. The Hungarian Budapest Jewish emigrants

Leo Szilárd (* 1898), Eugen Paul Wigner (* 1902, † 1995), John von Neumann (* 1903), and Edward Teller (* 1908) as Szilárd's Martians used Einstein's signature under the letter mailed to the USA president to throw their mostly undesired Japanese bombs. Like Vienna, after the collapse of the Danube monarchy Budapest also hosted and raised incredible talents: Budapest geniuses were predominantly Jewish and focused on physics even more than the emigrants from Vienna. Both cities irretrievably lost almost all those talents under the fascist boot: the collapsing Nazi empire forced them out for free. Like the literati of the sudden collapsed Byzantium, the Napoleonic empire, or the Soviet Union, they used their excellent education in their emigrational networks for the worldwide benefit of Westerners, at least for a while.

Sources and Literature

Archival Sources

Catholic Matriculas

Archiv der Diözese Gurk, Mariannengasse 69020 Klagenfurt am Wörthersee Matriculas (Škofija Krka, Celovec-Dom, Dioecesis Gurcensis, Diocese Gurk): Klagenfurt-Dom, Sterbebuchs (Books of Funerals.

Diözese Gurk, parish Klagenfurt-Dom Matriculas: Trauungsbuchs (Books of Weddings).

Bohinjska Bistrica parish Matriculas:
Book of Baptisms-Births of parish Bohinjska Bistrica no. 1 (Knjiga Rojstev in krstov Župnije Bohinjska Bistrica).

Direktions-Archiv der Sternwarte Kremsmünster

Derfflinger, Thaddäus; Koller, Marian or his collaborators. 1802-1824, 1848. Uibersicht der Sonnenmackeln, welche auf der Sternwarte zu Kremsmünster Seit dem 26. September 1802 beobachtet wurden. Up to 1824 inclusive; then again on July 26, 1848, Manuscript, Directions-Archive of the Sternwarte Kremsmünster.

Koller, Marian, Vorlesungen „aus der Naturgeschichte" an der philosophischen Lehranstalt zu Kremsmünster (1825-1830, Lectures on Natural history read at philosophical studies in Kremsmünster). Direktions-Archiv der Sternwarte Kremsmünster.

Koller, Marian, Vorlesungen „aus der gesamten Physik in der 8ten Classe des Lyzeum" an der philosophischen Lehranstalt zu Kremsmünster (1826-1839, Lectures about physics read at philosophical studies in Kremsmünster) von P. Marian Koller. Koller-Manuskripte, 15a (I. Abteilung: Chemie, Mechanik, Akustik on 694 pages together with later added title page) and 15b (II.

Abteilung: Optik, Magnetismus, Elektrizität, on 672 pages together with later added title page), Direktions-Archiv der Sternwarte Kremsmünster.

Koller, Marian, Abhandlung „Bestimmung des Weges eines homogenen Lichtstrahles durch mehrere brechende Flächen" (unpublished Treatise "Determination of the path of a homogeneous light beam through several refracting surfaces"). Direktions-Archiv der Sternwarte Kremsmünster.

Koller, Marian, Notes from the lectures of Dr. Josef Stefan, 1862-1864: a) "Theorie der Elasticität fester Körper". 1862, lectures in the winter semester 1862, altogether 36 folios (= 146 pages), following with few folios of annotations; the last 36 pages 177-213 (9 folios) were in fact devoted to the course of static and dynamics of elasticity probably connected with acoustic in new summer semester of 1861/62 began on 6 May 1862. Koller-Manuskripte, 18, Direktions-Archiv der Sternwarte Kremsmünster.

b) "Über die Theorie der Wärme". 1862/63, all fourteen lectures except the eighth and tenth in the winter semester of 1863, 35 Folios (=140 pages), Koller-Manuskripte, 31, Direktions-Archiv der Sternwarte Kremsmünster.

c) "Über Elektrodynamik und Theorie der Induction". 1863, lectures in the summer semester of 1863, 26 Folios (= 104 pages), Koller-Manuskripte, 32.

d) „Über die Theorie des Lichtes". 1863/64, lectures in the winter semester of 1864, 30 Folios (= 120 pages), Koller-Manuskripte, 26.

e) „Über Interferenz, Beugung, Polarisation", with red number 27 on the contemporary cover, the first five of Stefan's supposedly twelve lectures during the summer semester of 1864, which was also attended by L. Boltzmann, 10 Folios (40 pages), Koller-Manuskripte, 27, Direktions-Archiv der Sternwarte Kremsmünster.

Koller, Marian. Vortrag "über die Vorausberechnung von Sonnenfinsternissen" gehalten in der Conversation der Mitglieder des Doctoren-Collegiums der philosophischen Facultät v Wien, am 24. November 1855, von Dr. Marian Koller. Manuscript, Direkcije-Archiv der Sternwarte Kremsmünster.

Koller, Marian. Notes on the Lectures of (Vorlesungen von) Prof. Dr. Josef (Jožef) Stefan, manuscripts (reference codes) no 25 by Augustin Reslhuber's list of 1866, Directions-Archiv der Sternwarte Kremsmünster, Photo: Amand Kraml:

a) „Theorie der Elasticität fester Körper (Theory of elasticity of solids)", taught in winter semester 1861/62, 36 folios (= 144 pages), Koller-Manuskripte, 18.

b) "Über die Theorie der Wärme (On theory of Heat)", Koller-Manuskripte, 27, Direktions-Archiv der Sternwarte Kremsmünster, fourteen Stefan's lectures except the unnoted 8th and 10th lectures, taught during winter semester 1862/63, 35 folios (= 140 pages). L. Boltzmann attended those lectures a year and a half later in summer semester 1863/64 (Höflechner, 1994, 15). Related to Stefan's publications: *Bemerkungen über die Absorption der Gase*, Sitzungsberichte der Kaiserlichen Akademie der Wissenschaften in Wien (Wien.Ber.) 27: 375–430 (1858).

Über das Dulong-Petit'sches Gesetz, Wien.Ber. 36: 85–118 (1858).

Über der Erscheinungen der Gasabsorption, Programm der Ober-Realschule, Vienna (1859).

Über die specifische Wärme des Wasserdampfes, Ann. Phys. 110: 593 (1860).

Über die Fortpflanzungsgeschwindigkeit des Schales in gasförmigen Körpern, Ann. Phys. 118: 494 (1863).

Bemerkungen zur Theorie der Gase, Wien.Ber. 47, II: 81–97

c) "Über Elektrodynamik und Theorie der Induction (About electrodynamics and the theory of induction)", taught in summer semester 1862/63, 26 folios (= 104 pages).

d) „Über die Theorie des Lichtes (About the Theory of Light)", taught in winter semester 1863/64, 30 folios (= 120 pages).

e) „Über Interferenz, Beugung, Polarisation (On Interference, Diffraction, Polarization)", Koller-Manuskripte, 27, Direktions-Archiv der Sternwarte Kremsmünster, red number 27 at the modern hand's title-page, the first five of supposedly twelve Stefan's lectures taught in summer semester 1863/64 with L. Boltzmann also among the students after his matriculation in October 1863 (Höflechner, 1994, 15), 10 folios (= 40 pages (or 15 folios forming 60 pages) (Manuscript, Directions-Archive of the Observatory (Sternwarte) Kremsmünster).

Koller, Marian; his maternal 3rd cousin also Zois' protegee Franz Ksaverij Jožef Hanibal Count Hohenwart (nicknamed Jernej, 1771 Ljubljana-1844 at his castle Kolovec in Upper Carniola), correspondence with autobiographical content 1839-11-23 from Ljubljana and 1839-12-03 reply from Kremsmünster (Fellöcker, 1864, 247).

Ljubljana City Historical Archive
SI_ZAL SI LJ 184 Klasična gimnazija v Ljubljani (Classical Grammar School Ljubljana), Action fond 1, Individueller Ausweis, summer Semester April 1810, box 86, Fascicle 534

Vienna, AT-OeStA/AVA

Stefan Archive = AT-OeStA/AVA *Unterricht* UM *Unterrichtsministerium*, 1848-1940, AT-OeStA/AVA Unterricht UM (Ministerium für Kultus und Unterricht, OSTA, 1). Reference code: 4 Phil. Stefan. 1 6215/1858. Stefan, J. (1858–1878): Universitäts Akten, fol. 1–4, Ministerium für Kultus und Unterricht; Personalakts Jožef (Josef) Stefan: Four folios of documents on Stefan's career at the University of the Habsburg Ministry of Education and Worship (Kultus und Unterricht) including the documents dated: January 26, 1863, February 9, 1863 (pp. I-IV), September 20, 1866 (pp. 1-12).

Printed Literature

Airy, George Biddell. 1858. *Mathematical tracts on the lunar and planetary theories, the figure of the earth, precession and nutation, the calculus of variations, and the undulatory theory of optics : designed for the use of students in the university*, Cambridge: Macmillan and Co.

Alexiades, V.; Alan D. Solomon, 1992. *Mathematical Modeling of Melting and Freezing Processes*, CRC Press.

Bachelard, Gaston. 1928. *Étude sur l'évolution d'un problème de physique: la propagation thermique dans les solides,* Paris: J. Vrin. Reprint: 1973. Paris: J. Vrin.

Bartoli, Adolfo Guiseppe. 1879. Dimostrazione di un Theorem relativo alla teoria del raggiamento dato dal Professor R. Clausius: del dott. Adolfo Bartoli. Professore di fisica nell'instituto technico di Firenze. *Nuovo Cimento*. (3) 6: 265-276.

Baumgartner, Andreas. 1826. *Die Naturlehre*, Vienna: Heubner, January 1826. 3rd edition (Neue Auflage 3:) 1829.

Böhm, Josef Georg. 1852, Beobachtungen von Sonnenflecken und Bestimmung der Rotations-Elemente der Sonne; Beobachtungen von Sonnenflecken in den Jahren 1833, 1834, 1835 und 1836 (Beobachtungsort: Die k.k. Universitäts-Sternwarte in Wien). *Denkschriften der kais. Akademie der Wissenschaften. Mathem.Naturw.Kl.* 3/2: 39–42 with tables 8-11, 43-112. Separate academic print: the Academy also provided a joint a separate printed booklet on 74 pages with 4 tables in Vienna.

Brillouin, Louis Marcel. 1930. Sur quelques problèmes non résolus de la Physique Mathématique classique: Propagation de la fusion. *Annales de l'Institut Henri Poincaré*, Volume 1, No. (Fascicule) 3: 285-308 (published in 1932, Brillouin's lectures read at *Institut Henri Poincaré* on 13th and 16th May 1929). Reprint in 1931: Paris: Presses universitaires de France.

Brush, Stephen George. 1976. *The kind of Motion We Call Heat*. Amsterdam-New York-Oxford: North-Holland, 2nd part.

Bufon, Zmago. September 1970. Naravoslovec Marjan Koller. *Proteus: ilustriran časopis za poljudno prirodoznanstvo* (1970/71). 33/1: 40.

Caldwell, J. and Y. Y. Kwan. 2004. Numerical methods for one-dimensional Stefan problems, *Commun. Numer. Meth. Engng* 2004; 20:535–545 (DOI: 10.1002/cnm.691)

Christiansen, Christian. 1881. Einige Versuche über die Wärmeleitung, *Ann. phys.* (3) 14 (250)/9: 23-33.

Christiansen, Christian. 1 May 1883. II. Absolute Bestimmung des Emissions- und Absorptions- Vermögen für Warme, *Ann. phys.* (3) 19 (255)/6: 267-283.

Clément Desormes, Nicolas. 1827. Rapport fait à l'Académie des Sciences de l'Institut, sur un Mémoire relatif à un phénomène que présente l'écoulement des fluides élastiques, et au danger des soupapes de sûreté employées dans les appareils à vapeur; présenté; par M. Clément Desormes dans la séance du 4 décembre 1826 (Commissaires: Biot, Poisson & Navier, par rapporteur noted some historical reminiscences on D. Bernoulli's hydrodynamics and recommended Clément Desormes' paper in Paris on 10 September 1827 for publication in *Recueil des Savants étrangers (Mémoires présentés par divers savans à l'Académie royale des sciences de l'Institut de France))*. *Annales de chimie et de physique*, 1827, 36: 69-80.

Crank, Acton John. 1956. *The mathematics of diffusion*, Clarendon Press, Oxford.

Crank, Acton John. 1957. Two methods for the numerical solution of moving-boundary problems in diffusion and heat flow, *Quart.J.Mech.Appl.Math.*, 10 (1957), pp.220-231.

Crank, Acton John. 1984. *Free and moving boundary problems*, Clarendon Press, Oxford.

Desormes, Charles; Clément Desormes, Nicolas. 1819. Détermination expérimentale du zéro absolu de la chaleur et du calorique spécifique des gaz, Journal de physique, de chimie et d'histoire naturelle, tome 89/N° 7: p. 321-331-346, 428-455.

Crepeau, John C. 2007. Josef Stefan: His life and legacy in the thermal sciences. *Experimental Thermal and Fluid Science*, 31:795–803.

Crepeau, John C. 21 May 2008. Josef Stefan and Josef Loschmidt: Colleagues in Vienna, Contrasts of Stigler's Law. Lecture at Erwin Schrödinger Institute Vienna, Austria.

Crepeau, John C. (ed.). March 2013. *Jožef Stefan: His Scientific Legacy on the 175th Anniversary*. Sharjah (UAE): Bentham Science Publishers.

Cundrič, Ivo Janez. 2002. *Pozabljeno bohinjsko zlato*. Slovenj Gradec: Cerdonis/Maribor: Ma-tisk.

Cvirn, Janez. 1988. *Boj za Celje*. Ljubljana: Zveza zgodovinskih društev Slovenije.

Cvirn, Janez. 1997. *Trdnjavski Trikotnik*. Maribor: Obzorja.

Čermelj, Lavo. 1976. *Josip Stefan: življenje in delo velikega fizika*. Ljubljana: Mladinska knjiga.

Črnivec, Živka et all (ed.). 1999. *Ljubljanski klasiki 1563–1965*. Ljubljana: Maturanti klasične gimnazije.

Datsev, Asen Borisov (Assène Datzeff, Асен Борисов Дацев). 1950. Sur le probleme liniaire de Stefan, II, *Annuaire de l'Université de Sofia. I. Faculté Physico-Mathematique (Godišnik na Sofijskija Universitet "St. Kliment Ohridski" Fiziko-matematicheski fakultet)*. Livre 1 volume 46 (1950) pp. 271-325 (in French language with Bulgarian summary).

Datsev, Asen Borisov (Dacev, Assène Datzeff, Асен Борисов Дацев, Датцев). 1955. Sul problems bidimensionale di Stefan, *Doklady Akad. Nauk. SSSR (Proceedings of the USSR Academy of Sciences; Доклады Академии Наук СССР; DAN SSSR; Doklady Akademii Nauk Sojuza Sovetskich Socialističeskich Respublik = Comptes rendus de l'Académie des Sciences de l'Union des Rèpubliques Soviètiques Socialistes. Leningrad)*. Leningrad: Izd-vo Akademii nauk SSSR, 101: 441-444, 1955 (in Russan language).

Datsev, Asen Borisov (Дацев, Dacev, Assène Datzeff, Асен Борисов Дацев). 1955. О трехмерной проблеме Стефана (On Stefan's three-dimensional problem, Sul problema tridimensionale di Stefan), *Доклады Академии Наук СССР (Doklady Akad. Nauk. SSSR)*, 101: 629-632, 1955 (in Russan language).

Datsev, Asen Borisov (Assène Datzeff, Асен Борисов Дацев). 1963. Sur le problème de la propagation de la chaleur dans les corps solides (with an introduction of Louis de Broglie), Paris: Gauthier-Villars (*Върху проблема за разпространението на топлината в твърди тела*, Париж, изд. Готие-Вилар) (Stefan mentioned on p. 7).

Datsev, Asen Borisov (Assène Datzeff, Асен Борисов Дацев). 1970. *Sur le problème linéaire de Stefan (Върху проблема на Стефан)*, Paris: Gauthier-Villars, Mémoires de sciences physiques no 69.

Dewar, James. 1927. *The collected papers of Sir James Dewar* (edited by Lady Dewar (Helen Rose née Banks married Dewar); James Douglas Hamilton Dickson; Hugh Munro Ross; Ernest Charles Scott Dickson). Cambridge, UK: University Press.

Douglas, Jim Jr.; Gallie,
Thomas Muir Jr. 1955. On the numerical integration of a parabolic differential equation subject to a moving boundary condition, *Duke Math. J.* vol. 22 (1955) pp. 557-572.

Douglas, Jim jr. 1957. A (An) Uniqueness Theorem for the Solution of a Stefan Problem. *Proceedings of the American Mathematical Society,* Vol. 8, No. 2 (April 1957), pp. 402-408 (Presented to the Mathematical Society on 24 August 1956).

Dovjak, Martin (edited by Knoblehar & Koller). 1859. I. Metorologische Beobachtungen in Chartum, Ulibary und Gondokorò in den Jahren 1852 und 1853. *Jahrbücher der k.k. Centralanstalt für Meteorologie und Erdmagnetismus* 1854, Wien, 6: 499-523.

Dovjak, Martin; Kreil, Karl (commentator). 1859. Resultate aus der meteorologische Beobachtungen. *Jahrbücher der k.k. Centralanstalt für Meteorologie und Erdmagnetismus* 1854, Wien, 6: 523-527.

Evans, George W. II; Isaacson, Eugene; MacDonald, James Keene Lorne (sent from New York University, received 14 October 1949), Stefan-like problems, *Quarterly of Applied Mathematics*, Vol. 8/3 (October): pp. 312-319, 1950.

Evans, George W. II. 1951. A Note on the Existence of a Solution to a Problem of Stefan," *Quarterly of Applied Mathematics*, Vol 9 (1951) pp 185-193.

Fellöcker, Sigismund. 1864. *Geschichte der Sternwarte der Benediktiner-Abtei Kremsmünster*. Linz: J. Feichtinger's Erben. 1st of four parts.

Folk, Reinhard (* 1944 *Neuendettelsau* in Middle Franconia); Holovatch, Yurij (* 1957 Lviv). 2020. Crossing borders in the 19th century and now — two examples of weaving a scientific network, *Condensed Matter Physics* (Institute for Condensed Matter Physics of the National Academy of Sciences of Ukraine), Vol. 23, No 2, 23001: 1–15.

Font Martinez, Francesc (Departament de Física; Universitat Politècnica de Catalunya. BIOCOM-SC - Grup de Biologia Computacional i Sistemes Complexos). June 2014. *Beyond the classical Stefan problem,* PhD thesis

Supervised by Prof. Tim Myers. Universitat Politècnica de Catalunya, Barcelona, Spain.

Fresnel, Augustin Jean. 1866-1870. *Oeuvres completes*. Paris: Senarmont (Sénarmont), Verdet et Léonor Fresnel.

Frischauf J. (1865). Recenzija: K. Nejedli, Elementäre Ableitung der Budar-Horner'schen Auflösungsmethode Höherer Zahlengleichungen (22 strani), *Zeitschrift für die österreichischen Gymnasien.* 16: 865-866).

Graetz (Grätz), Leo. 1880. Ueber das Gesetz der Wärmestrahlung und das absolute Emissionsvermögen des Glases. *Ann.Phys.* (N.F.=3rd series, submitted in July 1880 from Strasbourg) 11 (247)/13: 913-930.

Graetz (Grätz), Leo. 1881. Ueber die Wärmeleitungsfähigkeit von Gasen und ihre Abhängigkeit von der Temperatur, *Ann.Phys.* (N.F.=3rd series) 14 (250)/10: 232-260 (on page 233 discussed just Stefan's papers published in 1872, 1873 and 1875, but not the one of 1879).

Gunderson, James R. July 1966. *A study of heat conduction with phase change*, Master Thesis at the University of Alberta in Edmonton under Lock's supervision.

Gupta, S.C., ed. *The Classical Stefan Problem: Basic Concepts, Modelling and Analysis*, Vol. 45, Elsevier, 2003.

Hachette, Jean Nicolas Pierre. 1827. De l'écoulement des fluides aériformes dans l'air atmosphérique et de l'action combinée du choc de l'air et de la pression atmosphérique, *Annales de chimie et de physique*, 35: 34-53.

Hayakawa, Hisashi; Bruno Philipp Besser; Tomoya Iju; Rainer Arlt; Shoma Uneme; Shinsuke Imada; Philippe-A. Bourdin, Amand Kraml. 8 Jan. 2020. Thaddäus Derfflinger's sunspot observations during 1802-1824, A Primary Reference to Understand the Dalton Minimum, *The Astrophysical Journal*, doi: 10.3847 / 1538-4357 / ab65c9

Hof- und Staats-Handbuch des Kaiserthumes Österreich (Hof- und Staatshandbuch des Österreichischen Kaiserthums). 1844-1868. Wien.

Höflechner, Walter (ed.). 1994. *Ludwig Boltzmann, Leben und Briefe* (The first part of a book: Ludwig Boltzmann, Dokumentation eines Professorlebens). Graz: Akademisch Druck und Verlagsanstalt.

Ingersoll, Leonard Rose; Zobell, Otto Julius; Ingersoll, Alfred Cajori, *Heat Conduction with Engineering and Geological Applications*, McGraw-Hill Book Co., Inc., New York, N.Y., 1948.

James, C. and K. Yuen-Yick. 2000. A brief review of several numerical methods for one-dimensional Stefan problems, *Thermals Science*, 13(2):61-72.

Južnič, Stanislav. 2005. Babbage's calculating machines, the proteus from Postojna cave, and the Carniolan museum society. *Acta Carsologica*. 34/1: 211-220.

Južnič, Stanislav. 2020. *The History of Vacuum and Vacuum Technologies*. Ljubljana: DVTS.

Kaiserseder, Markus. 2013. *Die österreichischen Missionsstationen im Sudan zur Mitte des 19. Jahrhunderts-Wegbereiter eines Kolonialismus?* Masters thesis at Viennese University.

Kieweg, Heinrich (junior in senior). 1998. Das ehrsame Handwerk der Messerer, Scharsacher, Klingenschmiede und Schleifer in Steinbach an der Steyr. Von den Anfängen bis um 1800, *Oberösterreichische Heimatblätter* (Linz: Land Oberösterreich). 52/1-2: 77-105.

Klemenčič, Drago (* 1938 Stomaž, † 2017 Ajdovščina) as editor-in-chief with the pseudonym -ič, Anonymous. 6 September 1981. Marian Koller (1792–1866), *Družina: slovenski katoliški tednik* (Ljubljana), volume 30: 15.

Klemenčič, I. (1885). *Experimentaluntersuchung über die Dielektricitätsconstante einiger Gase und Dämpfe. Sitzungsberichte der Mathematisch-Naturwissenschaftlichen Classe der Kaiserlichen Akademie der Wissenschaften.* Abteilung 2, volume 91: pp. 712–759 (session 19 March 1885 at *Wien.Anzeiger*). Translation: XLIV. Intelligence and miscellaneous articles. Experimental Research Upon the Determination of the Dielectric Constant of

Some Gases, by Klemenčič, a Privat Docent in Graz, *Philosophical Magazine*, 5th series, 19th Volume, number 120 (May 1885) pp. 393-395.

Klemenčič, I. (1890). Über die Untersuchung Elektrische Schwingungen mit Thermoelementen (Hertz'schen Schwingungen Antennenleitung geschalteten Thermoelement Hochfrequenz), *Wien.Berichte,* volume 99. Presented by Boltzmann at session of 17 July 1890, noted at *Wien.Anzeiger, Kaiserliche Akademie der Wissenschaften in Wien, Mathematisch-Naturwissenschaftliche Classe,* IIa, volume 27, no. 18, pp. 173-174. Reprint 1891: *Annalen der Physik.* Volume 278 (N.F. 4), No. of issue 3, pp. 416-424. Abbreviated translated report: XIV. Intelligence and miscellaneous articles. Investigation of Electrical Vibrations with Thermo-Elements. *Philosophical Magazine,* 5th series, 30th Volume, number 184 (September 1890) p. 284-284.

Klemenčič, I. (1892). *Über eine Methode zur Bestimmung der elektromagnetischen Strahlung.* In: *Sitzungsberichte der Mathematisch-Naturwissenschaftlichen Classe der Kaiserlichen Akademie der Wissenschaften.* Abteilung 2a, volume 101, 1892, pp. 310–318 (session 18 February 1892 at *Wien.Anzeiger).* Translation: XIV. Intelligence and miscellaneous articles. On a Method of Determining Electromagnetic Radiation. *Philosophical Magazine,* 5th series, 33rd Volume, number 104 (April 1892) p. 396-396.

Klemenčič, I. (1893). *Beiträge zur Kenntniss der Absorption und Verzweigung elektrischer Schwingungen in Drähten.* In: *Sitzungsberichte der Mathematisch-Naturwissenschaftlichen Classe der Kaiserlichen Akademie der Wissenschaften.* Abteilung 2a, volume 102, 1893, pp. 298–320 (session 16 March 1893 noted at *Wien.Anzeiger).* Abbreviated translated report noting data from *Wien.Berichte*: XIV. Intelligence and miscellaneous articles. On the Disengagement of Heat Occurring when Electrical Vibrations are Transmitted Through Wires. *Philosophical Magazine,* 5th series, 35th Volume, number 212 (June 1893) pp. 537-538.

Klun, Vinzenz Ferrerius. 1850. *Potovanje po Beli reki. Po izvirnem rokopisu Knobleharja.* Ljubljana 1850. Simultaneous print: 1850 *Reise auf dem weissen Flusse. Aus den Original-Maniscripen des General-Vicars von Central-Africa Ignaz Knoblecher, bearbeitet von Vinzenz Ferrerius Klun.* Laibach *(Journey along the White river. From the original maniscripts of the General Vicars of*

Central Africa Ignaz Knoblecher, edited by Vinzenz Ferrerius Klun. Laibach, 1850). Reprint: 1851. Laibach: Ignaz von Kleinmayer & Fedor Bamberg.

Knoblehar, Ignac; put in table by Koller, Marian. 1859. II. Wasserhöhen des Blues Nils in Jahre 1849. *Jahrbücher der k.k. Centralanstalt für Meteorologie und Erdmagnetismus* 1854, Wien, 6: 528 (with one table).

Knoblehar, Ignac; put in table by Koller, Marian. 1859. III. Tagebuch währen einer Reise auf den Weisen Nil non 13. November 1849 bis 16. Jänner 1850 (diary), *Jahrbücher der k.k. Centralanstalt für Meteorologie und Erdmagnetismus* 1854, Wien, 6: 529-533.

Koller, Marian. 1842. XVI. Extract from a letter of M. Marian Köller (sic!), Director of the observatory and Kremsmünsterto Francis Baily Esq., accompanying a catalogue of 208 Stars, *Memoirs of the the British Royal astronomical Society* (London: Moyes & Barclay, 12: pp. 373-380. Checked in preliminary form by Baily, London 7-8 August 1838, sent in an updated letter from Kremsmünster on 17 December 1838 with an introduction to French, read in London on 11 March 1842.

Koller, Marian. 18 July 1850. Comment of paper for Academy: Bericht über die vom Professor Dr. Böhm Abhandlung Beobachtungen von Sonnenflecken und Bestimmung der Rotations-Elemente der Sonne, *Sitzungsberichte der mathematisch-naturwissenschaftlichen Classe der kaiserlichen Akademie der Wissenschaften*, 5 (June-December)/7: 150-153.

Koller, Marian, Printed works on Astronomy, in: Astronomische Nachrichten (Astronomical News). Altona: Hammerich & Lesser:

August 1831, volume 8 No. 184: 315-316 Auszug eines Schreibens and der Herausgeber with obituary of Bonifacius Schwarzenbrunner, mailed on 18 May 1830.

1831. Prof. Koller Beobachtungen des Kometen und Sternbedeckungen von 1830 auf der Sternwarte in Kremsmünster (Observations of elements of orbit of comets in 1830 and 1831, Beobachtungen der Kometen v. J. 1830 and v. J. 1831). Bd. IX. *Astronomische Nachrichten*, volume 9, Issue 20, (Beobachtungen des von den Astronomen Nell de Bréauté am 8 Jänner 1830 entdeckten Kometen auf der Sternwarte Kremsmüster (agronomist-meteorologist-geographer Eléonore Suzanne Nell de Bréauté's (1794 Rouen (Seine-Maritime)-1855 La-Chapelle-de-Bourgay by Dieppe (Seine-Maritime)) comet discovered on 8 January 1831 which made him a corresponding member of astronomical section of Parisian academy after Koller's visit on 12 November 1838 instead of Johan Tobias Bürg who died three years earlier, published in *Comptes-rendus de l'Académie des sciences* including Note sur les *vérifications des glaces d'horizons artificiels* in 1842 14: 408-410) No. 207 pp. 283-284; Halley's comet (Beobachtungen des Kometen und Sternbedeckungen von 1830 auf der Sternwarte in Kremsmünster): Nr. 208, pp. 291-292)

Surveillance of the eclipses of stars. Bd. IX.

1833. Abhandlungen „über den Meridiankreis und das neue transportable Äquatoreale" der Sternwarte zu Kremsmünster. *Astronomische nachrichten* volume X, no. 225: 135-140, sent to the editor on 22 April 1832;

1833. Observations of Jupiter (Beobachtungen des Jupiters in den Jahren 1831 bis 1838) volume X, no. 225: 137;

1833. Observations of the opposition of Saturn in the years 1832 to 1838, sent 6 August 1832, volume X, No. 232: 265-266 compared to rectascensionen (Right ascension as the angular distance of a particular point measured eastward along the celestial equator from the Sun at the March equinox to the hour circle of the point in question above the earth) of Encke's *Berliner Astromisches Jahrbuch* as sent to the editor of *Astronomische nachrichten* on 22 April 1832; Observations of Uranus in the years 1831 to 1838; Observations of Vesta in the Years 1832 to 1838 (as the third largest objects in the asteroid belt, kilometres, discovered by H.W. Olbers in 1802) volume X, no. 225: 139-140 & volume 13 June-July: pp. 369f; Observations of the Juno in the Years 1832 to 1838 volume X, no. 225: 135-140; Observations of Pallas in the years 1832 to 1837 (as after Ceres the largest objects in the asteroid belt, with a mean diameter of 525 kilometres, discovered by H.W. Olbers in 1807); Observations der Ceres in den Jahren 1832 bis 1837; in the volumes X to XVI. Beobachtungen des Biela'schen Kometen im Jahre 1832 with the help of Wolfgang Danner by Bessel's formulas, volume XI, November 1832 no. 253 pp. 223-226 and continuation in next number. Koller observed from 20 November 1832 to 26 November 1832. 3D/Biela comet was rediscovered with its period of 6.6 years on 27 February 1826 in Bohemian Pevnost Josefov (Josefstadt fortress) by Martin Alois David's Prague student of astronomy, officer baron Wilhelm von Biela (* 1782 Rossla in Saxony-1856 Venice). Biela's Comet was the only comet known to intersect the Earth's orbit. Biela's Comet and Comet Encke's orbit were found to be shrinking in size, which was ascribed to the drag of an ether through which it orbited. On 14 January 1846 Biela's Comet was observed as split in several pieces;

Calculations of the meridian (latitude) difference between Kremsmünster and Altona from correspondent Meridian. Bd. XI.

Movements of the stars in the years 1832 to 1838. Volumes XI to XVI.

Observations of Mars (Beobachtungen des Mars) in den Jahren 1832 bis 1835. Volume XIII June-July: pp. 369f.

Observations of Halley's comet (Auffindung des Halley'schen Kometen) zu Kremsmünster in der Nacht vom 21. zum 22. August im Jahre 1835, und Observations of the same (Beobachtungen desselben) vom 21. August bis 15. October. Volumes XII to XIII (no. 11-12: Beobachtungen des Halley'schen Cometen auf der Sternwarte in Kremsmünster).

22 November 1835 pp. 143f, from 21 August 1835 to 15 October 1836 by Koller, Stampfer and Reslhuber pp. 153f & from 21 October 1835 to 13 November 1835 pp. 173f.

Observations of Encke's (Encke'schen) Comets in the Year 1838. Volume XVI (after Koller returned from his European tour);

Observations of solar eclipse (Beobachtung der Sonnenfinsterniss) am 15. März 1838. volume XVI.

Observations and calculations of elements of detected comets on December 2, 1839. Bd. XVII.

Height difference between (Höhenunterschied between) Krakow and Kremsmünster. Bd. XVII.

Beobachtungen u. Elements of the Comets discovered (entdecken) on 25 January 1840;

Observations of elements of orbit of comet detected on March 6, 1840 (Beobachtungen und Elements of March 6, 1840 entdecken Komets); Observations of elements of orbit of comet detected on 25 October 1840 (Beobachtungen und Elements of 26 October 1840 entdecken Comet), Bd. XVIII.

Beobachtung der grossen Sonnenfinsterniß am 7. Juli (astronom.) 1842 zu Kremsmünster. Bd. XX.

Beachtungen des grossen Kometen vom März 1843; Beob. des am 3. Mai 1843 von Mauvais entdeckten Kometen, volume XXI (C/1843 J1 comet discovered by Arago's assistant turned meteorologist Félix-Victor Mauvais (* 1809 Maîche, Doubs department, † 1854 Paris).

Results of meteorological observations of the Kremsmünster in the year 1845; General meteorological results from observations during several years in Kremsmünster;

Beobachtung der Sonnenfinsterniß am 25. April 1846 zu Kremsmünster. Bd. XXV.

Koller, Marian: In Gauß 'and Weber's Results of Magnetic observations (*Resultate aus den Beobachtungen des Magnetischen Vereins* based in Göttingen (Leipzig: Weidmann)):

Magnetic Observations. (Göttingen.)

Reconstructions of the variation of the magnetic declination in August and November terms 1839 to Kremsmünster; Year 1839

Vibrations of the variation of the magnetic declination in February, May, August terms; the declination and horizontal intensity in the November term of the year 1840; Year 1840

The movements of the variation of the magnetic declination and the horizontal intensity on the dates of February, May, August and November in the year 1841; Year 1841, 28.-29. Februar, 29.-30. Mai, 28.-29. August und 27.-28. November im Jahre 1840; Jahrgang 1841. 5: pp. 167, 194-197, 210-215, 228-233, 249, 251, 253, 255, 257.

Koller, Marian in: C. In Dr. J. Lamont's Annals of Meteorology and Earth Magnetism. (Munich.)

Magnetic termination conditions for Kremsmünster from October 1840 to December 1841; Hft. II.

About the horizontal intensity of the Kremsmünster; Hft. IV.

Terminsbeobachtungen vom Jänner und Februar 1842 zu Kremsmünster; Hft. VI.

The results of the majority of the Kremsmünster readings on the flammability of the atmosphere; Hft. X.

Magnetic termination rates for Kremsmünster from March 1842 to December

1843; Hft. IX.
Meteorological terms for the time of the Aquinoctien and the Solstice for Kremsmünster in the year 1844; Hft. XII.

Koller, Marian in: Annalen der k. k. Sternwarte zu Wien. Observations of stars at the Kremsmünster in 1839, 16th volume.

Koller, Marian and Baily in: *Memoirs of the British Royal Astronomical Society*. London: Moyes & Barclay, Published by Society and sold at their apartment in Somerset House:

Baily, Francis. 1826. New tables for determining the places of 2881 stars, *Memoirs R. Astr. Soc.* II: Appendix.

Baily, Francis. 1827. The Society's Gold medal presented to him for his "New tables for determining the places of 2881 stars", *Memoirs R. Astr. Soc.* III: 127f.

Baily, Francis. 12 May 1837. *An Address to Astronomical Observers, Relative to the Improvement and Extension of the Astronomical Society's Catalogue of 2881 Principal Stars*, London, 15 pages of 28 cm, noted at first page of Koller's article in 1842.

1838. D. Kaiser, A Calculation, observatory of Leyden, read 12. March 1837 Memoirs of the Royal Astronomical Society v.10: 303f, using Koller's observations of occultation of 117 Tauri on 19 March 1830 and occultation of α Sagittarii on 31 October 1830 at pp. 311-313, by Koller, *Astronomische nachrichten* No. 208.

1842. XVI. Extract from a letter of M. Marian Köller (sic!), Director of the observatory in Kremsmünster, to Francis Baily Esq., accompanying a catalogue of 208 stars, Volume XII: 373-380. Discussed in preliminary form with Baily in London on 7-8 August 1838, sent in French language from Kremsmünster on 17 December 1838, read in London on 11 March 1842 (A Catalogue of 208 fixed stars; The positions of 78 fixed stars contained in the A.S.C., represented by Baily as not determined with sufficient accuracy, *Monthly Notices of the Royal Astronomical Society*).

Koller, Marian in: *Monthly Notices of the Royal Astronomical Society* (by the vice-president Baily while the president was John Wrottesley, 2nd Baron

Wrottesley (1798-1867) in 1841-1843 after John Herschel's term in 1839-1841 and befoired Baily in 1843-1844) and *Phil. Mag.*:
1838. III The positions of 78 fixed stars contained in the A.S.C., represented by Baily as not determined with sufficient accuracy, deduced from Observations made with the meridian circle of the observatory of Kremsmünster. By M. Köller (Sic!, Koller) the director of the observatory. *Monthly Notices of the Royal Astronomical Society*, Volume 5 (1st volume printed in 1831), Issue 21, read on 11 March 1842, Page 173, indexed on p. 245 (Report of the Council of the Society to the Twenty-third Annual General Meeting, held this day). Reprint: *The London, Edinburgh and Dublin Philosophical Magazine and Journal of Science*, July-December 1842, volume 21: p. 483.

Koller, Marian in: *Jahrbuch des Oberösterreichischen Musealvereines*:
1841: Beiträge zur Landeskunde von Österreich ob der Enns und Salzburg. Über den Gang der Wärme in Oberösterreich. 5: 1-29.

Koller, Marian in: Jahresberichten des Museums "Francisco-Carolinum" in Linz.
Reports on meteorological and magnetic measurements at Kremsmünster from 1839 to 1845; in the years 1840 to 1846.
Posts for public observations from similar barometers.
Abhandlung "über den Gang der Wärme in Oberösterreich" aus zwanzigjährigen Temperaturbeobachtungen (1820-1839) zu Kremsmünster (send in July 1840, read on 3 August 1840), nebst einem Anhange „Untersuchung über die Temperatur des Quellenwassers zu Kremsmünster" Year (Jahrgang) 1841: 1-29;
Abhandlung "über die Feuchtigkeitsverhältnisse der Luft" aus, zehnjährigen Beobachtungen (1833-1842) zu Kremsmünster; Jahrgang 1843.

Koller, Marian in: Denkschriften der kais. Academy of Sciences in Vienna.
Abhandlung "über die Berechnung periodischer Naturerscheinungen", 1849, volume 1: 54-74 (as completely mathematical approach following Bessel, with some analysis of Kremsmünster data on the end; abstract in *Wien.Ber.* read on 13 April 1848 1: 147-149)

Koller, Marian in: *Wien.Ber.*:
Nachrichten über das am 8. Oktober in Kremsmünster gesehene Nordlicht (aus einem Schreiben der dortigen Astronomen P. Aufg. Reslhuber) *Wien.Ber.* 1848 1: 530-530, 606-623.

Koller, Marian in: Yearly of Natural History Association in Brünn (Verhandlungen des naturforschenden Vereines in Brünn, Brno): 1861 (Submitted on 10 December 1861, issued in 1862): Ueber das Passage-Instrument, 1: 3-37. Separate print in Brno: *Über das Passage-Instrument*, 1863. Noted in *Wien.Ber.*, 1863, 48 (January-July): 161. *Wien.Ber.*, 1864 49: 230; *Anwendung das Passage-Instrument zur Zeitbestimmung in einen belieben jedoch bekannten Azimuthe*, 5th session on 12th February 1863, *Wien.Ber.* 47, II: 116.
Abhandlung "Zur Theorie des August'schen Heliostaten". 1863 (Submitted at session on 14 October 1863, printed in 1864), 2: 3-14.. "Beitrag zur Theorie der Röhrenlibelle (Contribution to the theory of the tubular level)". 1864 (Read on 11 January 1865, printed in 1865), 3: 46-59. Abhandlung "Über die Änderungen, welche der Stundenwinkel eines Sternes in einem gegebenen Verticale durch die Fahler des Instrumentes erleidet". 1865 (Sent from Vienna in December 1865 as an update to Koller's paper about a Passage instrument, printed in 1866), 4: 271-277 (published as a 3rd paper after 1st paper of ordinary member professor at Real School in Brno Mendel, J. Gregor (1866). "Versuche über Pflanzenhybriden", Verhandlungen des naturforschenden Vereines in Brünn, Bd. IV für das Jahr, 1865, Abhandlungen: 3–47, while Mendel's meteorology including the ozone was 6th last paper on pages 318-330. The converted Jewish director of the Brno State Higher Real School where the meetings of Society were held and professor of arithmetic-accounting where Mendel was a supplement lower classes teacher of physics and Natural History in 1854-1868, editor of German liberals' daily in Moravia and city councillor in Brno in 1861-1886 Josef Adolf Auspitz (* 1812 Mikulov (Bře[]lav); † 1889 Brno-City), Obituary of Koller noting about his forthcoming article, 5: 40-46 (Peter J. van Dijk; T. H. Noel Ellis, Mendel's journey to Paris and London: context and significance for the origin of genetics, March 2020, PhilSci-Archive)

Kraml, pater Amand (Baptized Gerhard, * 1952 Waldkirchen am Wesen). 2012. Eine "physikalische Reise" durch Westeuropa im Jahr 1838. A talk at the 62nd Annual Meeting of the Austrian Physical Society at University of Graz, 18-21 September 2012. Graz: Karl-Franzens Universität. Pp. 103, 107.

Kraml, pater Amand. May 2013. Auswertung unserer Sonnenfleckenbeobachtungen durch Rudolf Wolf
http://www.specula.at/adv/monat_1305.htm

Kotschy, Theodor (Teodor Koczy). 1865. De plantis nilotico-aethiopicis Knoblecherianis, *Sitzungsberichte der mathematisch-naturwissenschaftlichen Klasse der kaiserlichen Akademie der Wissenschaften (mathematical-natural science class of the Imperial Academy of Sciences)* 50: 351–365.

Kreil, Karl. 1858. Resultate der fünfmonatlichen Beobachtungen in Chartum und dreimonatlichen Beobachtungen in Ulibary (Libo, Ulibo) und Gondokorò (by White Nil), *Denkschriften der Kaiserlichen Akademie der Wissenschaften. Mathematisch-naturwissenschaftliche Klasse* 15/1: 37-68.

Kreil, Karl. 1857. (summary of above by comparison of Sudan Dovjak's measurements with simultaneous data for Milano, Prague and Vienna) Über zwei Reihen meteorologischer Beobachtungen in den afrikanischen Missions-Stationen Chartum und Gondokoró (Gondokoro island 1,200 km south of Khartoum), *Sitzungsberichte der mathematisch-naturwissenschaftlichen Klasse der kaiserlichen Akademie der Wissenschaften* 25: 476-488.

Kyner, Walter Thomas. 1959. On a free boundary value problem for the heat equation, *Quart. appl. Math.* 17/3: 395-310 (received November 3, 1958 as the work done while was a temporary postdoc Member of the Institute of Mathematical Sciences, New York University).

Kyner, Walter Thomas. 1959. An existence and uniqueness theorem for a nonlinear Stefan problem, *J. Math. and Mech.* 8/4: 483-498.

Lamé, Gabriel; Clapeyron, Benoît Paul Émile. 1831. Mémoire sur la solidification par refroidissement d'un globe liquide. *Annales de Chimie et de Physique*, Volume XLVII, 1831, pp. 250-256. Read at Parisian Academy on 10 May 1830.

Lampreht, Rajmund (mentor Andrej Rahten), 2018. *Constitutional landowners in Carniola (Ustavoverni veleposestniki na Kranjskem).* doctoral dissertation, Maribor. Philosophical Faculty.

Landau, Hyman Garshin. 1950 (received on 19 November 1948). Heat conduction in a melting solid, *Quarterly of Applied Mathematics*, v8 n1 (19500401): 81-94.

Lecher, Ernst (1882). Über Ausstrahlung und Absorption, presented by the secretary Stefan on 2 March 1882 for *Wien. Anzeiger* 19/6: 57-59; *Wien.Ber.* volume 85/2: 412-454. Reprint from *Wien.Ber.* abbreviated by Lecher himself in 1882: *Ann. Phys. Chem.* 17 (Neue Folge): pp. 477-518.

Lecher, Ernst (1882). Ueber die Absorption strahlender Wärme in Wasserdampf und Kohlensäure, presented by the secretary Stefan on 9 June 1882 for *Wien. Anzeiger* 19/14: 131-132. Measured in Viennese Physics institute under supervision of Victor von Lang focussing on Tyndall and Röntgen's measurements.

Leibenzon, Leonid Samuilovich (Леонид Самуилович Лейбензон). 1931. *Руководство по нефтепромысловой механике (Rukovodstvo po neftepromyslovoǐ mekhanike, A guide to petroleum industry mechanics, Handbook on petroleum mechanics)*, Moscow: Гос. научно-техническое изд-во (Gos. nauchno-tekhnicheskoe izd-vo, GNTI). Reprint: 1951-1955. Собрание трудов (Collected works). Volume 3: Нефтепромысловая механика (Neftepromyslovaiā mekhanika, Oilfield Mechanics) Moscow: Izdat.Akad.Nauk SSSR, 1955, pp. 435-439.

Leibenzon, Leonid Samuilovich (Леонид Самуилович Лейбензон). 1939. К вопросу о затвердевании земного шара из первоначального расплавленного состояния (K voprosu o zatverdevanii zemnogo shara iz pervonachal'nogo rasplavlennogo sostoyaniya, On the question of the solidification of the globe from its original molten state, On the hardening of the sphere of the Earth from its original fused state). *Izv. Akad. Nauk SSSR, Ser. Geograf, i Geofiz.*, No. 6: pp. 625-660. Reprint: Collected works.

Leuckart, Friedrich S. (editor). 1840. (Amtlicher) Bericht über die Versammlung Deutscher Naturforscher und Ärzte, Freiburg 1838, Freiburg: Emmerling.

Lightfoot, Nicholas Morpeth Hutchinson, The Solidification of Molten Steels, *London Match Soc.*, series 2, Vol. 31/1750, 1929 (1930), pp. 97-116.

Lock, Gerald Seymour Hunter. 1969. On the Use of Asymptotic Solutions to Plane Ice-Water Problems. *Journal of Glaciology*, v8 n53 (1969): 285-300.

Lock, Gerald Seymour Hunter; Gunderson, James R.; Quon, Donald (Department of Chemical Engineering, University of Alberta); Donnelly, John K. (PhD in chemical engineering with Lock in Alberta in 1968, Department of Chemical Engineering, University of Calgary Canada). 1969. A study of one-dimensional ice formation with particular reference to periodic growth and decay, *International Journal of Heat and Mass Transfer*, v12 n11 (196911): 1343-1352.

Lorenz, Ludwig Valentin. 1881. Om Metallernes Ledningsevne for Varme og Elektricitet (About the Conductivity of Metals for Heat and Electricity). *Det Kongelige danske videnskabernes selskabs skrifter. Naturvidenskabelig og mathematisk afdeling* (Kjøbenhavn: B. Luno), Raekke 6, volume 2, 42 pages. Translation on 15 June 1881 with continuation on 15 July 1881: Ueber des Leitungsvermögen der Metallen für Wärme und Elektrizität, *Ann.Phys.* (3) 13/7: 422-447, 13/9: 582-606.

Loschmidt, J. (1865). Zur Grösse der Luftmoleküle. *Sitzungsberichte der Kaiserlichen Akademie der Wissenschaften Wien*. 52 (2): 395–413 (22nd session on 12 October 1865). Abbreviated reprint: Ueber die grösse der Luftmoleküle, *Schlömilchs Zeitschrift für Mathematik und Physik,* volume 10 (1865): 512f.

Lukas, Franz. 1859. Edited the anthology. *Jahrbücher der k.k. Centralanstalt für Meteorologie und Erdmagnetismus* 1854, Wien, 6: 499-527.

Lukas, Franz (Kreil's Prague assistant from December 1851). 1859. Geographische Skizze zum verschiedenen Tagebuche, *Jahrbücher der k.k. Centraelanstalt für Meteorologie und Erdmagnetismus* 1854, Wien, 6: 533–536 (with a map).

Lummer, Otto. 1900. Le rayonnement des corps noirs. Raports presentés au *Congrés International de physique réuni au Paris en 1900*. (ed. Lucien Poincaré; Charles-Édouard Guillaume). Paris. 41-99.

Lutstorf, Heinz Theo. 1893. *Professor Rudolf Wolf und seine Zeit 1816-1893.* Zürich: Schriftenreihe der ETH-Bibliothek, Band 31.

Maxwell, James Clerk. (1872). On Induction of Electric Currents, *Phil.Mag.* 43/289: 529-538.

Maxwell, James Clerk, On Loschmidt's Experiments on Diffusion in Relation to the Kinetic Theory of Gases, Nature, Vol. 8/198 (14 August 1873), pp. 298-300.

Maxwell, James Clerk, Molecules (British association (BAAS) lecture at Bradford o 22 September 1873 after the ill president James Joule was replaced by the Scottish chemist Alexander W. Williamson (1824-1904)), Nature, Vol. 8/204 (25 September 1873), pp. 437-441.

Mach, Ernst. 1860. Über die Änderung des Tones und der Farbe durch Bewegung, in: *Sitzungsberichte der kaiserlichen Akademie der Wissenschaften in Wien. Mathematisch-naturwissenschaftliche Classe*, Volume (Bd.) 41: 543-560. Reprint: *Annalen der Physik*, 2nd series (Folge), volume (Bd.) 112: 58-76.

Mach, Ernst. VI. Ueber die Controverse zwischen Doppler und Petzval, bezüglich der Aenderung (Änderung) des Tones und der Farbe durch Bewegung. *Schlömilchs Zeitschrift für Mathematik und Physik*, zvezek (volume) 6 (1861) 120-126.

Mach, Ernst. 1864. Vorläufige Bemerkung über das Licht glühende Gase. *Schlömilchs Zeitschrift für Mathematik und Physik*, volume 9 (1864): 69-70.

Meschutar, Andreas (Andrej Mešutar as a president). 1851. *Das Comité (Komité) des Marien-Vereins zur Unterstützung der Mission in Central-Africa*. Wien.

Meschutar, Andreas (Andrej Mešutar as a president). 10 October 1851. Ein Aufruf (A call), *Die Mission von Central-Afrika zur Bekehrung der Neger und der Marien-Verein (The mission of Central Africa to convert the Negroes and the Mary's Association)*. Wien: k.k. Hof- und Staatsdr.

Miranker, Willard L.; Keller, Joseph B. 1960. The Stefan problem for a nonlinear equation, *J. Math. and Mech. (Journal of Mathematics and Mechanics)* 9/1: 67-70.

Mitrović, Jovan. (University of East Sarajevo, 2012). Jožef Stefan and the dissolution-diffusion phenomena - Not only a historical note. *International Journal of Pharmaceutics*, 15 July 2012, 431/1-2: 12-15.

Moth, F.K. 1827. *Theorie der Differenzial-Rechnung und ihre Anwendung zur Auflösung der Probleme der Rectification, der Complanation und der Kubierung unabhängig von der Betrachtung der unendlich kleinen oder verschwindenden Grössen, der unendlichen Annäherung, oder der Grenzverhältnisse u. s. w.* Prague: Kronberger & Weber.

Moth, F.K. 1852. *Lehrbuch der Algebra für den Gymnasial-Unterricht bearbeitet von Fr. Moth,* Linz: F. Eurich.

Obermayer, Albert von. Fire of St. Elm. *Philosophical Magazine,* 5th series, 25th Volume, number 155 (April 1888) pp. 323-324.

Oersted, Hans Christian; Fourier, baron Jean-Baptiste-Joseph, On Sur quelques nouvelles Expériences thermo-électriques faites par M. le Baron Fourier et M. Oersted (Some New Experiments in Thermoelectricity Performed by Baron Jean-Baptiste-Joseph Fourier and Oersted), *Annales de chimie,* 22 (1823), 375-389.

Oersted, Hans Christian; Fourier, baron Jean-Baptiste-Joseph. 1823. Oersted et Fourier poursuivi les recherches de Seebeck thermoélectriques, *Journal de physique de chimie,* vol 96: 300-302 (Continued by Oersted's report on compressibility of water on 96: 302-303)

Pouillet, Claude Servais Mathias. 1828. *Élémens de physique expérimentale et de météorologie,* Pariz.

Prasad, B. S. N.; Mascarenhas, Rita. April 1978. A laboratory experiment on the application of Stefan's law to tungsten filament electric lamps. *American Journal of Physics.* 46/4: 420-423.

Provinzial-Handbuch von Osterreich ob der Enns und Salzburg für das Jahr 1843 (1848). Linz: Joseph Wimmer, izdajatelj Museum Francisco-Carolinum.

Puschl, Karl, Über das Entstehen progressiven Bewegungen durch Verbrauch lebendiger Kraft oscillatorischer Bewegungen, *Wien.Ber.* (1852) 9: 173-185.

Puschl, Karl, Über die Einwirkung von Licht- und Wärmewellen auf bewegliche Massenteilchen, *Wien.Ber.* (1855)

Puschl, Karl, Uber das Ursprung und die Gesetze der Naturkrafte nach mechanische Theorie der Wärme (in his Gesamtverzeichnis noted as: '...der Molekularkräfte nach dem Principe der Kräfterhaltung'), Program Melk 1861 pp. 3-44 and 1862 pp. 3-21

Puschl, Karl, Über den Wärmezustande der Gase, *Wien.Ber.* (1862) 45: 357-384

Puschl, Karl, Notiz über die Molekularbewegung in Gasen, *Wien.Ber.* (1863) 48: 35-39.

Puschl, Karl, *Das Strahlungwermögen der Atome als Grund der physikalischen und chemischen Eigenschaften der Körper*, Gerold, Wien 1869, V+324 pages.

Puschl, Karl, Über eine kosmische Anziehung, welche die Sonne durch ihre Strahlen ausübt, *Wien.Ber.* (1870) 59: 299-318 (not about the kinetic theory)

Puschl, Karl, Über Warmemenge und Temperatur der Körper, Wien.Ber. (1870) 62: 171-196; the continuation appeared at the same volume with equal title on additional 20 pages.

Puschl, Karl, Bemerkungen zur specifische Wärme der Kohlstoffs, Wien.Ber. (1874) 69: 142-147

Puschl, Karl, Über Körperwärme und Aetherdichte, Wien.Ber. (1874) 69: 324-333

Puschl, Karl, Über eine Modification der herrschenden Gastheorie, Wien.Ber. (1874) 70: 413-433.

Puschl, Karl, Über das Verhalten gesättigter Dämpfe, Wien.Ber. (1874) 70: 571-588.

Puschl, Karl, Über die Volumenveränderung der Kautschuks durch Wärme, Wien.Ber. (1875) 71: 95-98.

Puschl, Karl, Über den Einfluss von Druck und Zug auf die thermischen

Ausdehnungscoefficient der Körper und über das bezügliche Verhalten von Wasser und Kautschuk, Wien.Ber. (1875) 72: 245-256.

Puschl, Karl, Erniedrigung der Temperatur des Dichtigkeitsmaximums das Wassers durch Druck (1875) 72: 283-286.

Puschl, Karl, Grundzüge der Aktinischen Wärmetheorie, Wien.Ber. (1878) 77: 471-500

Puschl, Karl, Ueber die latente Wärme der Dämpfe, Wien 1879 (3: Wien 1883), comment by I.G.Wallentina (* 1852) in: Z.Math.Phys. 33: 228-229

Puschl, Karl, Der zweite Hauptsatz der mechanischen Wärmetheorie und das Verhalten des Wassers, Wien.Ber. (1884) 89: 631-635

Puschl, Karl, Über den Wärmezustand der Gase, Wien.Ber. (1902) 111: 187-214

Puschl, Karl, Über das Gesetz von Dulong und Petit, Wien.Ber. (1903) 112: 1230-1245

Reden, Friedrich Wilhelm Otto Ludwig baron. 2 June 1857. Ueberblick (Überblick) der bisherigen Thätigkeit (Tätigkeit) und Erfolge österreichischer geistlicher und weltlicher Sendboten in Central-Afrika. *Mitt. d. Geogr. Ges. (Mitteilungen der k.k. Geographischen Gesellschaft in Wien)* Wien: M. Auer 1857, 1: 150–160, 169.

Redozubov, Dmitrij Vasil'evič (Редозубов Дмитрий Васильевич (1904—1978)), *Закономерности температурного поля вечной мерзлоты на Воркуте.* Труды Института мерзлотоведения АН СССР, т. I. Изд-во АН СССР, 1946.

Redozubov, Dmitrij Vasil'evič (Редозубов Дмитрий Васильевич (1904—1978)), "The Stefans Problem for a Linear Initial Temperature Distribution in a Semi-Infinite Medium," Bull.
Acad. Sci., USSR Geophys. Ser.
No. 4, April 1962, p. 364.

Reich, Karin & Elena Roussanova. 2018. Karl Kreil und der Erdmagnetismus; Seine Korrespondenz mit Carl Friedrich Gauß im historischen Kontext. Paragraph 4. Kreil als Mitglied des Göttinger magnetischen Vereins (after his innitial studies in Kremsmünster), str. 89 – 114. Vienna: Verlag der Österreichischen Akademie der Wissenschaften

Reitlinger, Edmund, On the stratification of the electric light, *Philosophical Magazine*, 4th series, 1863, 25th Volume, number 168 pp. 317-318

Reslhuber, Pater Augustin, 1866: Eine Lebensskizze von Dr. Marian (Wolfgang) Koller. *Zeitschrift der Österr. Gesellschaft für Meteorologie*, Wien (eds. Jelinek and J. Hann) 1/23 (11 December 1866): 353-361, 1/24 (21 December 1866): 375-382. Modified Reprint: 1866/67. Die Feierliche Sitzung der Kaiserlichen Akademie der Wissenschaften am 31 Mai 1867, Wien Hof- und Staatsdruckerei, 1867, 15: 105–143. Wien: Carl Gerold's Sohn.

Robida, Karl (Karel Lucas, Klagenfurt 10 November 1863). 1864. XV. Zur Theorie der Gase. *Schlömilchs Zeitschrift für Mathematik und Physik,* volume 9/3 (1864): 218-221. Against scientifically meaningless Clausius' atmosphere with aether (Robida 220-221) before Robida published his Höhenbestimmungen der Erdatmosphäre in Klagenfurt program in 1866. Robida also disliked Krönig's Ann.Phys. papers & Stefan's (218) paper in *Zeitschrift für Mathematik und Physik* 1863 8: 355f after Grailich criticized Robida in 1858 and 1859 (Grailich, *Zeit.öst.Gymn.* 10: 319-321). Robida proclaimed Stefan-Krönig-Clausius' theory as self-contradictory, scientifically meaningless and false; therefore, Robida formulated early form of Loschmidt's reversibility paradox on p. 219 as the trajectory of molecules under gravity must be parabolic to prevent their escape into cosmos. Robida supported his fellow Benedictine Puschl. Robida also praised Johann August Grunert's (1797–1872) student who took over his *Archiv der Mathematik und Physik* in 1872 Berliner university professor from 1870 Ernst Reinhold Eduard Hoppe (1816 Naumburg in Southern Saxony-1900. Hoppe published Ueber die Wärme als Aequivalent der Arbeit, *Annalen der Physik und Chemie*, v173 n1 (1855): 30-34 against Clausius. Hoppe also published: Ueber Bewegung und Beschaffenheit der Atome, *Annalen der Physik und Chemie*, v180 n6 (1858): 279-292). Robida also supported the Berliner Grammar school professor Emil Carl Georg Gustav Jochmann (1833-1871). Robida was probably unable to publish that same article in *Wien.Ber.* where his former student Stefan was already a corresponding member but Robida was not).

Robida, Karl. 1866. Höhenbestimmungen der Erdatmosphäre. *Klagenfurt Grammar School program.*

Rubinstein, Lev Isakovič (Rubinštein, Лев Исакович Рубинштейн). 1947. О решении задачи Стефана (On the solution of Stefan Problem). *Известия АН СССР. Серия: География и геофизика.* 11 (Numbers 1 and 6.): pp. 37-54 (MR 8 516).

Rubinstein, Lev Isakovič. 1947. Об определении положения границы раздела фаз в одномерной задаче Стефана (Stefan Problem). *Докл. АН СССР (Doklady Akademii Nauk SSSR, Acts of Academy of Sciences of USSR)*. Сер. (series) A. Volume 58, No. 2: pp. 217-220 (On the determination of the portion of the boundary which separates two phases in the one-dimensional problem of Stefan, *Acts of Academy of Sciences of USSR*).

Rubinstein, Lev Isakovič. 1951. On the uniqueness of solution of the homogeneous problem of Stefan in the case of a single-phase initial condition of the heat conducting medium, *Докл. АН СССР (Doklady Akademii Nauk SSSR, Acts of Academy of Sciences of USSR)*. Сер. (series) A. Volume 79, No. 2: 45-47.

Rubinstein, Lev Isakovič. 4 April 1953. *Докл. АН СССР (Doklady Akademii Nauk SSSR, Acts of Academy of Sciences of USSR)* 90: pp. 987-990. Sent from Turkmenian branch of the All-Union Research institute in west Turkmenian center of oil industry Balkanabat (Балканабат, Neftedag, Nebit-Dag) on 4 April 1953. Presented to the academy on 24 April 1953 by one of the leading designers of Soviet nuclear bomb project the academician head of Moscow computer department Sergei Lvovich Sobolev (Сергéй Львóвич Сóболев, 1908 Petersburg-1989 Moscow). Translation in December of the same year 1953: *On the dynamics of evaporation* (Verdunstung) *of ideal multi-component liquid mixtures*. Oak Ridge, Tennessee: U.S. Atomic Energy Commission, Technical Information Service.

Rubinstein, Lev Isakovič. 21 May 1953. *Докл. АН СССР (Doklady Akademii Nauk SSSR, Acts of Academy of Sciences of USSR)* 91: pp. 767-769. Sent from Turkmenian branch of the All-Union Research institute in west Turkmenian center of oil industry Balkanabat (Балканабат, Neftedag, Nebit-Dag) on 21 May 1953. Presented to the academy on 23 May 1953 by one of the leading designers of Soviet nuclear bomb project the academician Sergei Lvovich Sobolev (Сергéй Львóвич Сóболев, 1908 Petersburg-1989 Moscow). Translation in December of the same year 1953: *On the Dynamics of Evaporation of Polycomponent Solutions in a Nonvolatile Solvent*. Oak Ridge, Tennessee: U.S. Atomic Energy Commission, Technical Information Service.

Rubinstein, Lev Isakovič. 1958. К вопросу о численном решении интегральных уравнений задачи Стефана, *Изв. вузов. Матем. (Известия высших учебных заведений. Математика,* On question of numerical solving of integral equation of Stefan Problem, *Reports of Higher Schools of Mathematics*), No. 4, pp. 202-214.

Rubinstein, Lev Isakovič. 1967. *Проблема Стефана*. Рига (Riga: Звайгзне (Latvijskij Gosudarstvennyj universitet imeni Petra Stučki, Vyčislitel'nyj centr), 458 pages. Translation of Alan D. Solomon: 1971. *The Stefan Problem, Translations of Mathematical Monographs*, vol. 27, Providence, R.I.: American Mathematical Society.

Rubinstein, Lev Isakovič. 1971. *The Stefan problem*, Trans.Math.Monographs-Vol. 27, Amer.Math.Soc., Providence.

Friedman, T. S.; Rubinstein, Lev Isakovič; H. M. Geiman. 1972. On the Transfer of Non-electrolytes Across Deformable Semipermeable Membranes. *Heat and Mass Transfer*. Vol. 3. Iykoff, A. V., ed. Minsk: Nauka I Technika.

Geiman, H. M.; S. V. Krumin; Rubinstein, Lev Isakovič; T. S. Friedman. 1972. IV International Biophysics Congress; Abstracts of Contributed Papers; Sections IX–XV; EXV b 1/6. August 7–14; Moscow.

Rubinstein, Lev Isakovič. 1974. Passive transfer of low-molecular nonelectrolytes across deformable semipermeable membranes. - I: Equations of convective diffusion transfer of non-electrolyteas across deformable membranes of large curvature. *Bull. of Math. Biol.* 36, 365–379. Russian translation in 1999: *Tsitologia*, 41, 506–511 entitled: separating subdomain containing the solution of an enzyme and subdomain.

Geiman, H.M.; Rubinstein, Lev Isakovič (both went from Riga to School of Applied Science and Technology, The Hebrew University of Jerusalem in 1974 and were still there in 1982). 1974. Passive transfer of low-molecular nonelectrolytes across deformable semipermeable membranes. II: Dinamics of a single muscle fibre swelling and shrinking and related changes of the T-system tubule form. *Bulletin of Mathematical Biology*, 36, 379–401.

Rubinstein, Lev Isakovič (Jerusalem); Antonio Fasano (Firenze); Mario Primicerio (Firenze). December 1980. Remarks on the analyticity of the free boundary for the one-dimensional Stefan problem, *Annali di matematica pura ed applicata*, volume 125, number 1, pages 295-311.

Rubinstein, Lev Isakovič; Geiman, H.M.; Shachaf, M. (all at School of Applied Science and Technology, The Hebrew University of Jerusalem in 1982). 1982. Heat Transfer with a Free Boundary Moving Within a Concentrated Thermal Capacity, *IMA Journal of Applied Mathematics*, Volume 28, Issue 2, March 1982, pp. 131–147.

Rubinstein, Isaak; Charach, Ch. (Ben-Gurion University of the Negev). 1992. Pressure-temperature effects in planar Stefan problems with density change. *Journal of Applied Physics*, 71:1128.

Rubinstein, Lev Isakovič; Rubinstein, Isaak. 1995. *Partial Differential Equations in classical mathematical physics*. New York: Cambridge.

Rubinstein, Lev Isakovič; Martuzāns, Bruno Janovičs (Martuzans from Latvia). 1995. *Free boundary problems related to osmotic mass transfer through semipermeable membranes,* (Volume 4 of Gakuto international series: Mathematical sciences and applications), Tokyo: Gakkōtosho Co.

Rubinstein, Lev Isakovič. 1999. On the disturbance of osmotic equilibrium as a possible cause triggering plasmalemma invagination or protrusion at the initial stage of phagocytosis. *Theory Biosci.* 118, 141–160.

Rubinstein, Lev Isakovič. 2003. A mathematical problem related to the deformation of membranes, bounding the intracellular tubules, at the initial stage of micropinocytosis. *Theory in Biosciences* (Urban & Fischer), 121, no. 4 (4 March 2003): 395-417.

Ruoff, Arthur Louis. 1958. An alternate solution of Stefan's problem, *Quart. appl. Math.* 16: 197-201.

Salvatori, L. and N. Tosi, Stefan Problem through Extended Finite Elements: *Review and Further Investigations, Algorithms* 2(3), 2009, DOI: 10.3390/a2031177

Schneebeli, Heinrich. 1884. Untersuchungen im Gebiet der strahlenden Wärme, *Ann. Phys.* (3) 22/7: 430-438.

Schrötter, Anton von Kristelli. 1849. Report about academic meteorological commission at session on 15 March 1849 (Sitzung vom 15. März 1849). *Wien.Ber.*, 1: 169-176 (Mathematisch-naturwissenschaftliche Classe. Jahrgang 1849, Erste Abtheilung, Die Monate Jänner – Mai). Koller and Doppler were named as the additional members of commission on final page 176.

Schrötter, Anton von Kristelli, 1863, General-Secretary of Academy after 1851 Obituary of Kreil. *Die feierliche Sitzung der kaiserlichen Akademie der Wissenschaften*, Vienna on 30 May 1863 pp. 87, 118-152. On pp. 145-147 Schrötter included Koller's about Kreil reminiscences reprinted from *Wien. Ber.* of 15 May 1863 pp. 427–428 (Koller, Marian, Bericht über ein nahezu vollendetes handschriftliches Werk der Verstorbenes Akademikers Dr. K. Kreil (died 21 December 1862) über die Klimatologie in Böhmen, 13th session on 15 May 1863, *Wien.Ber.* 47, II: 391, 427–428. Koller's reminiscences on Kreil were reprinted in: *Die feierliche Sitzung der kaiserlichen Akademie der Wissenschaften*, Vienna on 30 May 1863 pp. 145-147 as a part of General-Secretary of Academy after 1851 Anton Schrötter von Kristelli obituary of Kreil on pp. 87, 118-152. There, Koller especially highlighted Kreil's research of the history of Bohemian meteorology focused on the Jesuits Joseph Stepling (1716-1778) and Antonin Strandt (Strnad, 1746-1799), the Premonstratensian Aloys Martin David (1757 Dřevohryzech u Teplé-1836 Teplé) and the Piarist Cassiano Hallashka (Hallaschka, Franz Ignaz Cassian, František Ignác Kassián Halaška, 1780 Budišov nad Budišovko (Bautsch)-1847 Prague). Stefan's work about translation of heat was published just before Koller's obituary of Kreil).

Schuster, Peter Maria (* 1939); Krumpel, Helmut (* 1941, ur.). 2006. *Schöpfungswoche – Tag drei: Josef Stefan zur Huldigung, Zvezek 3*, Vienna: Living Edition, STARNA Ges.m.b.H.

Sekulić, Martin. „Beziehung zwischen der elektromotorischen Kraft und der chemischen Wärmetönung." *Anzeiger der Kaiserlichen Akademie der Wissenschaften*, 16, 15 (1878): 129. Read by the secretary of mathematical-Natural History class of the academy (Sekretär der mathematisch-naturwissenschaftlichen Klasse) j. Stefan on 21 June 1878.

Senčar-Čupovič, Ilinka. 1983. Vitalizam u južnoslavenskoj kemijskoj literaturi 19. stoljeća. *Građe za Povijest znanosti u Hrvatskoj (Discussions for history of science in Croatia)*, Zagreb: JAZU, 73-118.

Selig (Selling), Franz. 1956. Bemerkungen zum Stefanschen Problem, *Oester. Ingenieur Arch. (Österreichische Ingenieur- und Architekten, Österreichische Ingenieurzeitschrift: ÖIZ)*. 10: 277-280.

Sestini, Benedetto. 1845, 1847. *Memoria sopra i colori delle stelle del Catalogo di Baily, osservati dal P.B. Sestini*, Roma: Marini, 1845: pp. IV, VI & *Memoria seconda*. 1847: pp. 12-14.

Sestini, Giorgio. 1952. Esistenza di una soluzione in problemi analogli a quello di Stefan, *Rivista di Matematica della Universita di Parma*, series 1 volume 3 (1952) pp. 3-23.

Sestini, Giorgio. 1952. Esistenza ed unicità del problema di Stefan relativo a campi dotati di simmetria, *Rivista di Matematica della Universita di Parma*, series 1 volume 3 (1952); pp. 103-113.

Sestini, Giorgio. 1953. Esistenza ed unicità in problemi analoghi a quello di Stefan, *Proc.of the Eighth Inter.Congress on Theoretical and Applied Mechanics, Istanbul* (1953), pp. 439-440.

Sestini, Giorgio. 1957. Sopra un teorema di unicità in problemi unidimensionali analoghi a quello di Stefan (About the uniqueness theorem in one-dimensional problems analogous to Stefan's), *Bollettino dell'Unione matematica italiana (Bulletin of the Italian Mathematical Union)*, series 3, volume 12, pp. 516-519.

Sestini, Giorgio. 1959. Ancora su di un teorema di unicità in problemi unidimensionali analoghi a quello di Stefan, *Bollettino dell'Unione matematica italiana (Bulletin of the Italian Mathematical Union)*, series 3, volume 14, pp. 373-375.

Sestini, Giorgio. 1960. Problemi di diffusione lineari e non lineari analoghi a quelli di Stefan, *Conferenze del Seminario di matematica dell'Università di Bari*, Bologna: Zanichelli, no. 55-56; pp. 1-26.

Sestini, Giorgio. December 1960. Sul problema unidimensionale non lineare di Stefan in uno strato piano indefinito, *Annali di matematica pura ed applicata,* series 4, 1960, vol. 51, pp. 203-224.

Sestini, Giorgio. *Lezioni di meccanica razionale,* Firenze 1961.

Sestini, Giorgio. December 1961. Sul problema non lineare di Stefan in strati cilindrici o sferici, *Annali di matematica pura ed applicata,* series 4, 1961, vol. 56/1: pp. 193-207.

Sestini, Giorgio. Su un problema non lineare del tipo di Stefan, *Atti della Accademia nazionale dei Lincei. Rendiconti Lincei. Scienze fisiche e naturali,* series 8, 1963, vol. 35, pp. 518-523.

Sestini, Giorgio. Problemi analoghi a quello di Stefan e loro attualità, *Rendiconti del Seminario matematico e fisico di Milano,* 1967, vol. 37, pp. 39-50.

Sitar, Sandi. 1993. *Jožef Stefan pesnik in fizik; ob stoletnici smrti.* Ljubljana: Park.

Stefan, J. (1862). Ueber (Über) die Integralsinus und einige verwandte bestimmte Integrale, *IV Jahres-Bericht der Öffentlichen Ober-Realschule in der Inneren Stadt (Bauernmarkt Nr. 11) zu Wien (Öffentlichen Ober-Realschule auf dem Bauernmarkte (in der Inneren Stadt) zu Wien.* Wien: Anton Schweiger). 41-48. Abstract published in *Schlömilchs Zeitschrift für Mathematik und Physik,* volume 5 (1862); Ueber (Über) das bestimmte Integral v *Schlömilchs Zeitschrift für Mathematik und Physik,* zvezek 8, 1863, pp. 229-231.

Stefan, J. (1864). Review: Ad. Wüllner. Lehrbuch der Experimentalphysik mit theilweiser benutzung von Jamins cours de physique de l'École Polytechnique (first two parts: mechanics, optics. Leipzig: B.G. Teubner, 1863, *Zeitschrift für die österreichischen Gymnasien,* 15: 158-166.

Stefan J. (1865). Review: Schellbach, K. H. 1864. Die Lehre von den elliptischen Integralen und der Thetafunctionen (Funktionen). Berlin: Reimer. *Zeitschrift für die österreichischen Gymnasien.* 16: 747-748.

Stefan J. (1865). Book review: Trappe, Albert Ferdinand. 1865. Die Physik für den Schulunterricht bearbeitet von Albert Ferdinand Trappe, 3rd edition on 296 pages. Breslau (Wrocław): Ferdinand Hirt. *Zeitschrift für die österreichischen Gymnasien.* 16: 670-673

Stefan J. (1865). Book review: Neumann, Carl. 1863. Die Magnetische Drehnung des Polarisationebene des Lichtes: Versuch Einer Mathematischen Theorie. Halle: Waisenhausen, VIII + 82 pages. *Zeitschrift für die österreichischen Gymnasien.* 16: 305-306. On page 472 about Stefan academician.

Stefan, J. (1871). Über den Einfluß der Wärme auf die Brechung des Lichtes in festen Körpern, *Wien.Ber.*, 1871, 63, pp. 223–245 (1871).

Stefan, J. (1872). Über die mit dem Soleil'schen Doppelquarz ausgeführten Interferenzversuche, *Wien.Ber.* 66, II, pp. 325–354. About the use of optical apparatus of Jean-Baptiste-François Soleil (1798-1878).

At least a dozen of English abridgements of Stefan's papers published in *Philosophical Magazine* were mostly translated from *Wien. Anz. (Anzeiger der (Kaiserlichen) Österreichischen Akademie der Wissenschaften, Mathematisch-Naturwissenschaftliche Klasse)*:

Stefan, J. (1873). *Poskusi o izparevanju (Versuche über die Verdampfung),* Wien.Ber., 68 (session on 23 October, printed in November 1873), II, pp. 385–423. Abstract: LXIV. Intelligence and miscellaneous articles. Experiments on evaporation. *Philosophical Magazine*, 4th series, 46th Volume, number 308 (December 1873) pp. 483-484.

Stefan, J. (1874). *K teoriji magnetnih sil (Zur Theorie der magnetischen Kräfte),* Wien.Ber. 69 (12. 2. 1874), II, pp. 165–210 (1874). Abstract: Contribution to the theory of magnetic forces. *Philosophical Magazine*, 4th series, 47th Volume, number 312 (April 1874) pp. 318-319. In his three-part discussion, Stefan discussed magnetic force according to Ampère and magnetic induction according to Koller's acquaintance Denis Poisson. In support of Maxwell's new field theory, Stefan raised doubts against Wilhelm Weber's (1804-1891) competing theory by results of measurements of the ferromagnetism published by Helmholtz and Maxwell's protégé Henry Augustus Rowland (1848-1901 Baltimore), Gustav Kirchhoff (1824-1887), his

Russian doctoral student Alexander Stoletow (Aleksander Grigorievich Stoletov, Алекса́ндр Григо́рьевич Столе́тов, 1839-1896 Moscow), and Gustav von Quintus-Icilius (1824–1885 Hannover).[34]

Stefan, J. (1874). *Poskusi o navidezni adheziji* (*Versuche über die scheinbare Adhäsion*), Wien.Ber. 69 (30. 4. 1874), II, pp. 713–728; tudi PA 154, pp. 316 (1874). Abstract: Experiments on apparent adhesion. *Philosophical Magazine*, 4th series, 47th Volume, number 314 (June 1874) pp. 465-466. J. Stefan argued that the adhesion of two marble slabs does not involve forces between invisibly small components that were already for two centuries a means of proving the existence of a vacuum. Stefan was able to reject the old idea of a vacuum between two compressed planes, as he had many other more appropriate tools at his disposal to prove his kinetic atomism. Stefan experimented with glass plates immersed in water, salt water, alcohol, and air proved that their mutual force decreased with the fourth power of the distance. He linked the problem to capillary hydrodynamics and no longer to statics. Stefan determined the coefficients of internal friction (viscosity) based on the results of the student of Arago and Ampère at the Parisian Polytechnic, Jean Léonard Marie Poiseuille (1797 Paris – 1869 Paris), based on the explanations of Oscar Emil Meyer and Maxwell. A similar "Huygens effect" was explained by Huygens' confidant, the poet-critic Jean Chapelain (1595–1674), with the assumption of pyramid-shaped atoms supposed to take a leading role in water under depleted air. Of course, Huygens rejected that idea as he saw in the phenomenon the pressure of a tiny invisible substance, just like Hooke. Later, Laplace tried to explain Huygens' effect with capillarity. In the 19th century, these phenomena interested the Belgian Jean-Jacques-Daniel Dony (Abbot Dony, 1759 Liège – 1819), the Englishman Charles Frederick Partington (Wartington, † 1857) as lecturer to various institutions and assistant librarian to London Institution established in 1806. After them, the German Helmholtz continued similar research. In the 20th century, research was continued by Huygens' Dutch colleagues Kamerlingh-Onnes, his student Willem Hendrik Kees (1876 Texel–1956 Leiden), and Casimir at the Philips Research

---

[34] J. Stefan, Zur Theorie der magnetischen Kräfte, *Wien. Ber.* (12 February 1874), II, **1874**, 69, pp. 165–210. Translated abstact: Contribution to the theory of magnetic forces. *Philosophical Magazine*, 4th series, April **1874**, 47th volume, no. 312, pp. 318–319; M. Koller, Notices on lectures of Prof. Dr. Jozef Stefan under the title „Über Elektrodynamik und Theorie der Induction" v letnem semestru **1862/63**, 26 folios (= 104 pages), Direktions-Archiv der Sternwarte Kremsmünster, Koller-Manuskripte *32*, pp. 52, 55, 60, 79-81, 94, 96, 98.

Laboratories in 1948. The electromagnetic causes of the Casimir effect suggested that Newton may have been among the closest to truth, but the mechanical forces of adhesions today complement the intermolecular forces that cause the adhesion of different particles and the cohesion of similar particles. Therefore, despite Stefan's experiments, the vacuum is still involved in modern explanations.[35]

Stefan, J. (1879). On the Diffusion of Liquids, *Philosophical Magazine*, 5th series, 7th Volume, number 40 (January 1879) pp. 74-75 (abbreviated from: Über die Diffusion der Flüssigkeiten (1: Über die optischen Beobachtungsmethoden (*O difuziji tekočin I. O optičnih opazovalnih metodah*)), *Wien.Ber.* II, Volume 78 (5 December 1878) pp. 957-975). J. Stefan relied on a concept introduced in Zurich in 1855 by the anatomical processor Adolf Eugen Fick (1829 Kassel-1901 Blankenberge in Flanders, Belgium) as Fick's laws of diffusion governing the diffusion of gas through a liquid membrane. Fick studied in Marburg with Stefan's later Viennese mentor Carl Ludwig (1816-1895), so all three worked closely together. In 1870, Fick first measured cardiac output using what is now called Fick's principle. In addition to Fick, Stefan also cited the experiments of Ernst Voit (1838 Speyer-1921 Munich), who later in 1883 became secretary and head of the photometry department of the Vienna Exhibition under Stefan's technical leadership.[36]

Stefan, J. (1879). On the Diffusion of Liquids, *Philosophical Magazine*, 5th series, 7th Volume, number 43 (April 1879) pp. 295-297 (abbreviated from Über die Diffusion der Flüssigkeiten, 2 Abhandlung (2: Berechnung der Graham'schen Versuche (*O difuziji tekočin II. Preračun Grahamovih poskusov*)), *Wien.Anzeiger*, volume 16, number 3 (1879), pp. 24-27. Full text of that article: Über die Diffusion der Flüssigkeiten, 2 Abhandlung, *Wien.Ber.* II. Volume 79, number 1 (3rd session on 3 January 1879) pp. 161-214). In the second part of his discussion of diffusion, Stefan referred in addition to Fick to

[35] J. Stefan, Versuche über die scheinbare Adhäsion, *Wien. Ber.* II, 30 April **1874**, *69*, pp. 713–728; Abbreviated reprint: *Ann. Phys.*, **1875**, *154* (230/2), pp. 316–318. Abstract: Experiments on apparent adhesion. *Philosophical Magazine*, 4. series, June **1874**, *47th volume*, no. 314, pp. 465–466.
[36] J. Stefan, Über die Diffusion der Flüssigkeiten (1: Über die optischen Beobachtungsmethoden), *Wien. Ber.* II, **1878**, volume *78*, no. 10 (26. session, 5. december 1878) pp. 953, 957–975. Translation of abstract: On the Diffusion of Liquids, *Philosophical Magazine*, Series 5, January **1879**, 7th volume, no. 40. pp. 74–75; A. Fick, Ueber Diffusion, *Ann.Phys.* **1855**, volume *170 (94)*, no. 1, pp. 59-86; Boncelj, 1960, 33.

the measurements of Thomas Graham (* 1805; † 1869), who a few years before his measurements became the last master of the mint as a successor of Newton and John Hershel.[37]

Stefan, J. (1879). Über die Beziehung zwischen der Wärmestrahlung und der Temperatur, *Wien.Ber.* II. Volume 79, number 1, (8th session on 20 March 1879): pp. 391-428. Abstract: *Wien.Anzeiger*, volume 16, number 8 (1879), pp. 87f. No English Abstract was printed in *Philosophical Magazine* because the editors there preferred more experimental research like in *Ann.Phys*).

Stefan, J. (1879). On the deviations of Ampère's theory of Magnetism from the Theory of the Electromagnetic Forces, *Philosophical Magazine*, 5th series, 8th Volume, number 46 (July 1879) pp. 83-84 (abbreviated from: Über die Abweichungen der Ampèreschen Theorie des Magnetismus von der Theorie der elektromagnetischen Kräfte, *Wien.Anzeiger*, volume 16, number 10 (1879), pp. 110-111. Celotna razprava: Über die Abweichungen der Ampèreschen Theorie des Magnetismus von der Theorie der elektromagnetischen Kräfte, *Wien.Ber.* II. Volume 79, number 2 (10th Session on 17 April 1879) pp. 659-679). J. Stefan used the measurements of Julius Plücker doctoral student and collaborator the Jew August Beer (1825 Trier–1863 Bonn) because Plücker was the main backer of Faraday's ideas at the European continent. Stefan also used the results of Boltzmann's equally Jewish doctoral student of electrodynamics at Stefan's Viennese laboratory, Max Margules (1856 Brodi by Lvov in Ukraine-1920 Perchtoldsdorf at the east of Lower Austria). Margules excelled in meteorology of storms at Viennese Meteorological institute established by M. Koller but

---

[37] J. Stefan, Über die Diffusion der Flüssigkeiten, 2 Abhandlung (2: Berechnung der Graham'schen Versuche), *Wien. Anzeiger*, **1879**, volume *16*, no. 3, pp. 24–27. Entire paper: Über die Diffusion der Flüssigkeiten, 2 Abhandlung, *Wien. Ber.* II, 3rd session, 3 January **1879**, volume *79*, no. 1, pp. 161–214. Translation of abstract: On the Diffusion of Liquids, *Philosophical Magazine*, Series 5, April **1879**, 7th volume, no. 43, pp. 295–297; T. Graham, Liquid diffusion applied to analysis, *Philosophical Transactions*, **1861**, *151*, str, 183-224. Reprint: *Philosophical Magazine*, Series 4, **1862**, *23*rd volume, pp. 204-223, 290-306, 368-380. Translation: Mémoire sur la diffusion moléculaire appliquée à l'analyse, *Annales de Chimie et de Physique*, Series 3, **1862**, *65*, 129-207.

abandoned his Berliner academic career to keep his Jewish faith.[38] Stefan also referred to his own decade earlier work.[39]

Stefan, J. (1880). *O nosilnosti magnetov* (*Über die Tragkraft der Magnete*), Wien.Ber. 81, pp. 89–116 (15.1.1880 *Anzeiger* pp. 14-15). Translation: On the carrying-power of magnets, *Philosophical Magazine*, 5th series, 9th Volume, number 55 (March 1880) pp. 232-233. Seven years after his successful heading of the Viennese electrical exhibition, Stefan relied on Kirchhoff's decade earlier research of magnetic inductions (G. Kirchhoff, Zur Theorie des in einem Eisenkörper inducirten Magnetismus, *Ann.Phys.*, **1871**, *218* (new series volume 1): 1-15) and Rowland's research at his post of assistant professor of physics at Rensselaer at Troy, New York on 2 June 1873 (H.A. Rowland (communicated by J.C. Maxwell), On Magnetic Permeability, *Phil.mag.*, 4th series, July-December 1873, 46, pp. 140-159, Stefan quoted page 158). Rowland then still supported Edison but criticized his industry as foreign to science in 1883. Stefan also praised Adalbert Carl Ritter von Waltenhofen zu Eglofsheimb (Karl, * May 14, 1828 castle Admontbichl west of Graz in Styria, † 5 February 1914 Vienna). Waltenhofen was then the professor of physics of the Prague German Technical University. In 1867-1883, Waltenhofen was the first professor and later the director of the Institute of Electrical Engineering at the Viennese University of Technology from 1883 to 1899, nominated during Stefan's Viennese Internationale Elektrische Ausstellung of 1883. In Vienna, Waltenhofen's successors also taught the main Slovenian successor of Stefan's electrical engineering, Milan Vidmar who was promoted at the Viennese University of Technology (Technical High School) on 16 July 1910. (A.C. Waltenhofen, Über elektromagnetische Tragkraft, *Wien.Ber.*, (sent from Prague on 4 April 1870) **1870**, 61: 739-754, Stefan quoted p. 746). Right after Stefan's article his student Boltzmann published about the new Maxwell's theory of viscosity of gases.

---

[38] A. Beer, Ueber das Verhältniss des Laplace-Biot'schen Gesetzes zu Ampère's Theorie des Magnetismus; Vergleich der von Neumann und Plücker aufgestellten Theorien der magneto-elektrischen Induction, *Ann.Phys.* **1855**, *94* (170/2), pp. 177-192; M. Margules, Über Theorie und Anwendung der elektromagnetischen Rotationen. Vorgelegt in der Sitzung am 23 Mai 1878. *Wien.Ber.* II, **1878**, 77, 805-818. Reprint: *Annalen der Physik*, **1878**, *242/1*, pp. 59-72; M. Margules, Bemerkung zu den Stefan'schen Grundformeln der Elektrodynamik: Vorgelegt in der Sitzung am 17 October 1878, *Wien.Ber.*, II, *78*, **1878**, 779-788.

[39] J. Stefan, *Über die Grundformeln der Elektrodynamik*, *Wien.Ber.* II, **1869**, *59*, pp. 693–769.

Stefan, J. (1881). O izhlapevanju iz posod s krožnim ali eliptičnim robom (Über die Verdampfung aus einem kreistförmig oder eliptisch begrenzten Becken), *Wien.Ber.* 83 (junij 1881), II, pp. 943–954; Reprint: Wiedmann, *Annalen der Physik und Chemie* 17, pp. 550 (1881). Not published in *Phil. Mag.*

Stefan, J. (1888). *O produkciji močnih magnetnih polj (Über die Herstellung intensiver magnetischer Felder)*, Wien.Ber. 97, II a, pp. 176–183 (9 February 1888). Translation: On the production of intense magnetic fields, *Philosophical Magazine*, 5th series, 25th Volume, number 155 (April 1888) pp. 322-323.

Stefan, Josef. 1889. On a theory of ice-forming, particularly in Polar Seas (Über die Theorie der Eisbildung, insbesondere über die Eisbildung im Eismeere), *Sitzungsberichte der Kaiserlichen Akademie der Wissenschaften in Wien*, II a, **98**: 965-983. Reprint: *Ann.Phys.* (1891) 42: 269-286. Not translated in *Phil.Mag.* Abridged in: *Monatshefte für Mathematik und Physik* (Vienna: Ministerium für Cultus und Unterricht) 1890, 1: 1-6.

Stefan, J. (1889-1890). O izhlapevanju in raztapljanju kot pojavih difuzije (Über die Verdampfung und die Auflösung als Vorgänge der Diffusion), *Wien.Ber.* II a, 98 (21. november 1889), pp. 1418–1442; Reprint: WA 41/12 (1890), pp. 725-747. Translation: On evaporation and solution as processes of diffusion, *Philosophical Magazine*, 5th series, 29th Volume, number 176 (January 1890) pp. 139-140.

Stefan, J. (1890). O električnem nihanju v premih vodnikih (Über die elektrische Schwingungen in geraden Leitern), *Wien.Ber.* 99, II a (9 January 1890 & 16 January 1890), pp. 319–339; Reprint: WA 41 (1890), pp. 400. Translation: On electrical vibrations in straight conductors, *Philosophical Magazine*, 5th series, 29th Volume, number 179 (April 1890) pp. 373-374 & number 180 (May 1890) 450-452.

Stefan, J. (1890). O teoriji nihanja razelektrenja (Über die Theorie der oscilatorischen Entladung), *Wien.Ber.* 99 (2.6.1890), II a, pp. 534–548; Reprint: WA 41/11 (1890), pp. 421-434. Translation: On the theory of oscillation discharge. *Philosophical Magazine*, 5th series, 30th Volume, number 184 (September 1890) pp. 373-374 & number 180 (May 1890) 282-283.

Stefan, J. (1891). O Wheatstonovi določitvi hitrosti elektrike (Über (Ueber) Wheatston's Bestimmung der Geschwindigkeit der Elektrizität), *Akademischer*

Anzeiger, pp. 106 (23 April 1891). *Wien.Ber.*, **1891**, *100/4*: 469 (just a note about the title even if *Phil.Mag.*, cited *Wien.Ber.*). Translation: On Wheatstone's determination of velocity of electricity. *Philosophical Magazine*, 5th series, 31st Volume, number 193 (June 1891) pp. 519-520 & 32nd Volume, number 198 (November 1891) pp. 480-480. Stefan's last paper published in *Phil.Mag.* as he died 14 months later. Kirchhoff, in his paper on the Motion of Electricity in Wires, first showed in 1857 that, under certain conditions, electricity moves in a thin wire according to the laws of wave-motion, and with a velocity which may be put equal to that of light.[40] The agreement between the velocities of electricity and light is only attained when the first travels in a straight wire stretched in the air. Kirchhoff has restricted his investigation to this case. If the bases of his calculation are applied to other cases, for instance to a wire which is wound in a zigzag, or is coiled in a spiral, it is found that electricity travels in such a wire with far greater velocity. In Wheatstone's well-known experiment[41] a wire was used which was coiled in twenty straight windings, and the velocity of electricity was found to be half as great again as that of light. I think that in the preceding I have given the right explanation of this result. I have, however, attempted to give this explanation an experimental support, and have employed the method given by Hertz of producing stationary waves in wires. I used a circuit like that in Wheatstone's experiment, but on a smaller scale, connected it with a couple of long straight wires, and compared the length of the wave in the circuit with the length of the same wave in the straight wires. The wave in the circuit is considerably longer, and in conformity with this the velocity of electricity in the circuit is greater than in the straight wires, and, according to my experiments, in a ratio which exceeds that found by Wheatstone.

Stenographic minutes of the sessions of the Provincial Assembly of the Duchy of Styria - Verhandlungen des steiermärkischen Landtages, 1867–1914, stenographic minutes of the 25th session of the Provincial Assembly of Styria.

Strachan Richard. 1873; 1879-1888. *Contributions to our knowledge of the meteorology of the arctic regions*, Meteorol. Office. Official, 18 (467 pages) and 34: 1/5. London: Stat. Off., 1879-1888.

---

[40] G. Kirchhoff, Über die Bewegung der Elektricität, *Ann. Phys.*, **1857**, *178 (12)*, pp. 529–544.

[41] C. Wheatstone, An account of some experiments to measure the velocity of electricity, and the duration of electric light, Phil.Trans, **1834**, *124*, 583-591.

Strachan, Richard. 1879. The barometer and its uses, wind and storms. *Modern meteorology A series of six lectures delivered under the auspices of the Meteorological Society in 1878.* London: E. Stanford. Translation: 1882. Vorlesung 3. Das Barometer und seine Anwendung, Winde, Sturme. *Die moderne Meteorologie: sechs Vorlesungen gehalten auf Veranlassung der Meteorologischen Gesellschaft zu London.* Braunschweig : Vieweg, pp. 81–118.

Strakosch-Grassmann, Gustav. 1905. *Geschichte des österreichischen Unterrichtswesens ... Mit 95 Porträts und 29 Abbildungen im Texte und zwei Beilagen,* Wien: A. Pichler's Witwe & Sohn.

Strnad, Janez. 1986. *Jožef Stefan, Ob stopetdesetletnici rojstva (Josef Stefan, By his 150th anniversary of birth)*, DMFA SRS, Ljubljana.

Šarler, Božidar. 1995. Stefan's work on solid-liquid phase changes. In anthology: Stefan and Moving Front Problems (guest editor Alain Jacques Kassab of the Department of Mechanical and Aerospace Engineering at the University of Central Florida (UCF) in Orlando). *Engineering analysis with boundary elements*, 16/2: 83-92.

Šarler, Božidar. 2011. On Stefan's Research of Multiphase Systems (O Stefanovih raziskavah večfaznih sistemov), Lecture at the Institute Jožef Stefan, 14 December 2011. unpublished.

Šubic, Ivan. 1902. Dr. Josip Stefan. *Zbornik znanstvenih in poučnih spisov* (Ljubljana: Slovenska Matica), 4: 62–85.

Šubic, Simon, 1864, Innere Arbeit und specifiche Warme, *Wien. Anz.*, 1: pp. 22-25.

Šubic, Simon, 14 July 1864 Über die specifische Warme, die innere Arbeit und das Dulong-Petit'sche Gesetz, Wien.Ber.mat.-nar.Kl. I 50: 169 in II 50: 168. Stefan's expert opinion issued on 21 July 1864, publication refused, Šubic's manuscript lost, published abstract: Wien.Anz. 1864 1: 134-136.

Šubic, Simon, 8. 2. 1872 Ueber die Temperatur-Constante, Wien.Ber.mat.-nat.Kl. 65: 27. Loschmidt's expert opinion on 20 February 1872 (returned on 7 March 1872) publication refused. Abstracts: Wien.Anz. (1872) 9: 26, Chem.C.Bl. (20. 3. 1872) 3: 177 (reprint from Wien.Anz. 9, the surname misspelled as S.Subié; J.chem.soc. (2) 9: 591 (H. W.'s translation from Chem.C.Bl. 3, also an error in spelling of a surname), Inst. 1872 175. Orinted in Pogg.Ann. at the end of the year 1872, 147: 452-468. Abstract: Fortschr. 1872, 17: 490-491 (avtor Nn).

Tarzia, Domingo Alberto. 1979. *Sur le problème de Stefan à deux phases,* PhD at the Université Pierre et Marie Curie in Paris.

Tarzia, Domingo Alberto. 1988. *A Bibliography on Moving-Free Boundary Problems for the Heat-Diffusion Equation. The Stefan Problem.* Progetto Nazionale M.P.I., Equazioni di evoluzione e applicazioni fisico-matematiche, 1st. edition.

Tarzia, Domingo Alberto. 1991. *Problèmes à frontière libre du type elliptique et parabolique; le problème de Stefan,* Habilitation à diriger des recherches at the Université Pierre et Marie Curie in Paris.

Tarzia, Domingo Alberto. 2000. A Bibliography on Moving-Free Boundary Problems for the Heat-Diffusion Equation. The Stefan and Related Problems. *MAT-Series A,* 2, 1-297. Rosario, July 2000 (Manuscript submitted at Rosario in October 1999).

Tikhonov, Andrey Nikolayevich (Tychonoff, Tihonov, Андре́й Никола́евич Ти́хонов). 1936. *О функциональных уравнениях* типа *Вольтерра и их приложениях к уравнениям математической физики (On functional equations of Volterra type and their applications to equations of mathematical physics).* PhD in Moscow University (MGU). Reprinted: 1938a. О функциональных уравнениях типа Вольтерра и их приложениях к уравнениям математической физики, *Бюллетень Московского государственного университета имени М. В. Ломоносова (Бюлл. МГУ),* секция А, серия: математика и механика, Т. 1, вып. 8. С. 1-25 (Functional equations of Volterra type and their applications to certain problems of mathematical physics, *Bulletin of Lomonosov University in Moscow),* section A, series of mathematics and mechanics, volume=Issue 1, no. 8: 1-25.

Tikhonov, Andrey Nikolayevich. 1938b. Об уравнении теплопроводности для нескольких переменных, *Бюллетень Московского государственного университета имени М. В. Ломоносова (Бюлл. МГУ)*, секция А, серия: математика и механика, Т. 1, вып. 9. С. 1-49 (On equation of thermal conduction for several variables, *Bulletin of Lomonosov University in Moscow*, section A, series of mathematics and mechanics, volume=Issue 1/9, 1-49). Simultaneous abbreviated French translation published in Moscow: Theoremes d'unicité pour l'équation

de la chaleur, *Матем. сб.* (Математический *сборник*, Российская академия наук, Математический институт им. В.А. Стеклова Российской академии наук (МИАН), Москва) 42, No 2 (1938), 199—216.

Trench, William F. June 1959. On an explicit method for the solution of a Stefan problem, *J. Soc.Indust. Appl. Math.* (*Journal of the Society for Industrial and Applied Mathematics, SIAM*) 7/2: 181-204 (abridged doctoral thesis in Mathematics at University of Pennsylvania defended in 1958, sent on 1 May 1958).

Truesdell, Clifford Ambrose. 1980. *Tragicomical History of Thermodynamics 1822-1854*. New York-Heidelberg-Berlin: Springer-Verlag.

Vaniček, Alois (ed.). 1860. *Schematismus der österreichischen Gymnasien und Realschulen für das Schuljahr 1859-60*. Prag: F. Tempsky.

Vertovec, Matija. 1847. *Kmetijska kemija, to je, natorne postave in kemijske resnice obernjene na člověško in živalsko življenje, na kmetijstvo in njegove pridelke*. Ljubljana. Bettered reprint wit somewhat modified title: *Kmetijska kemija ali natorne postave in kemijske resnice obernjene na člověško in živalsko življenje, na kmetijstvo in njegove pridelke*. 1856. Ljubljana: J. Blaznik.

Vidmar, Luka. 2004. *Correspondence of Žiga Zois between 6 April 1812 and 27 June 1813 (Korespondenca Žige Zoisa med 6.4.1812 in 27.6.1813)*. Ljubljana: ZRC SAZU.

Violle, Jules. 1879. Sur la radiation de platine incandescent, *Comptes Rendus*, 88: pp. 171-173.

Violle, Jules. 1881. Review: Stefan. Ueber die Beziehung zwischen der Wärmestrahlung und der Temperatur (Sur la relation entre le rayonnement calorifique et la température); Sitzungs berichte d. K. Akademie d. Wissenschaften in Wien, p. 84. *J. Phys. Theor. Appl.*, 1881, 10 (1), pp. 317-319.

Viskanta, Raimondas. 1988. "Heat Transfer During Melting and Solidification of Metals," *J. Heat Transfer*, **110**, pp.1205-1219.

Voronina, Margarita Mihajlovna. 1987. *Gabriel' Lame, 1795-1870*, Leningrad: Nauka.

Vuik, Cornelis (professor in Numerical Analysis at the Delft University of Technology, Zuid-Holland, Netherlands). 1993. Some historical notes about the Stefan problem. *Delft University of Technology Report*, 93-107. Simultaneous Reprint: 1993. Some historical notes about the Stefan problem, *Nieuw Archief voor Wiskunde*, 4e serie, 11 (2): 157–167.

Wisniak, Jaime. 2006. Josef Stefan – Radiation, Conductivity, and other Phenomena. *Revista CENIC (Centro Nacional de Investigaciones Científicas) Ciencias Químicas (Ciudad de La Habana, Cuba)*, Vol. 37, No. 3, January 2006, 188-195.

Wolf, Rudolf; Larcher, Verena (Ed.). 1993. *Rudolf Wolfs Jugendtagebuch 1835 - 1841 (Rudolf Wolf's juvenile diary, 1835 - 1841)*. Zürich: Schriftenreihe der ETH-Bibliothek, Volume (Band) 30.

Wurzbach, Constantin knight Tannenberg. 1864. *Biographisches Lexikon des Kaiserthums Oesterreich, enthaltend die Lebensskizzen der denkwürdigen Personen, welche 1750 bis 1850 im Kaiserstaate und in seinen Kronländern gelebt haben*, Wien: L.C. Zamarski Kaiserlich-königlichen Hof- und Staatsdruckerei, volume 12.

Yao, L.S. and J. Prusa, 1989, "Melting and Freezing," *Adv. Heat Transfer*, **19**, pp.1-95.

Summary

How did an illegitimate suburban son Stefan make it to his top scientific positions? The merits belonged to his mentor, the Slovenian compatriot Maran Koller. Dr. Ernst Wilhelm ritter von Brücke (1819 Berlin-1892 Vienna) recommended Stefan to an adviser at the Ministry of Education and Worship, Marian Koller, who even personally attended Stefan's classroom at least during six semesters. He noted almost the entire cycles of Stefan's university lectures between 1862 and 1864, on his last occasion together with the young student L. Boltzmann. As a Carniolan, Koller particularly loved the Slovene Stefan. Stefan graduated from the Benedictine Grammar School in Klagenfurt during the mayoralty of M. Koller's brother, Andreas, who was there in the office from 10 November 1850 to 13 April 1852. Stefan used to study the textbooks of Koller's travel companion Kunzek. Stefan was a Grammar School student of Andreas's son-in-law, the Slavist Anton Janežič. The Benedictine Karl Robida was Stefan's class teacher, which was especially appreciated by the Benedictine Marian Koller even if Robida later publicly criticized Stefan's and Clausius' kinetic theory of gases. Koller's recommendation enabled his Slovenian compatriot Stefan's steep way up to the very top of European science and engineering. By Koller's recommendation, Stefan became the youngest professor of mathematical physics at the University of Vienna, a co-director of the Physical Institute on 9 March 1863, a full member of the Viennese Academy of Sciences as the first recipient of the Ignaz Lieben's Prize on 27 April 1865 and director of the Viennese Physical Institute after Ettingshausen's retirement by imperial order on 1 October 1866, just a dozen days after Koller's sudden death. Soon, Stefan published his major breakthroughs now called Stefan Problem, Stefan Number and Stefan(-Boltzmann) Law. Most of Stefan's experimental works were translated and published with almost no delay in Berlin, Leipzig, London, Paris, and Geneve, which is the other main novelty of this study. The aim is to illustrate how the dying Habsburgian monarchy topped the worldwide erudition by Koller, Franz Liszt, Stefan, E. Mach, Boltzmann, S. Freud's two months younger N. Tesla, Gustav Mahler, Gustav Klimt, both Johann Strauss, Stefan Zweig. After the Habsburgian monarchy collapsed and following Einstein's Prague semesters from April 1911 through July 1912, the Viennese Erwin Schrödinger built quantum physics which mostly the Jewish Budapest naturalized Americans developed into a still reigning mainstream. In that context, the swansong of Habsburgian monarchy is not just the story of our past erudition, but about our future, especially if we have another collapsing empire(s) onboard today.

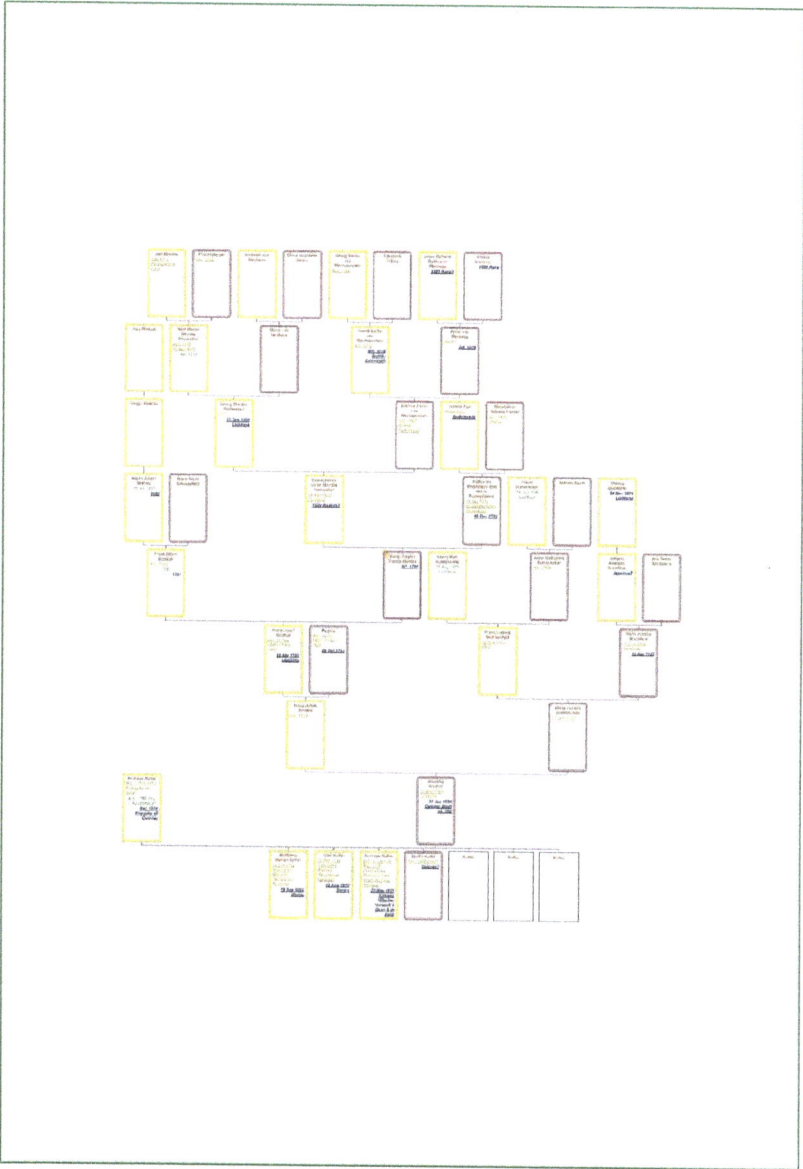

Figure 15: Koller's ancestors by all branches

www.ingramcontent.com/pod-product-compliance
Lightning Source LLC
Chambersburg PA
CBHW060813220326

41598CB00022B/2605